W9-CNB-312

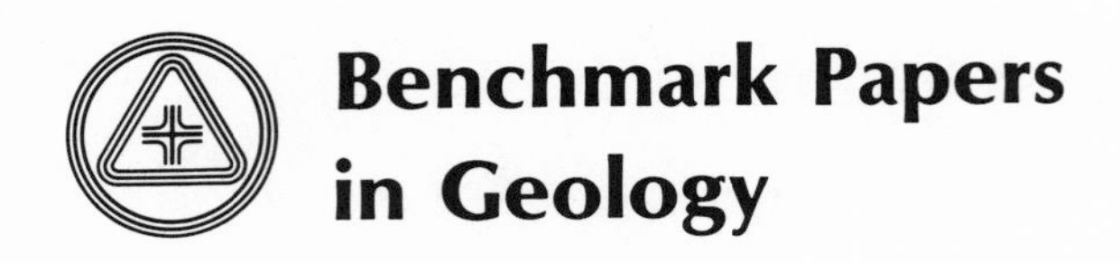

Benchmark Papers in Geology

Series Editor: Rhodes W. Fairbridge
Columbia University

Volume

1 ENVIRONMENTAL GEOMORPHOLOGY AND LANDSCAPE CONSERVATION, Volume 1: Prior to 1900 / *Donald R. Coates*
2 RIVER MORPHOLOGY / *Stanley A. Schumm*
3 SPITS AND BARS / *Maurice L. Schwartz*
4 TEKTITES / *Virgil E. Barnes and Mildred A. Barnes*
5 GEOCHRONOLOGY: Radiometric Dating of Rocks and Minerals / *C. T. Harper*
6 SLOPE MORPHOLOGY / *Stanley A. Schumm and M. Paul Mosley*
7 MARINE EVAPORITES: Origin, Diagenesis, and Geochemistry / *Douglas W. Kirkland and Robert Evans*
8 ENVIRONMENTAL GEOMORPHOLOGY AND LANDSCAPE CONSERVATION, Volume III: Non-Urban / *Donald R. Coates*
9 BARRIER ISLANDS / *Maurice L. Schwartz*
10 GLACIAL ISOSTASY / *John T. Andrews*
11 GEOCHEMISTRY OF GERMANIUM / *Jon N. Weber*
12 ENVIRONMENTAL GEOMORPHOLOGY AND LANDSCAPE CONSERVATION, Volume II: Urban Areas / *Donald R. Coates*
13 PHILOSOPHY OF GEOHISTORY: 1785–1970 / *Claude C. Albritton, Jr.*
14 GEOCHEMISTRY AND THE ORIGIN OF LIFE / *Keigh A. Kvenvolden*
15 SEDIMENTARY ROCKS: Concepts and History / *Albert V. Carozzi*
16 GEOCHEMISTRY OF WATER / *Yasushi Kitano*
17 METAMORPHISM AND PLATE TECTONIC REGIMES / *W. G. Ernst*
18 GEOCHEMISTRY OF IRON / *Henry Lepp*
19 SUBDUCTION ZONE METAMORPHISM / *W. G. Ernst*
20 PLAYAS AND DRIED LAKES: Occurrence and Development / *James T. Neal*
21 GLACIAL DEPOSITS / *Richard P. Goldthwait*
22 PLANATION SURFACES: Peneplains, Pediplains, and Etchplains / *George F. Adams*
23 GEOCHEMISTRY OF BORON / *C. T. Walker*
24 SUBMARINE CANYONS AND DEEP-SEA FANS: Modern and Ancient / *J. H. McD. Whitaker*
25 ENVIRONMENTAL GEOLOGY / *Frederick Betz, Jr.*
26 LOESS: Lithology and Genesis / *Ian J. Smalley*

27 PERIGLACIAL PROCESSES / *Cuchlaine A. M. King*
28 LANDFORMS AND GEOMORPHOLOGY: Concepts and History / *Cuchlaine A. M. King*
29 METALLOGENY AND GLOBAL TECTONICS / *Wilfred Walker*
30 HOLOCENE TIDAL SEDIMENTATION / *George deVries Klein*
31 PALEOBIOGEOGRAPHY / *Charles A. Ross*
32 MECHANICS OF THRUST FAULTS AND DÉCOLLEMENT / *Barry Voight*
33 WEST INDIES ISLAND ARCS / *Peter H. Mattson*
34 CRYSTAL FORM AND STRUCTURE / *Cecil J. Schneer*
35 OCEANOGRAPHY: Concepts and History / *Margaret B. Deacon*
36 METEORITE CRATERS / *G. J. H. McCall*
37 STATISTICAL ANALYSIS IN GEOLOGY / *John M. Cubitt and Stephen Henley*
38 AIR PHOTOGRAPHY AND COASTAL PROBLEMS / *Mohamed T. El-Ashry*
39 BEACH PROCESSES AND COASTAL HYDRODYNAMICS / *John S. Fisher and Robert Dolan*
40 DIAGENESIS OF DEEP-SEA BIOGENIC SEDIMENTS / *Gerrit J. van der Lingen*
41 DRAINAGE BASIN MORPHOLOGY / *Stanley A. Schumm*
42 COASTAL SEDIMENTATION / *Donald J. P. Swift and Harold D. Palmer*
43 ANCIENT CONTINENTAL DEPOSITS / *Franklyn B. Van Houten*
44 MINERAL DEPOSITS, CONTINENTAL DRIFT AND PLATE TECTONICS / *J. B. Wright*
45 SEA WATER: Cycles of the Major Elements / *James I. Drever*
46 PALYNOLOGY, PART I: Spores and Pollen / *Marjorie D. Muir and William A. S. Sarjeant*
*47 PALYNOLOGY, PART II: Dinoflagellates, Acritarchs, and Other Microfossils / *Marjorie D. Muir and William A. S. Sarjeant*

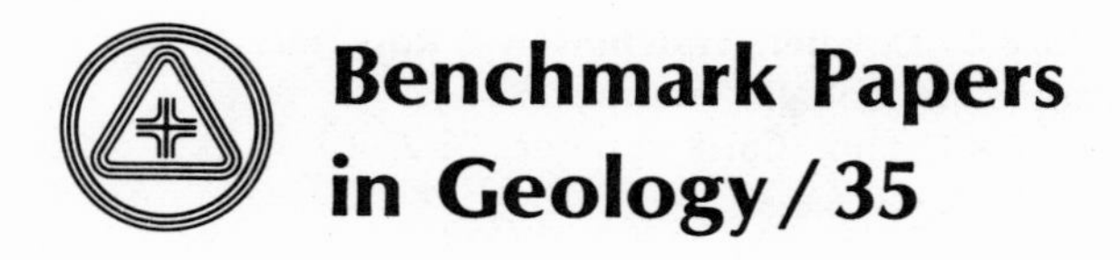

A BENCHMARK®Books Series

OCEANOGRAPHY
Concepts and History

Edited by

MARGARET B. DEACON

To my father
G. E. R. Deacon

Benchmark Papers in Geology, Volume 35
Library of Congress Catalog Card Number: 76-27682
ISBN: 0-87933-202-6

80 79 78 1 2 3 4 5
Manufactured in the United States of America.

LIBRARY OF CONGRESS CATALOGING IN PUBLICATION DATA
Main entry under title:
Oceanography : concepts and history.
(Benchmark papers in geology ; 35)
Bibliography: p.
Includes index.
1. Oceanography—Addresses, essays, lectures. I. Deacon, Margaret.
GC26.027 551.4'6 76-27682
ISBN 0-87933-202-6

Distributed world wide by Academic Press,
a subsidiary of Harcourt Brace Jovanovich,
Publishers.

SERIES EDITOR'S FOREWORD

The philosophy behind the "Benchmark Papers in Geology" is one of collection, sifting, and rediffusion. Scientific literature today is so vast, so dispersed, and, in the case of old papers, so inaccessible for readers not in the immediate neighborhood of major libraries that much valuable information has been ignored by default. It has become just so difficult, or so time consuming, to search out the key papers in any basic area of research that one can hardly blame a busy man for skimping on some of his "homework."

This series of volumes has been devised, therefore, to make a practical contribution to this critical problem. The geologist, perhaps even more than any other scientist, often suffers from twin difficulties—isolation from central library resources and immensely diffused sources of material. New colleges and industrial libraries simply cannot afford to purchase complete runs of all the world's earth science literature. Specialists simply cannot locate reprints or copies of all their principal reference materials. So it is that we are now making a concerted effort to gather into single volumes the critical material needed to reconstruct the background of any and every major topic of our discipline.

We are interpreting "geology" in its broadest sense: the fundamental science of the planet Earth, its materials, its history, and its dynamics. Because of training and experience in "earthy" materials, we also take in astrogeology, the corresponding aspect of the planetary sciences. Besides the classical core disciplines such as mineralogy, petrology, structure, geomorphology, paleontology, and stratigraphy, we embrace the newer fields of geophysics and geochemistry, applied also to oceanography, geochronology, and paleoecology. We recognize the work of the mining geologists, the petroleum geologists, the hydrologists, the engineering and environmental geologists. Each specialist needs his working library. We are endeavoring to make his task a little easier.

Each volume in the series contains an Introduction prepared by a specialist (the volume editor)—a "state of the art" opening or a summary of the object and content of the volume. The articles, usually some thirty to fifty reproduced either in their entirety or in significant extracts, are selected in an attempt to cover the field, from the key papers of the last century to fairly recent work. Where the original works are in foreign languages, we have endeavored to locate or commission translations. Geologists, because of their global subject, are often acutely aware of the oneness of our world. The selections cannot, therefore, be restricted to any one country, and whenever possible an attempt is made to scan the world literature.

To each article, or group of kindred articles, some sort of "highlight commentary" is usually supplied by the volume editor. This commentary should serve to bring that article into historical perspective and to emphasize its particular role in the growth of the field. References, or citations, wherever possible, will be reproduced in their entirety—for by this means the observant reader can assess the background material available to that particular author, or, if he wishes, he, too, can double check the earlier sources.

A "benchmark," in surveyor's terminology, is an established point on the ground, recorded on our maps. It is usually anything that is a vantage point, from a modest hill to a mountain peak. From the historical viewpoint, these benchmarks are the bricks of our scientific edifice.

RHODES W. FAIRBRIDGE

PREFACE

The intention of this book is to introduce oceanography by tracing the historical development of some of the ideas central to the science and of some of its principal fields of activity. It does not attempt to give an account of the scope of present-day oceanography, which many excellent books already do, but tries in the thirty-nine papers or extracts included, to illustrate some of the most important milestones in its development, though not quite all of these milestones could be said to represent movement in a forward direction. Reasons of space have meant that not all branches of marine science, such as marine optics or acoustics, which are important today but have a relatively short history, are represented, nor, for the same reason, could the work of all the leading oceanographers today be included as would have been desirable had more than a single volume been permitted.

Oceanography was recognized as a science in the late nineteenth century, during a period of expanding interest and involvement in work at sea by scientists of many disciplines and many nations. The voyage of H.M.S. *Challenger,* 1872–1876, the centenary of which has recently been celebrated, was one of the first major consequences of this interest. The voyage is therefore often referred to as the foundation of oceanography; but, in fact, its history began much earlier. A convenient starting point is the scientific revolution of the seventeenth century when, as far as we know, observation began to supplement speculation about such phenomena as the tides, depth, and saltness of the ocean, for the first time since the decline of Greek science. The subsequent development of the science of the sea was interrupted, however, more than once by long periods of inactivity. The causes of these fluctuations may be sought primarily in the many factors that militated against research at sea, in particular the financial and practical difficulties that remained almost insurmountable until the development of machine power in the last century and of large-scale finance in the present. The need to make accurate observations at sea has always forced oceanography to make capital of discoveries in other branches of science and technology, as in navigation and engineering, from the marine chronometer to the computer and from the steam engine to modern electronics.

The difficulties of oceanography largely arose because it almost entirely lacked the first stage common to most other sciences, in which great advances could be made by relatively few people working with relatively simple apparatus in an observatory or laboratory. Accordingly, only in the last thirty to thirty-five years, since continuing financial support for science on a large scale became government policy in democracies, has oceanography been able to grow freely.

These vicissitudes of fortune also interrupted the theoretical development of the science. It is not uncommon for ideas to be rejected and then revived as new evidence comes to light. But for many years in the middle of the nineteenth century it was believed that the depths of the ocean were both motionless and lifeless and previous evidence to the contrary was either disputed or forgotten. One cannot help suspecting that perhaps Victorian scientists were unconsciously justifying their lack of interest in the deep ocean, or inability to reach it, by saying that nothing was going on there anyway.

Has oceanography left behind the uncertainties of its early history or will its career continue to be chequered? We are now perhaps emerging from a period of intense growth and whether it is followed by an era of consolidation or will turn out, like earlier periods of activity, to have been an episode, remains to be seen. Much depends on the attitude of society as a whole. Activity could lessen if, for example, funds were curtailed, perhaps because of a reaction of public opinion against science that might result from the disillusionment caused by the naive belief held in some quarters that knowledge somehow confers political power *per se*. Again, if oceanographers' freedom to pursue research were too severely curtailed by national and international legislation on territorial waters and the exploration of the sea bed, they might find other subjects more rewarding.

The history of oceanography has always attracted interest and a considerable amount has been written on it. References to some of this historical literature will be found in the bibliography, which also includes works cited in the commentaries, some general works on oceanography, and references for those extracts from before 1900 that are without bibliographies of their own, omitting one or two items without direct bearing on marine science.

Readers unfamiliar with the scientific journals of the early nineteenth century may be dismayed at the prospect of trying to obtain local publications. They should bear in mind the convenient contemporary custom of almost immediately reprinting or summarizing articles, in translation where necessary, in a number of different periodicals. These additional references have been given where known.

I should like to thank everyone who has helped in the preparation of this volume, in particular the authors and institutions who have given permission for work to be reprinted. A special mention must be made of Dr. Adrian Gill of the University of Cambridge, for kindly making available Mrs. Hudson's translation of Marsigli's essay on the currents of the Bosporus; to Signorina Tilde Pedenovi of SACLANT ASW Research Centre, La Spezia, who kindly overlooked and added to the translation; and to Dr. and Mrs. Tom Allen who enlisted her aid. I am very grateful to the series editor for his helpful suggestions, to Mrs. Valerie West and Mrs. Jacqueline Skidmore for help with typing, and finally to my father for his assistance throughout and especially with regard to the twentieth-century work.

Some of the papers reprinted here have been in existence over 300 years, making it impossible to locate good quality original material to print from. In spite of this fact, the papers have been printed facsimile in order to preserve their original flavor. I hope that the reader will understand.

MARGARET B. DEACON

CONTENTS

PART III: TIDES

PART IV: WAVES

PART V: CHEMISTRY OF SEA WATER

PART VI: DEPTHS OF THE OCEAN AND THE SEA BED

PART VII: MARINE BIOLOGY

CONTENTS BY AUTHOR

OCEANOGRAPHY

INTRODUCTION

Oceanography as an established science is still less than a hundred years old. Indeed, it may be said to have become well established only during the period of expansion following World War II. Looking back, many present-day oceanographers have seen the voyage of HMS *Challenger* as the beginning of their science, a view shared by at least one member of the expedition (Buchanan 1895). However, it was the events of the 1880s and 1890s which consolidated the work of this and of similar expeditions making scientists begin to think of the science of the sea as "oceanography" (e.g., Boguslawski 1884). Yet, as the *Challenger Report* (Thomson and Murray 1895, "Summary: Vol. 1") itself made clear, on several occasions during the previous centuries, attempts had been made to establish a science of the sea. For various reasons they had failed and for long periods the subject enjoyed a curious half-life, dominated by no less curious half-truths (Deacon 1971).

Scientific interest in the sea can be traced back to philosophers of Greek and Roman times (Deacon 1971, ch. 1) Foremost among them was Aristotle who, in the fourth century B.C., discussed the cause of the saltness of the sea, described the evaporation–precipitation cycle between the ocean and the atmosphere, and studied the marine fauna of the Aegean Sea (Aristotle 1908–1952). During the next two or three hundred years, a considerable knowledge of tides was acquired (see Part III).

The learning of the Middle Ages was not geared to the study of natural phenomena. Writers continued to show an interest in tides and the saltness of the sea (Sarton 1927–1948); but in an age in which biblical revelation was seen as the ultimate source of knowledge, argument from literary sources, in these as in other subjects, was the standard method of procedure. In the memorable words of Edward Gibbon (1776), "A cloud of critics, of compilers, of commentators, darkened the face of learning, and the decline of genius was soon followed by the corruption of taste." There were a few exceptions, people who owed their knowledge to those whose livelihood depended on practical knowledge of the sea, the sailors and fishermen of the coasts of Europe and the Mediterranean. In an age when literacy was confined to a few, these people, like Chaucer's shipman in the *Canterbury Tales*, relied on their own knowledge and the accumulated experience handed down by word of mouth.

At the Renaissance, literacy spread; and with the introduction of printing this accumulated knowledge, together with new discoveries, began to appear in treatises on navigation and in sailing directions. Arab writers had described the currents of the Indian Ocean and their twice-yearly reversal with the monsoons as early as the ninth century (Warren 1966; Aleem 1967). As Renaissance voyagers traveled farther from home, the movements of the other oceans were discovered. An early description of currents, particularly in the Atlantic, is found in William Bourne's *Treasure for Travellers* (1578) together with information on tides and the effect of the sea on coastlines (Deacon 1971, ch. 3).

Soundings began to appear on charts in the late sixteenth century (Destombes 1968). These soundings, made on the continental shelf in water less than 100 fathoms deep, formed an important part of the navigator's knowledge in the waters of Northern Europe. In bad weather conditions, familiarity with the depths of the sea and the composition of the sea bed could replace the sights and bearings normally used to fix a ship's position (Waters 1958).

Several scientific writers of the sixteenth century described an apparatus they thought might be more accurate for measuring great depths than a lead and line (Multhauf 1960). The lineless sounder basically consisted of a weight to which a float was attached, in such a way that it would be released when the weight struck the sea bed. The idea was even older and had already been described by several fifteenth-century precursors of the scientific movement, including Cardinal Nicholas of Cusa (1650). None of these people, however, appear to have actually used the device. Nor, with a few exceptions, did the numerous writers discussing tides make observations for themselves (see Part III).

The development of the scientific movement in the seventeenth century restored scientific observation to its rightful place, alongside argument from propositions; in fact the pendulum for a time swung to the opposite extreme, following Sir Francis Bacon's idea that, by collecting sufficient information, the laws of nature would be revealed. This was a strong influence in the early years of the Royal Society of

London and one of the subjects to which the members applied it was the study of the sea. At that time (the early 1660s), the sea was of rapidly increasing importance, both politically and economically, in the life of the British people, and the Royal Society hoped that their work would not only increase the stock of scientific knowledge but also practically assist the sailor (Deacon 1971, ch. 4).

Robert Boyle (1666) made a list of subjects he thought should be studied. It included currents, waves, depth, salinity, temperature, the effect of storms, and the possible medical properties of sea water. He omitted tides as they had already been dealt with in a previous article (see Paper 13). The Royal Society succeeded in obtaining some cooperation from travelers and sailors, for example Sir Robert Holmes (Deacon 1971, p. 409); and its work certainly influenced a wider circle, as shown, for example, by the interest in the Strait of Gibraltar (see Part II). However, in spite of this good start, as the century drew to a close activity tailed off. This was partly due to a change in direction of the scientific movement, away from the study of crafts and trades and toward the theoretical sciences. In the case of marine science, it was also due to the particular difficulties involved, especially the need to rely on other people to do work at sea and the problems connected with the apparatus they had to use. The greatest single difficulty was that most of the instruments depended on a wooden float to bring them to the surface. While these floats worked well in the shallow water where they were tested, in deep water, unknown to the researchers, they would become waterlogged and sink (see Part VI). Unfortunately, the sea trials of these instruments were never sufficiently numerous to show that something was fundamentally wrong with the design. There were always other factors which could have been responsible for their failure.

The most interesting results of the Royal Society's work, apart from the discoveries about tides (see Part III), were the four essays by Boyle (1671, 1673) on the depth, temperature, salinity, and plant life of the sea (Deacon 1971, ch. 6), and the later work of Hooke (Deacon 1971, ch. 8). Hooke devised numerous new devices for work at sea, including instruments to measure depth by finding pressure or with a clockwork gauge, a self-registering thermometer for taking deep-sea temperatures, and a water sampler that could work at any depth (Derham 1726, pp. 225–248). When Hooke at last made them public in 1691 he was bitter about the lack of interest and of opportunities for making use of his inventions.

Italy was another country in which a practical interest in the sea developed at about the same time. This interest is reflected in two studies by L. F. Marsigli, remarkable both in concept and results. The first was an examination of the currents of the Bosporus (Paper 3);

the second, a study of the Gulf of Lions. In 1706–1707, Marsigli (1725) was living at Cassis, near Marseilles (Olson and Olson 1958). He studied depths out to the edge of the continental shelf, temperature, and specific gravity, and included serial temperature measurements, the movements of the water, and the marine life of the area. Marsigli was critical of the Royal Society for trying to study the whole ocean at once and not attempting a limited project, but he found that he could not adequately observe the water movements of even this relatively small area by himself and concluded that marine research would only flourish with the support of the state.

With the revival of the natural sciences in the mid-eighteenth century interest in marine science was renewed (Deacon 1971, ch. 9–10). Scientists like Stephen Hales, who was inspired by Hooke's work (see Parts II and VI), worked on new apparatus for making measurements and, as the voyages of discovery sent by the maritime nations to the Pacific and Australasia and to the polar regions became increasingly scientific in their nature, more and more observations were made at sea. On Cook's second voyage, 1772–1775, the astronomers William Wales and William Bayly measured surface and some subsurface temperatures. The French astronomer Chappe d'Auteroche (1772) measured specific gravities for Lavoisier on a voyage across the Atlantic in 1768–1769. Constantine Phipps (1774) on his unsuccessful voyage to the Arctic in 1773 made deep soundings of 683 and 780 fathoms out of a spirit of scientific curiosity. The ill-fated expedition of La Pérouse took with it two instruments for measuring the depth of the sea and an apparatus for finding temperature and salinity below the surface, presumably a kind of water sampler (Milet-Mureau 1799). Expeditions in the early nineteenth century made long series of observations of surface temperature and specific gravity and some important subsurface observations (see Part II).

In all but a few cases, however, the scientists themselves were still dependent on others to collect their information, apart from one or two like William Scoresby, the younger, who was a whaling captain. Nevertheless, with data from the expeditions and other sources more progress toward founding a science of the sea was made in the first years of the nineteenth century than ever before.

The development of chemistry had already made possible the first analyses of sea water by Lavoisier (Paper 25) and Bergman (1779, vol. 1, pp. 179–183) and these were refined and extended by chemists in many countries. The Swiss chemist Alexander Marcet also studied the distribution of salinity and first fixed the temperature of maximum density of sea water (Paper 26). James Rennell, the geographer, mapped the surface currents of the Atlantic (see Paper 4). At the same time Rumford (1800), Humboldt (1814 vol. 1, pp. 72–75), and other scientists were speculating about polar currents flowing toward the equator in the depths of the sea, as a result of density differences in the ocean. The existence of such a polar flow was convincingly demonstrated by the Russian physicist Lenz (Paper 5). The French mathematician Laplace had revolutionized the study of tides and instituted long-term observations at Brest (see Part III). The foundations of marine biology were also laid during this period (see Part VII).

Yet, in spite of this wide range of talent, marine science failed to establish itself. The underlying reason was economic. A scientist without private means then earned his living either as an academic or in a profession such as medicine. The state provided a few limited opportunities for research at sea on the voyages of exploration, but once the voyage was over there was no chance of going back to test a theory or examine a new phenomenon. It was not surprising, therefore, that someone such as Lenz should turn for the greater part of his career to studies of electricity and magnetism. The scientists did not yet see themselves as contributing to a new science, but even if they had, neither the state nor the universities at that time would have seen it as their job to provide the funds needed to undertake the work and to attract the pupils who could carry it on further.

Nevertheless, many of the ideas and methods that have since proved significant were first developed in the period 1770 to 1830 (Deacon 1971, ch. 11). They included the internal circulation of the ocean and its causes and the dynamical study of tides. Among the instruments first used at that time were the self-registering thermometer (Six 1794), the messenger (see Paper 26) and the self-registering tide gauge (see Paper 16). The reversing thermometer and current meters for use in deep water were developed slightly later by the extraordinarily inventive French scientist Georges Aimé (1845), who died in 1846.

In the absence of a coherent drive to establish marine science, and as the original scientists involved had either died or turned to other studies, the achievements of the early nineteenth century made very little impact on the science of the early Victorian period. Marine biology continued to develop unhindered, since at this time it was still largely confined to the shore and coastal waters and at a stage of development when many people, from country ministers to retired businessmen, could make contributions by collecting and describing new species. Tides, too, continued to be studied, having aroused interest within the established framework of astronomy and mathematics. Chemists continued to add to the list of elements detected in sea water as their analytical techniques improved. Paradoxically, it was physical oceanography, which had earlier made the most rapid progress, and the vast body of the oceans themselves, which now ceased to arouse the scientists' imagination.

This did not mean that physical observations were no longer made on voyages of discovery but the naval officers who carried out these routine operations were often not aware of work such as Lenz's on thermometers and Marcet's on the behaviour of sea water at low temperatures. They saw nothing odd in obtaining temperatures of about 4°C in deep water since they had no idea that sea water differed from fresh water in its behavior and assumed that its temperature of maximum density was the same. Many scientists also accepted this idea unquestioningly and either dismissed or did not know of the circulation model proposed by Lenz and others. The 4°C fallacy was endorsed by most of the voyages of the mid-nineteenth century, in particular the American, French, and British voyages of the late 1830s and early 1840s, led by Wilkes (1845), Dumont d'Urville (1842–1854), and Sir James Clark Ross (1847), respectively (Deacon 1971, ch. 13).

A comparable fallacy, also disregarding earlier work, had taken hold among the biologists. Forbes's idea of an azoic zone from 300 or 400 fathoms downwards (see Paper 37) was regarded by many as an established truth, although Aimé (1845) and others had published evidence to the contrary. While the deep ocean was beyond their reach, such a point of view did not significantly hinder research; but, as new opportunities arose, it caused much controversy.

In the 1840s there was something of a vogue for deep-sea sounding (see Part VI). But instruments had so far been used only to about 1000 fathoms, and the labor involved in retrieving them was so enormous and the results so open to doubt that some captains, such as Parry (1821), were deterred from it altogether. Such decisions, it may safely be assumed, found whole-hearted support among the crew.

With the development of steam power during the 1850s and 1860s, the undertaking of observations in the depths of the ocean became more feasible. This was partly as a result of the greater manoeuvrability of the steamship but even more because of the technology based on steam power being developed in connection with submarine telegraph cables, which culminated in the laying of the first successful trans-Atlantic cable in 1866 (Dibner 1964). The art of making deep-sea soundings with lead and line was now perfected (Paper 33) and apparatus for recording depth and for bringing up samples of the sea bed, such as Brooke's sounding machine (Maury 1855) were developed. It might have been supposed that scientists would have been eager to try to make use of these new opportunities; but physical oceanography had lost the appeal it possessed in the 1820s, since it was still widely believed that the ocean depths were all at the same temperature and presumably more or less static, in spite of the efforts of Maury (1855) and others to shake the entrenched belief in their constant temperature of 4°C.

Many biologists, too, still believed that the deep sea held nothing of interest, in spite of the growing amount of evidence to the contrary. It was, however, the biologists who, eventually, took up the challenge of this unknown region. The Norwegian zoologist Michael Sars had soon ceased to believe in the azoic zone (Sivertsen 1968). His work and that of his son G. O. Sars, the dredgings of Count Pourtalès in 1867 with the U.S. Coast Survey (Scheltema 1972), and the cruise of HMS *Lightning* in 1868 with W. B. Carpenter and Wyville Thomson, showed that the 400-fathom limit of life was imaginary. In 1869, Thomson (1873) in HMS *Porcupine* brought up living creatures from nearly 2500 fathoms (Deacon 1971, ch. 14).

It was Carpenter who succeeded in persuading the British government to agree to a large-scale expedition (Burstyn 1968; 1972). HMS *Challenger* (Linklater 1972) was captained by G. S. Nares, a surveyor, and several of his officers later became eminent in the hydrographic service. Thomson headed a team of civilian scientists containing three biologists—H. N. Moseley, R. von Willemoes-Suhm, and John Murray—and J. Y. Buchanan as chemist. The *Challenger* sailed late in 1872 and did not complete her voyage round the world until May 1876 (Deacon 1971, ch. 15). Large collections of marine creatures were made and it was established beyond question that the depths of the oceans supported life. The *Lightning* and *Porcupine* voyages finally exposed the 4°C fallacy and reopened discussion on ocean circulation. The *Challenger* was equipped to obtain serial temperature measurements and water samples. The voyage laid the foundation for the study of the abyssal sediments and answered questions that had arisen from the work of the cable surveyors (see Parts VI and VII).

Wyville Thomson's principal achievement was, perhaps, the determination with which he settled the future of the work when the expedition returned. He persuaded the government to finance the writing and publication of the report under his own direction and carried his point, against some bitter opposition, that experts should be employed on the work irrespective of nationality. He had in mind the history of previous expeditions, such as that of Wilkes whose report was still coming out when the *Challenger* sailed and was never entirely published (Haskell 1942). When Thomson died, in 1882, the work of the *Challenger* report was taken over by John Murray; it was finished in 1895.

Meanwhile other countries were organizing expeditions to do marine research (Wüst 1964; G. E. R. Deacon and Marr 1964). Some of the voyages were long circumnavigations on the *Challenger* model—among them were the USS *Enterprise*, the Italian ship *Vettor Pisani*, and the Russian *Vitiaz*. Others took the form of shorter cruises in selected areas over a number of years, for example the Norwegian ship *Vøringen* which worked in the North Atlantic between 1876 and 1878 and the French ships *Travailleur* and *Talisman* in the Atlantic in the early 1880s. Like the *Challenger,* these expeditions generally worked in several fields; the German Plankton Expedition in the *National* in 1889 was rather an exception. In this further activity Britain alone was not represented, successive governments apparently feeling that their continuing contribution to analysis of the *Challenger* material was sufficient. Paradoxically, in other countries, scientists found the example of the *Challenger* of assistance in persuading their governments to support marine science (Deacon 1971, ch. 16).

The expense of work at sea still presented a formidable barrier to the development of marine research. Governments still did not regard continuous investment in pure science as one of their duties. One person not hampered by these considerations was himself a head of state, Prince Albert of Monaco. In his yachts named *Hirondelle* and *Princesse Alice,* carrying international teams of experts, he explored the oceanography of the Mediterranean and of the Atlantic north into the Arctic from the 1880s onward. He established two centers for marine science, the Musée Océanographique at Monaco, and an institute at Paris.

There were also opportunities to obtain data and samples from marine surveyors. The work of naval hydrographers was extended to the deep ocean, where they made use of new methods and apparatus, such as the Kelvin sounding machine and the reversing thermometer, and sometimes themselves made new contributions. Samples of the sea bed from the Indian and Pacific oceans, provided by naval surveyors, were used by Murray and Renard (1891) in their *Report on Deep-Sea Deposits.* The American zoologist Alexander Agassiz (1888) made three cruises in the U.S. Coast Survey Ship *Blake* between 1877–1880. He introduced wire rope for dredging, to complement the general introduction of wire for sounding. Two of the captains of the *Blake* were Sigsbee (1880), who improved on the Kelvin sounding machine, and Pillsbury (1891) who invented a current meter and used it to study the Gulf Stream. Captain Magnaghi of the Italian ship *Washington* introduced an improved frame for reversing thermometers. Soundings by survey and cable ships as well as from oceanographic vessels were collected for inclusion in the Carte Générale Bathymétrique des Océans, compiled under Prince Albert's direction (Viglieri 1968).

Hydrography came into the category of "useful" sciences on which expenditure was thought to be justified. In the late nineteenth century, governments also began finding money for fisheries research because of concern among fishermen about their declining catches. The United States Fish Commission was set up in 1871 under the direction of Spencer F. Baird (Galtsoff 1962). Alexander Agassiz made several cruises on the Commission's ship, the *Albatross* (Hedgpeth 1945). The Scottish Fisheries Board was reconstituted with a scientific section in

1882. In Norway, government-sponsored fisheries research, which had begun in the 1850s, was directed by G. O. Sars and later by Johan Hjort (Solhaug and Saetersdal 1972). The fisheries vessel *Michael Sars* was also used by Norwegian physical oceanographers.

In Britain, funds for fisheries work helped swell the meagre budgets of marine biological laboratories at Plymouth, St. Andrews, and elsewhere in their early days. And, some years later, stations at Aberdeen and Lowestoft were established primarily for fisheries work. For those who were not biologists, there was no such assistance; and for those who, unlike Prince Albert and Alexander Agassiz, were not rich enough to finance their own expeditions the outlook was poor. John Murray, Britain's leading oceanographer before World War I, though he became wealthy enough to maintain a small research establishment, could not afford ocean-going expeditions. During the 1880s and 1890s, he led teams in studies of the sea- and fresh-water lochs of Scotland. In 1910, four years before he died, he financed a cruise in the *Michael Sars* (Murray and Hjort 1912).

Murray soon recognized the possibilities for marine science in the renewed interest in polar exploration. If oceanography was not sufficiently exciting to capture the imagination of the public, exploration of the last unknown areas of the globe had that power; and, since ships had to be used, they could perform oceanographic work as well. Among the Antarctic expeditions that made notable contributions in this field were the *Belgica* expedition of 1897–1899; the Swedish South Polar Expedition led by O. Nordenskjold in 1901–1903; the German expeditions in the *Gauss,* 1901–1903, and the *Deutschland,* 1910–1912; the Scottish National Antarctic Expedition of 1902–1904 in the *Scotia,* led by W. S. Bruce; and Jean Charcot's voyages in the *Français,* 1903–1905, and the *Pourquoi Pas?,* 1908–1910 (G. E. R. Deacon 1968). Nansen (1897) and his companions in the *Fram* made observations in the Arctic Ocean during their drift in the pack ice.

In spite of the continuing difficulties of ways and means, the last decades of the nineteenth century had seen a critical change in marine science. Instead of thinking of themselves firstly as chemists, mathematicians, etc., scientists working on the sea were beginning to call themselves oceanographers. This was due, in part, to greater opportunities for research and the expansion of science as a movement and as an occupation. But the opportunities, in physical oceanography especially, were still limited and had to be vigorously pursued. The development of common view points and a common technical language among oceanographers from different countries was also of great importance. The multiplication of scientific journals as well as improvements in transport meant that marine scientists were able to develop a new reference group, in addition to that of their original discipline, which, to a much greater degree, ignored national barriers. Joining in common projects such as preparation of the *Challenger* Report; meeting at the International Geographical Congresses, British Association, and similar gatherings; and, in the 1890s, participating in early international collaboration at sea, the nineteenth-century oceanographers established their collective identity.

It would be a mistake to assume that the development of modern oceanography was then assured by some infallible law of evolution, and that there was no longer any danger that the new science might suffer the same fate as the earlier movements in marine science. The strength of the new movement, however, lay in its very dispersal and while circumstances were unfavourable to it in some countries, it gained new life in others.

In the 1890s and early 1900s, oceanography became particularly strong in Scandinavia and Germany. In Scandinavia, Otto Pettersson, a professor at the "High School" in Stockholm, was responsible for much of the initiative in arranging collaboration between nations in marine research. As a result of his proposals (Pettersson 1894) the Scandinavian countries cooperated with Germany and the United Kingdom to make joint observations in the North Atlantic and in 1899 a conference was held that led to the formation of the International Council for the Exploration of the Sea. Pettersson's colleague Vilhelm Bjerknes, who developed a mathematical theory of ocean circulation; the explorer Nansen, who became professor of oceanography at Oslo in 1908; and Martin Knudsen, who perfected the technique of measuring salinity by titration, all held university posts and were able to gather students and assistants around them. From 1902 until 1908 the I.C.E.S. Central Laboratory was in operation in Christiania (Oslo) with Nansen as director and V. W. Ekman as his deputy (Went 1972b). Its work included improving apparatus and the manufacture of "standard sea water," a task later transferred to the Council's central bureau, under Knudsen, in Copenhagen. Ekman is particularly remembered for his work on dead water (1906) and on ocean circulation (Welander 1968). In Germany, knowledge of ocean circulation was furthered by Schott, Brennecke, Merz, Wüst, and others. The Institut für Meereskunde was established at Berlin in 1900.

In France, Britain, and the United States, physical oceanography did not do so well. In Britain, the *Titanic* disaster of 1912 led to a single expedition the following year. The U.S. Coast Guard began mounting the international ice patrol shortly afterwards. W. S. Bruce had founded the Scottish Oceanographical Laboratory but failed to put it on a secure footing before the outbreak of World War I, and it closed in 1920. In the same year, however, the first chair of oceanography in Britain was founded at Liverpool by W. A. Herdman, a former student of Wyville Thomson. It was held for many years by Proudman

(1968) who persuaded the University to establish the Tidal Observatory, under A. T. Doodson, at Birkenhead. Five years later, in 1925, the *Discovery* Committee, a government-financed body, began sending oceanographic expeditions to the Antarctic to investigate the whale population, anticipating its severe decline due to reckless killing.

In the United States, oceanography fared better in the 1920s and 1930s. The Scripps Institution of Oceanography was founded in 1925 by the University of California as a successor to an existing marine biological station (Vaughan 1937). The Bingham Oceanographic Laboratory at Yale and the Woods Hole Oceanographic Institution (Lillie 1944) were both founded in 1930. Other universities in the coastal states were setting up research laboratories, though some of these were confined to biological work.

It was World War II that finally involved governments in supporting oceanography as a whole, thus making possible the expansion that has since taken place. Both in the United States (Schlee 1973) and in the United Kingdom scientists were drawn into the war-time effort to devise such things as submarine and mine detectors and wave forecasts for military landings. Electronics had already come to the assistance of oceanography with the development, in the 1920s, of the echo sounder; but it was during the war that its full potential began to be realized and systematic work began on the subsidiary branches of physical oceanography, such as acoustics, optics, and electrical and magnetic conditions.

War-time needs and postwar expenditure on science benefited oceanography and reestablished the balance of the field, which had been heavily tilted on the biological side. Increased funds were provided for existing institutes and for the setting up of new laboratories, such as the National Institute of Oceanography in Britain. There was a corresponding increase in the size and number of research vessels and the number of scientists, so that oceanography entered a period of unprecedented prosperity and activity.

In recent history, as in earlier times, marine scientists have made good use of developments in other sciences. Sound waves have been employed to map the layers of sediment below the sea bed, to give sideways as well as vertical profiles of the sea floor, and to trace schools of fish and also to transmit instructions to or record information from instruments deep in the ocean. Radio waves are used to receive signals from satellites that enable ships' positions to be determined with great accuracy. Perhaps the most important development has been the introduction of computers. In oceanography, they have helped to improve navigation and made possible continuous recording of variables such as salinity and temperature and analysis of the enormous amounts of data now available.

Oceanography is still a very expensive science but not, like space research, limited to the superpowers. There can be much useful activity at various levels of sophistication. Collaboration has remained a strong tradition and numerous international projects have been undertaken during the last ten years. Even the largest nations find it convenient to pool their resources so that a more thorough investigation of particular topics can be made.

The picture of the ocean that has emerged from this work is very different from that held by the average early Victorian scientist. To begin with, theories of sea-floor spreading have shown that, instead of being permanent, the ocean basins are continually being modified. The deep water of the ocean is not uniform, nor motionless, as it was once thought to be; recent work has shown that in addition to the long-term movements of the great water masses, there are smaller and much more variable movements going on all the time. Far from being lifeless or inhabited only by a few simple organisms, the depths of the ocean support a very varied range of creatures, highly adapted to their existence. What was thought, little over a 100 years ago, to be a dead world, has turned out to be alive and continually changing.

Part I

SOME SEVENTEENTH CENTURY INSTRUCTIONS FOR MAKING SCIENTIFIC OBSERVATIONS AT SEA

Editor's Comments on Papers 1 and 2

1 *Directions for Sea-men, Bound for Far Voyages*

2 *An Appendix to the Directions for Seamen, Bound for Far Voyages*

The great snag in carrying out a program for studying the sea, such as that proposed by Boyle (1666), was the small likelihood of the scientists' being able to go to sea. To overcome this difficulty the Royal Society of London ordered Lawrence Rooke, 1622–1662, to draw up the *Directions for Seamen* (published in a slightly amended form in 1666) which its members hoped would encourage sailors to make the observations that scientists could not make themselves. As well as information about the sea, the *Directions,* Paper 1, gave instructions for observations in the fields of astronomy, geography, magnetism, and meteorology.

Two early presidents of the Royal Society, Sir Robert Moray and Lord Brouncker, led the way in testing apparatus for use at sea, including the lineless sounder, a water sampler, and a thermometer (Birch 1756, vol. 1). In 1663, following some failures with their apparatus, Robert Hooke produced improved versions of the lineless sounder and the water sampler. The designs were published in *An Appendix to the Directions for Seamen,* Paper 2, and in the extended "Directions for Observations and Experiments to be made by Masters of Ships . . ." (1667).

1

Reprinted from *Royal Soc. London Philos. Trans.* 1:140–143 (Jan. 1666)

DIRECTIONS FOR SEA-MEN, BOUND FOR FAR VOYAGES

It being the Design of the *R. Society*, for the better attaining the End of their Institution, to study *Nature* rather than *Books*, and from the Observations, made of the *Phænomena* and Effects she presents, to compose such a History of Her, as may hereafter serve to build a Solid and Useful Philosophy upon; They have from time to time given order to several of their Members to draw up both *Inquiries* of things Observable in forrain Countries, and *Directions* for the Particulars, they desire chiefly to be informed about. And considering with themselves, how much they may increase their *Philosophical* stock by the advantage, which *England* injoyes of making Voyages into all parts of the World, they formerly appointed that Eminent Mathematician and Philosopher Master *Rooke*, one of their Fellowes, and *Geometry* Professor of *Gresham Colledge* (now deceased to the great detriment of the Common-wealth of Learning) to think upon and set down some *Directions* for *Sea-men* going into the *East* & *West-Indies*, the better to capacitate them for making such observations abroad, as may be pertinent and suitable for their purpose; of which the said Sea-men should be desired to keep an exact *Diary*, delivering at their return a fair Copy thereof to the *Lord High Admiral* of *England*, his Royal Highness the *Duke* of *York*, and another to *Trinity-house* to be perused by the *R. Society*. Which *Catalogue* of *Directions* having been drawn up accordingly by the said Mr. *Rook*, and by him presented to those, who appointed him to expedite such an one, it was thought not to be unseasonable at this time to make it publique, the more conveniently to furnish Navigators with Copies thereof. They are such, as follow;

1. To observe the Declination of the *Compass*, or its Variation from the *Meridian* of the place, frequently; marking withal, the *Latitude* and *Longitude* of the place, wherever such Observation is made, as exactly as may be, and setting down the *Method*, by which they made them.

2. To carry *Dipping Needles* with them, and observe the Inclination of the Needle in like manner.

3. To remark carefully the Ebbings and Flowings of the Sea, in as many places as they can, together with all the Ac-

dents, Ordinary and Extraordinary, of the Tides; as, their precise time of Ebbing and Flowing in Rivers, at *Promontories* or *Capes*; which way their Current runs, what Perpendicular distance there is between the highest Tide and lowest Ebb, during the Spring Tides and Neap Tides; what day of the *Moons* age, and what times of the year, the highest and lowest Tides fall out: And all other considerable Accidents, they can observe in the Tides, cheifly neer Ports, and about Ilands, as in St. *Helena*'s Iland, and the three Rivers there, at the *Bermodas* &c.

4. To make Plotts and Draughts of prospect of Coasts, Promontories, Islands and Ports, marking the Bearings and Distances, as neer as they can.

5. To sound and marke the Depths of Coasts and Ports, and such other places nere the shoar, as they shall think fit.

6. To take notice of the Nature of the Ground at the bottom of the Sea, in all Soundings, whether it be Clay, Sand, Rock, *&c.*

7. To keep a Register of all changes of Wind and Weather at all houres, by night and by day, shewing the point the Wind blows from, whether strong or weak: The Rains, Hail, Snow and the like, the precise times of their beginnings and continuance, especially *Hurricans* and *Spouts*; but above all to take exact care to observe the *Trade-Wines*, about what degrees of *Latitude* and *Longitude* they first begin, *where* and *when* they cease, or change, or grow stronger or weaker, and how much; as near and exact as may be.

8. To observe and record all Extraordinary *Meteors*, Lightnings, Thunders, *Ignes fatui*, Comets, *&c.* marking still the places and times of their appearing, continuance, &c.

9. To carry with them good Scales, and Glasse-Violls of a pint or so, with very narrow mouths, which are to be fill'd with Sea-water in different degrees of *Latitude*, as often as

they pleafe, and the weight of the Vial full of water taken ex-actly at every time, and recorded, marking withall the degree of *Latitude*, and the day of the Month : And that as well of water near the Top ; as at a greater Depth.

2

Reprinted from *Royal Soc. London Philos. Trans.* 1:147–149 (Feb. 1666)

PHILOSOPHICAL *TRANSACTIONS.*

Munday, Feb. 12. 166 5/6.

The Contents.

An Appendix *to the* Directions *for Seamen, bound for far Voyages.*

Hereas it may be of good use, both *Naval* and *Philosophical*, to know, both how to sound depths of the sea *without a Line*, and to fetch up water from any depth of the same; the following waies have been contrived by Mr. *Hook* to perform both; (which should have been added to the lately printed *Directions for Seamen*, if then it could have been conveniently done.)

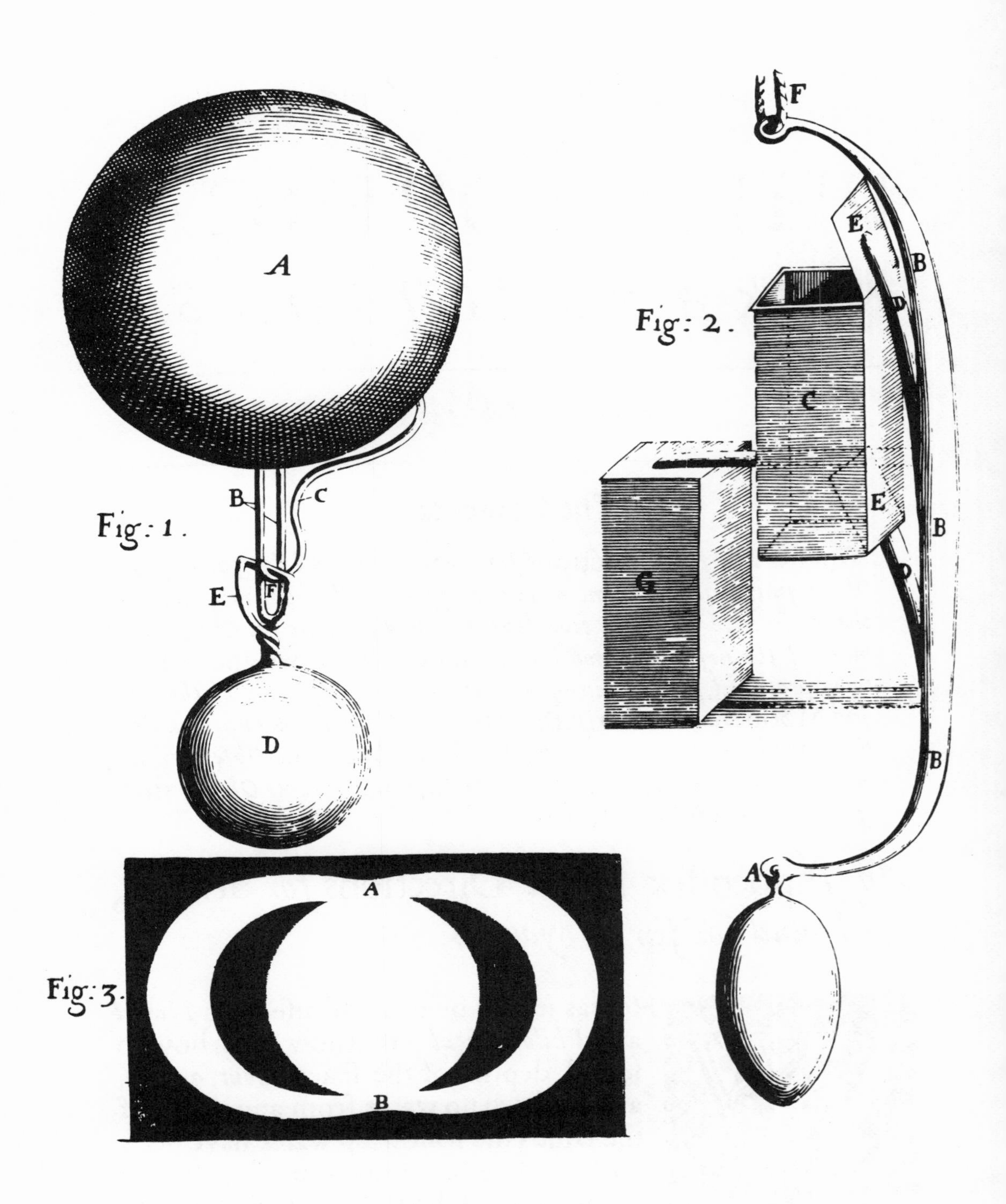
A
B
C
Fig: 1.
E
F
D
F
E
B
Fig: 2.
D
C
E
B
G
D
B
A
Fig: 3.
A
B

Firſt, for the ſounding of depths without a Cord, conſider ***Figure*** 1, and accordingly take a Globe of *Firr*, or *Maple*, or other light Wood, as A: let it be well ſecured by Verniſh, Pitch, or otherwiſe, from imbibing water; then take a piece of Lead or Stone, D, conſiderably heavier then will ſink the Globe: let there be a long Wire-ſtaple B, in the Ball A, and a ſpringing Wire C, with a bended end F, and into the ſaid ſtaple, preſs in with your fingers the ſpringing Wire on the bended end: and on it hang the weight D, by its ring E, and ſo let Globe and all ſink gently into the water, in the poſture repreſented in the firſt *Figure*, to the bottom, where the weight D touching firſt, is thereby ſtopt; but the Ball, being by the *Impetus*, it acquired in deſcending, carried downwards a little after the weight is ſtopt, ſuffers the ſpringing wire to fly back, and thereby ſets it ſelf at liberty to reaſcend. And, by obſerving the time of the Ball's ſtay under water (which may be done by a Watch, having minuts and ſeconds, or by a good Minut-glaſs, or beſt of all, by a Pendulum vibrating ſeconds) you will by this way, with the help of ſome *Tables*, come to know any depth of the ſea.

Note, that care muſt be had of proportioning the weight and ſhape of the Lead, to the bulk, weight, and figure of the Globe, after ſuch a manner, as upon experience ſhall be found moſt convenient.

In ſome of the Tryals already made with this Inſtrument, the Globe being of Maple-wood, well covered with Pitch to hinder ſoaking in, was $5\frac{11}{16}$ inches in diameter, and weighed $2\frac{1}{2}$ pounds: the Lead of $4\frac{1}{2}$ pounds weight, was of a *Conical* figure, 11. inches long, with the ſharper end downwards, $1\frac{9}{16}$ inches at the top, and $\frac{1}{16}$ at the the bottom in diameter. And in thoſe Experiments, made in the *Thames*, in the depth of 19. foot water, there paſſed between the Immerſion and Emerſion of the Globe, 6. ſeconds of an hour; and in the depth of 10. foot water, there paſſed $3\frac{1}{2}$ ſeconds or thereabout: From many o which kind of Experiments it will likely not be hard to find out a method to calculate, what depth is to be concluded from any other time of the like Globes ſtay under water.

In the ſame Tryals made with this Inſtrument in the ſaid River of *Thames*, it has been found, that there is no difference in time, between the ſubmerſions of the Ball at the greateſt depth, when it roſe two Wherries length from the place where it was let fall (being carried by the Current of the *Tide*) and when it roſe within a yard or ſo of the ſame place where it was let down.

The *other* Inſtrument, for Fetching up water from the depth of the ſea, is (as appears by *Figure* 2.) a ſquare woodden *Bucket* C, whoſe bottoms *EE*, are ſo contrived, that as the weight A, ſinks the Iron B, (to which the Bucket C, is faſtned by two handles DD, on the ends of which are the moveable bottoms or Valves EE,) and thereby draws down the Bucket; the reſiſtance of the water keeps up the Bucket in the poſture C; whereby the water hath, all the while it is deſcending, a clear paſſage through; whereas, as ſoon as the Bucket is pulled upwards by the Line F, the reſiſtance of the water to that motion beats the Bucket downward, and keeps it in the poſture G, whereby the Included water is preſerved from getting out, and the Ambient water kept from getting in.

By the advantage of which Veſſel; it may be known, whether ſea water be Salter at and towards the bottom, then at or near the top: Likewiſe, whether in ſome places of the ſea, any ſweet water is to be found at the bottom; the *Affirmative* whereof is to be met with in the *Eaſt Indian* Voyages of the induſtrious *John Hugh Van Linschoten*, who page 16 of that Book, as 'tis *Engliſhed*, records, that in the *Perſian Gulph*, about the Iſland *Barem*, or *Babarem*, they fetch up with certain Veſſels (which he deſcribes not) water out of the ſea, from under the ſalt-water, four or five fathom deep, as ſweet, as any Fountain water.

Part II

CIRCULATION OF THE OCEAN

Editor's Comments on Papers 3 Through 12

3 MARSIGLI
Excerpt from *Internal Observations on the Thracian Bosphorus, or True Channel of Constantinople, Represented in Letters to Her Majesty, Queen Christina of Sweden*

4 RENNELL
On the Levels of the Caribbean and Mexican Seas, and the Origin of the Florida or Gulf-Stream

5 LENZ
On the Temperature and Saltness of the Waters of the Ocean at Different Depths

6 *The General Oceanic Circulation*

7 *A New Thermometer*

8 WÜST
The Origin of the Deep Waters of the Atlantic

9 DEACON
The Northern Boundaries of Antarctic and Subantarctic Waters at the Surface of the World Ocean

10 SWALLOW
A Neutral-Buoyancy Float for Measuring Deep Currents

11 CREASE
Velocity Measurements in the Deep Water of the Western North Atlantic

12 NAMIAS
Large-Scale Air-Sea Interactions over the North Pacific from Summer 1962 through the Subsequent Winter

Scientific studies of currents at sea can be traced back to the late seventeenth century. In the 1660s and 1670s, British engineers at Tangier were attempting to explain the water movements in the Strait of Gibraltar (Deacon 1968; 1971, ch. 7). In 1661, the Royal Society had asked the Earl of Sandwich, admiral of the fleet sailing for the Mediterranean, to undertake a series of oceanographic observations, including a search for the undercurrent in the Strait (Birch 1756, vol. 1, pp. 29–30; reprinted in Deacon 1971, pp. 407–408). How the idea of such an undercurrent originated is uncertain. Seamen may have been responsible. Sir Henry Sheeres (1703), one of those who made a study of the problem, wrote that they were "very positive in their Assertions affirming, that there is a Disemboguing, and that it is performed by a counter Current." As proof, he continued, they cited "a vulgar experiment they make by letting fall an anchor, and veering but two or three cables." This had the effect of slowing the ship's motion so that it stemmed the surface current instead of being carried along by it. Clearly, this phenomenon should rather be seen as one of increased resistance (as the weighted rope entered layers of water that were not moving as fast as the surface current); the westward movement now known to exist in the depths of the Strait, would have been much deeper. However, in the Sound, at the entrance to the Baltic, it was reported about the same time that a boat was towed against the surface current, which was only 4 or 5 fathoms deep, by a bucket weighted with a cannon ball and lowered into the undercurrent (Smith 1684).

The problem that focussed attention on the Strait of Gibraltar was the difficulty of explaining the equilibrium of the Mediterranean Sea, which had no other visible outlet. Instead of running out into the Atlantic, as might be expected, the current in the Strait ran eastward, into the Mediterranean. Various explanations were put forward to explain this during the seventeenth century. Sheeres, for example, concluded some years before Edmond Halley (1687) published his paper on the same theme that evaporation by the sun used up so much of the water of the Mediterranean that the quantity received from rainfall and from rivers was insufficient to maintain it at the general level of the ocean.

Richard Bolland (1704), who was working under Sheeres on the harbor at Tangier, charted the surface current and showed how it occupied the center of the Strait, whereas there were areas subject to tidal influence on either side. He believed in the existence of the undercurrent, though he was unable to suggest how it worked, and devised instruments to observe it, though it seems unlikely that he was able to use them before his death in 1678.

In the following year, a young Italian military engineer, Count Luigi Ferdinando Marsigli, 1658–1730, made a comprehensive survey of the Bosporus, which was remarkable no less for its time than for the youth of the author (Olson and Olson 1958). Marsigli was a pupil of Geminiano Montanari, who himself wrote on the water movements of the Adriatic, and, through the writings of Robert Boyle, he knew of work done in Britain. Part of his account published in 1681, *Osservazioni intorno al Bosforo Tracio,* is given as Paper 3 in this section. Marsigli addressed his work to ex-Queen Christina of Sweden in the form of a letter.

Marsigli began by explaining the surface movements of the water. He had measured the principal surface current (the Maestra) with a current meter. He observed the tides and winds, and the barometric pressure in the area, and determined the specific gravity of the waters from different parts of the Strait and from adjacent seas with the aid of a hydrostatic balance. He applied several important principles to his observations, including what is now known as the continuity principle—that the speed of the current increases where the channel narrows. Last but not least, he observed the difference in density between the Black Sea, which is relatively fresh because of the number of rivers emptying into it, and the more saline water of the Mediterranean, and showed that in the Bosporus a layer of incoming Mediterranean water lay below the fresher water of the outflowing surface current. The situation is therefore the reverse of that obtaining in the Strait of Gibraltar.

The existence of an undercurrent in the Bosporus was affirmed by local fishermen. Marsigli made tests of the undercurrent itself and demonstrated experimentally how the presence of adjacent bodies of water of differing salinity would set up and maintain a circulation of this kind. In the concluding part of the work, not given here, he tries to explain the salinity of the sea and says something about the marine life of the area.

Ocean currents had been reported by explorers and described by writers such as William Bourne (1578); and, in the seventeenth century, the first attempts at world maps of ocean currents appear. At first the distinction between tides and currents was not clearly understood (Burstyn 1966). But as the theory of tides developed and explanations were given by Halley for the patterns of the winds and by Hadley for the effect of the earth's rotation on them, it was assumed that as the major currents followed roughly the same pattern, they were dependent on winds for their motive power.

One of the principal exponents of the wind-driven current system was Major James Rennell, 1742–1830. He had been a surveyor in the Royal Navy and then with the East India Company before returning to England in 1778. His *Investigation of the Currents of the Atlantic,* from which Paper 4 is taken, was based on half a century of research into data from logbooks and from observations made at sea by travelers and seamen. Rennell (Paper 4) believed that winds blowing over the sea surface created drift currents that might be converted into stream currents running independently of the wind. This occurred where a current met the land and its waters piled up so that a new current, in a different direction, would be set in motion by the difference in level. Rennell believed that this was how the Gulf Stream originated.

Matthew Fontaine Maury undertook similar work on a larger scale during the 1840s and 1850s. As Superintendent of the Naval Observatory in Washington he had access to the logbooks of American ships and, with the data he collected, compiled wind and current charts which enabled sailing ships to make faster ocean passages (Williams 1963). Maury organized an international conference at Brussels in 1853 at which participating nations agreed that their naval and merchant ships should keep meteorological registers. Maury (1855) was also one of the first to write a book of general interest about marine science. This book, *The Physical Geography of the Sea,* is one of the classics of oceanography and ran to many editions.

Like many nineteenth century scientists interested in the sea, from Humboldt and his contemporaries onward (see Paper 26 and the editor's comments in Part V), Maury believed in the existence of oceanic circulation due to density differences. This was suggested by information on the distribution of temperature and salinity in the sea collected in the late eighteenth and early nineteenth centuries and by growing understanding of the physical properties of sea water (Deacon 1971, chs. 9–11.)

Temperature measurements of the subsurface layers of the ocean had already been attempted in the eighteenth century, by hauling up water from the depths of the ocean and measuring its temperature at the surface (Hales 1751; Ellis 1751). Later methods included lowering a well-insulated thermometer and leaving it for sufficient time to adjust to the temperature of the water (Saussure 1779–1796 vol. 3, pp. 195–201) and using a self-registering thermometer (Cavendish 1757). The first such thermometer to prove effective at sea was the self-registering maximum and minimum thermometer specially designed by James Six (1794). This was apparently first used on the voyage of circumnavigation of the globe made by the Russian ships, *Nadeshda* and *Neva,* 1803–1806 (Krusenstern 1813), at a time when the ships of the other European countries were engaged in the Napoleonic wars. Observations showed that in middle and low latitudes the temperature of the sea decreased with depth, to a point not far above freezing and that, in the tropics, it fell below the lowest atmospheric temperature experienced during the year, in the deeper layers.

The polar origin of the deep water of the ocean was most ably argued, in the first half of the nineteenth century, by Emil von Lenz,

1804–1865, physicist on the second Russian expedition under Otto von Kotzebue in 1823–1826. Lenz was one of the few people before the late 1860s to realize the limitations of the Six thermometer. In collaboration with G. F. Parrot (1833), he made observations showing that the effect of pressure at great depths overcame the protective glass recommended by Six and made the reading appear higher than it should be. At sea, Lenz used a version of Hales's apparatus for bringing up water and applied a correction for the warming that would take place as it came to the surface.

Lenz published two papers on the results of his work (1831, 1847). A contemporary summary of the first is given here (Paper 5). Points to notice include his reference to salinity maxima north and south of the equator, with a minimum slightly south of the equator between them; the discovery of the saline water mass in the North Atlantic; and the sudden change at about 1000 fathoms to colder, less saline water which Lenz ascribed to a polar current. He developed these ideas in his second paper and suggested that the polar flow existed in both hemispheres and that low temperatures and salinities detected near the equator were due to polar water rising toward the surface.

Ideas such as these, however, failed to be generally accepted at the time. It was not until much later that there was an effective challenge of the idea that the depths of the ocean were occupied by water at 39°F or 4°C (see the editor's comments for Parts I and V). The voyages of the *Lightning* and *Porcupine* played an important part in this new approach. W. B. Carpenter and Wyville Thomson had previously shared the common assumptions that the depths of the ocean were at a constant temperature of 4°C and that the movements of the sea, apart from tides, were due to winds. In this respect their expeditions caused more rethinking of old ideas that they had anticipated. In a revealing letter to Sir John Herschel (1869) Carpenter wrote:

> The *Hydra* temperature soundings which came to the Admiralty just before we started, were "pooh-pooed" both at the Admiralty and by the President of the Royal Society; and it was only the chance of my meeting Dr. Frankland, and learning from him the fact of the continuous contraction of *sea*-water down to its freezing point that prepared me for what we actually met with.

The *Hydra* was a survey ship that had been sounding a telegraph-cable route in the Indian Ocean. Edward Frankland was a leading British chemist.

During the four years before the *Challenger* expedition sailed, a committee of scientists cooperated with instrument makers to produce thermometers better protected against pressure and to establish corrections to be applied to their results. They also investigated other ways of measuring temperature in the sea and an electrical-resistance thermometer by C. W. Siemens was tried out (Deacon 1971, ch. 14). Then, while the *Challenger* was still away, the firm of Negretti and Zambra (1874) announced a reversing thermometer. The idea of using a messenger to turn the thermometer upside down and break off the thread of mercury at the desired depth had been employed by Aimé (1845) but his work had not been taken up. The Negretti and Zambra thermometer is a more direct ancestor of the thermometers used by oceanographers today. The extract given here (Paper 7) is from a summary of their paper, printed in *Nature.*

The data obtained by the *Lightning* and *Porcupine* led Carpenter to take a keen interest in the distribution of temperature and he concluded that density differences in the ocean must cause cold water to move from the polar regions toward the equator at depth while a compensatory poleward movement of warm water took place at the surface. His ideas created opposition among those who believed that winds alone were responsible for ocean currents (Paper 6; Deacon 1971, chs. 14 and 15). Carpenter therefore made two expeditions to the Strait of Gibraltar and was the first to publish facts and figures about the undercurrent of dense Mediterranean water that flows out into the Atlantic (Carpenter 1872; Carpenter and Jeffreys 1870; Nares 1872). However, finding his critics unconvinced by this demonstration, he pursued his idea of one great expedition to investigate all the oceans of the world and bring back data to cut the ground from under his opponents' feet. This was the real origin of the *Challenger* expedition.

In one way the *Challenger* achieved what Carpenter had hoped it would. It showed (Thomson 1877) how water only a few degrees above freezing point, spread northward from Antarctic regions in the depths of the Atlantic and how its influence in certain areas is modified by the presence of submarine ridges which interrupt the deep flow. Similarly, it was found in the East Indies that marginal seas, which, like the Mediterranean, are cut off from the open ocean below a certain level, remain at the same temperature from that depth to the bottom, because the deep polar flow cannot surmount the barrier (Tizard 1875).

However, the *Challenger* temperature and current observations were not among the most successful part of the expedition's work and matters were not improved by the slight attention that was paid to them in the Report (Buchan 1895). Carpenter's original enthusiasm had done good by ventilating the subject and stimulating further work, both new and retrospective, such as Prestwich's monumental paper on observations made during the previous 125 years (1875). Unfortunately he was not successful in getting British physicists and mathematicians to work on it, although they were actively studying waves and tides. It was in Scandinavia and Germany that the study of ocean circulation emerged as a scientific discipline during the 1890s and early 1900s.

In Sweden the new movement was headed by Otto Pettersson, who suggested that ocean circulation was generated by melting ice (1904, 1907), and his colleague Vilhelm Bjerknes, who developed a mathematical theory of ocean circulation. This was applied to the ocean by J. W. Sandstrom and B. Helland-Hansen (Welander 1968). Their work was carried out in a series of voyages in the Norwegian fisheries vessel *Michael Sars* and later in the *Armauer Hansen* (Schlee 1973). Fridtjof Nansen was also closely involved in this work, though originally a zoologist. (In 1901 he became director of the Central Laboratory of I.C.E.S. at Christiania, where he was professor of zoology. In 1908 he returned to be professor of oceanography, at the renamed University of Oslo.) V. W. Ekman (1905) described "Ekman spirals"—the increasing deflection of movements in successively deeper layers of fluids caused by friction and the effect of the rotation of the earth. He too worked extensively on the theory of ocean circulation.

The particular contribution of German oceanographers was the mapping of the circulation of the Atlantic. The work was done by researchers on successive voyages, in which the Antarctic expeditions were of great importance, among them Gerhard Schott in the *Valdivia,* W. Brennecke in the *Planet* and *Deutschland,* and A. Merz and Professor Georg Wüst of the *Meteor* expedition (Wüst 1968). Brennecke showed that saline water sinks in the North Atlantic and flows southward above the Antarctic bottom water. Meinardus of the *Gauss* expedition showed how low salinity, subantarctic, intermediate, water flows northward above the North Atlantic deep water after sinking from the surface at the Antarctic Convergence. The line of the convergence was traced round the southern ocean by the ships of the Discovery Committee, the *Discovery,* the *Discovery* II, and the *William Scoresby,* in the 1920s and 1930s (G. E. R. Deacon 1937). Paper 9 is a report of this work.

So deeply engrained in the imagination of oceanographers was the model after Lenz's work with deep water rising to the surface at the equator, that it was still represented in Brennecke's diagram of water movements in the Atlantic. Merz and Wüst constructed meridional sections, based on the results of expeditions from the *Challenger* onward, which finally established that the Atlantic ocean had to be regarded as a single entity and not as two more or less independently circulating cells. The model proposed by Merz on the basis of the evidence existing then was verified by the work of the *Meteor* expedition described in the paper by Wüst, Paper 8. Nansen had established in 1912 the Arctic origin of deep water in the North Atlantic. The formation of the dense Antarctic bottom water has been explained more recently by Foster and others, as shown by Gill (1973), and found to take place mainly in the Weddell Sea, from which it spreads northward beyond the equator in all the oceans.

Since World War II new ways of making measurements have revolutionized and stimulated work on the theory of ocean currents. The development of the bathythermograph and other instruments for continuously recording the properties of the water and the use of computers have made much more information available about the distribution of water masses. Current meters of the Ekman type, which had to be lowered from a ship, have been advantageously replaced by floats and moored current meters, which can provide information for periods ranging from days to several months.

Neutrally buoyant floats equipped with signaling devices were invented and first used by John Swallow, of the National Institute of Oceanography, in 1955 (Paper 10). They enable oceanographers to track actual movements of deep layers in the ocean while moored current meters record changes in direction and speed of water movements at a single point. Whereas scientists used to think that movement in the depths of the ocean was slow and uniform in nature, measurements with these new instruments show that it is in fact highly variable—as argued in Paper 11 by Crease—both in speed and direction and that smaller movements or eddies within the larger water masses are continuously taking place. They appear to be analogous to movements in the atmosphere, where turbulent air currents are created around high- and low-pressure areas. The cooperative MODE (Mid Ocean Dynamic Experiment) studies in the Atlantic in 1973 were designed to throw new light on the working of these eddies.

Topics that have attracted special attention during the past thirty years include the behavior of currents on the eastern and western boundaries of the ocean and the equatorial currents. Oceanographers have studied the intensification of currents on the western sides of the great oceans. The phenomenon has been observed not only in surface currents such as the Gulf Stream (Stommel 1958) but in deep-water movements as well. On the eastern sides of the oceans, and in particular regions of offshore winds in places such as the Arabian peninsula, the movement of surface layers of water away from the land by trade winds and monsoons leads to the upwelling of deep cold water which, more rich in nutrients, supports plentiful marine life.

One of the most surprising discoveries was the finding of the equatorial undercurrents. Nineteenth-century scientists who accepted the idea of a thermohaline circulation expected an upward movement in the subsurface layers at the equator and attributed to it the area of low salinity found there at the surface. While working with the cable ship *Buccaneer* in 1885–1886, J. Y. Buchanan (1886; reprinted (1919) discovered the equatorial undercurrent in the Atlantic. He wrote:

> In connection with the absence of easterly currents off the coast may be taken the very remarkable under-current which is found setting in a

> south-easterly direction with a velocity of over a mile per hour at three stations almost on the equator, and to the northward of the island of Ascension. For the double purpose of examining the currents and of obtaining a large specimen of the bottom, the "Buccaneer" was anchored in 1800 fathoms of water by means of an ordinary light anchor fitted with a canvas bag to receive the mud which would otherwise fall off the flukes on its being weighed. While the ship was lying thus at anchor, the surface water was found to have a very slight westerly set. At a depth of 15 fathoms there was a difference, and at 30 fathoms the water was running so strongly to the south-east that it was impossible to make observations of temperature, as the lines, heavily loaded, drifted straight out, and could not be sunk by any weight the strain of which they could bear. (Buchanan 1919, pp. 99–100)

The theory of the time was not adequate to incorporate this finding and the incident was forgotten for seventy years, until the discovery of the Cromwell current in the Pacific by a group from the Scripps Institution of Oceanography at La Jolla (Cromwell, Montgomery, and Stroup 1954). A similar current was later observed in the Indian Ocean but there it has been found to be seasonal and to flow only during the north-east monsoon.

The understanding of ocean currents and of the way in which water masses are distributed has proved to be of great importance in other fields, for example in understanding the distribution of marine life and in learning more about the weather and its changes. In Paper 12, Namias shows how, in the summer preceeding the severe winter of 1962–1963, climatic anomalies over the North Pacific allowed warmer water than usual to accumulate in the central and eastern parts of the ocean. He suggests that the heat given off by this water during the ensuing autumn and winter may have provided the energy needed to power the cyclones which brought the Arctic air abnormally far south over both the United States and Europe during the ensuing winter.

3

OSSERVAZIONI
INTORNO
AL
BOSFORO TRACIO
OVERO
CANALE DI CONSTANTINOPOLI
Rappresentate in Lettera
ALLA SACRA REAL MAESTA
DI
CRISTINA
REGINA DI SVEZIA
DA
LVIGI FERDINANDO
MARSILII.

In Roma, Per Nicolò Angelo Tinassi 1681.

Con Licenza de' Superiori.

INTERNAL OBSERVATIONS ON THE THRACIAN BOSPHORUS, OR TRUE CHANNEL OF CONSTANTINOPLE, REPRESENTED IN LETTERS TO HER MAJESTY, QUEEN CHRISTINA OF SWEDEN

Count Luigi Ferdinando Marsigli

This extract was translated by Mrs. E. Hudson, from Osservazioni intorno al Bosforo Tracio, *originally published in Rome in 1681; reprinted in* Boll. Pesca Piscic. Idrobiol. 11:*734–758 (1935)*

A denotes the Maestra which starts from the Black Sea, and which mostly vanishes in the Sea of Marmora. B is another current which begins in the fresh waters in the place called Chiattana and finishes at Arnagiui, in a path, as you see, contrary to the Maestra. C is another current, but quite small, which goes the opposite way to the Maestra all along the Scutari shore, which is made in the shape of a crater. D is the last current, also contrary in motion to the Maestra, which it overlaps at the Tower of Leandros situated above a little reef, having its imperceptible beginnings in Chalcedon.

Before considering the reasons for these currents, I feel it is necessary for me to indicate to Your Majesty the examination I made to determine the speed of the main Maestra Current, in itself and also the variations occurring in different locations. So, the water comes in from the Black Sea, with a not very violent motion, which then receives the first change perhaps from the shock it receives in the Gulf of Sciarria in Europe, which deflects it to Asia at Gibugli Bacgesi, from which it continues into the strait of the Castles, where it passes at maximum strength, more than in all the rest of the strait, augmented by multiplied shocks, and by the narrowness of the location which does not allow any vessel to stop, being, if not greater, at least equal to any swift River. For the rest of the Canal, it reduces that violence, in some part, and goes on to aim at the point of the Seraglio, or ancient Byzantium, almost at a right angle. In the afore-mentioned place of the Castles, I believe, was the Temple of Mercury of which Polybius writes, which was situated above a rock, which in the shape of a headland came out into the sea from the European side, in the narrowest part of the Canal, which is the same place, he also asserts, where the greatest impetus in its currents begins. As I wanted to measure the velocity at the Tower of Leandros, I had made a wooden machine, with a six spoked wheel, each measuring one palm and four inches and two quarters of Roman measure; this wheel I put in a wooden frame of 7 palms and on the top I put a scale of a palm and two quarters, to show the turns of the wheel of the lower part moved, by the course of the water striking at a right angle, staying at the same inclination to the horizontal, and therefore during a hundred vibrations of a Pendulum eight inches and six-eighths long, the scale made 38 turns,

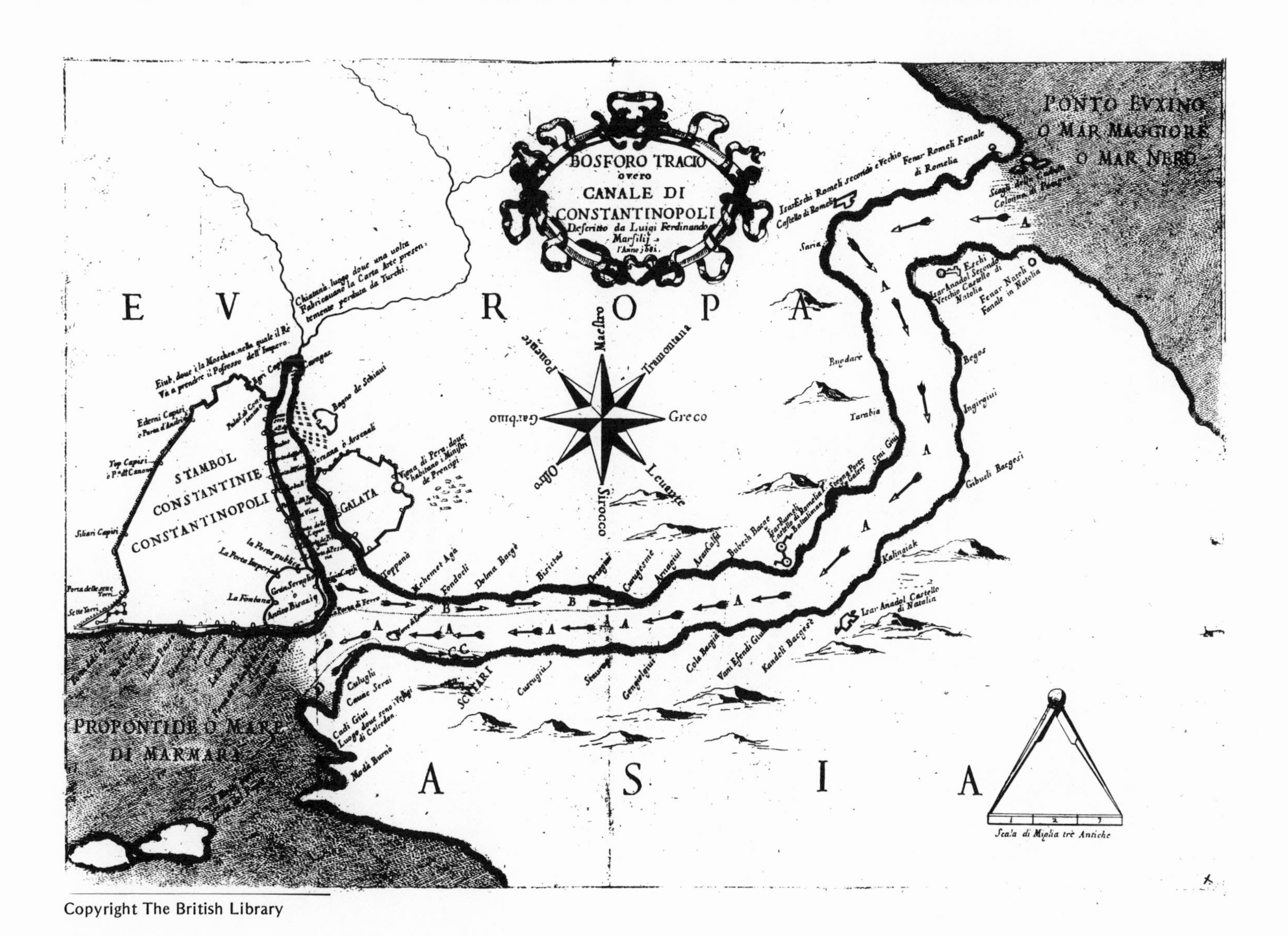
PONTO EUXINO
O MAR MAGGIORE
O MAR NERO
BOSFORO TRACIO
overo
CANALE DI
CONSTANTINOPOLI
Descritto da Luigi Ferdinando
Marsilij
EUROPA
ASIA
Maestro
Tramontana
Greco
Levante
Sirocco
Ostro
Garbino
Ponente
STAMBOL
CONSTANTINIE
CONSTANTINOPOLI
GALATA
SCUTARI
Bagno de Schiaui
Fenar Romeli Fanale
di Romelia
Fenar Natoli
Fanale in Natolia
Isar Anadol Castello
di Natalia
Kalingiak
Bugdare
Tarabia
Begos
Ingirgiui
Gibuli Bacgesi
Kandeli Bacgesi
Cola Bacgie
Bisictas
Dolma Bacgè
Fondocli
Mehemet Aga
Toppanà
Porta di Ferro
La Fontana
Cadi Giui
Moda Burnò
PROPONTIDE O MARE
DI MARMARA
Scala di Miglia trè Antiche

with the loss of one wing, which broke after the first four turns. This was a mischance that greatly reduced the velocity, which I could have observed better without that incident; at that time a light Northerly wind was blowing which was just enough to ripple the water but not enough to alter it from its characteristic motion. That experiment could easily give encouragement to anyone to compare the speed of any river to that of this place, which observed in the afore-mentioned way, I determined that the current at the Castles, which is the greatest in the channel, may be almost a third more than that which I observed at the Tower of Leandros.

Knowing the effects which are produced in this channel, I believe that it would not be difficult for me to investigate the reasons with some solid principles. Therefore considering the Maestra, as it empties out of the Black Sea into the White, I say that it goes with the described velocity by reason of the narrowness of the Canal where it passes. It being known that the nature of fluids is that the velocity builds up in proportion to the diminution of the cross-section through which it must pass, and also by reason of the slope of the level, which here I do not know to be necessary, for what reason I will tell you later; one could attribute to other reasons the low readings of the Mercury Barometer in the Black Sea, in comparison to that of the Adriatic; that is, to the quality of the air and not of the quantity in the larger or smaller cylinders.

Therefore, it remains to talk about the lateral opposing currents B, C and D. About B one must consider, as a reason for its motion, the streams of so-called sweet or fresh waters, which in proportion to their strength maintain it as far as Arnagiui, where it is overcome by the Maestra.

C is not difficult to explain if one only considers the shape of the shore at Scutari which is a hollow where the current beats against itself, reflecting back in proportion to the size of the crater.

D may arise from the same sort of reflection that the Maestra makes on some rocks which come up before Chalcedon. In comparison with the other two this is not very fast, as the Maestra here lacks speed as compared with other places because of its width. It is at any rate true that in each riverbed the slope is not necessary every time there is a driving out force, which itself might have been drawn out and stopped the slope, by chance, not by necessity. With all this, even though I think little of the supposed slope, I will bring to Your Majesty the Levelling which the Barometer, or Torrecillian Tube, may give.

This Barometer shows the Black Sea to be higher than the Adriatic, as one can distinctly see from...Table [1], which contains the observations made by me in Constantinople and on Mount Olympus in Bithynia, whose peaks I could not ascend because of the deep snow, with some others from an interested Englishman, and others from the Adriatic taken from the diary of His Excellency Signor Girolamo Coraro, who applied himself even among public affairs, with the greatest effect to natural studies.

The difference therefore of the Mercury, in these two seas taken according to the average heights, constituted by the series of various observations is very revealing: whereas in Constantinople it shows 39¼ in., in Venice it is 40½ in. of a Roman foot in measurement; that is, in the Bosphorus it is less than in the Adriatic by 1¼ in. and at the halfway mark of Mount Olympus 5¼ inches less; that is less than at sea-level, which it dominates. The other measurements will serve again for comparison with those of other countries in order to examine the effect of heat and cold.

Table 1.
Table of the various levels of mercury observed in Constantinople, in the Black Sea, and in Venice in the Adriatic, and in Bithynia on Mount Olympus.

Place	Year	Month	Height
Observations made in Galata and in Pera in the Bailiff's Residence, not far above sea level, and those of 77 and 78 I had from an English merchant, Signor Nort, who observed them for his English friends.	1677	7 December	42 in.
	1677	28 July	37 1/4 in.
	1678	-----	41 in,
	1679	15 December	41 3/4 in.
At the Bailiff's house, before going to the sea it was Stavros, and it was the place (or, in place of?) of the sea or channel.	1680	30 July 6 August	36 1/2 in.
	1680	4 June	40 in.
In the Black Sea, where is the Rock said to be of Pompey's Column.	1680	4 June	39 1/2 in.
	1680	4 August	40 2/7 in.
At the Bailiff's I did the usual experiment.	1680	8 August	39 8/9 in.
Halfway up Mount Olympus in the aforementioned place.	1680	2 May	34 in.
In the City of Bursia at the roots of that mountain in Bithynia.	1680	3 May	39 in.
Observations made in Venice over several years by His Excellency Signor Girolamo Coraro show the major, minor and medium heights.	----	-----	39 in.
	----	-----	40 1/2 in.
	----	-----	42 1/2 in.

Before passing on to the consideration of that movement, which is universal to all seas, that is that of the ebb and flow of the tides, I deem it necessary to tell you about the nature of the Winds of the Channel which alter their movement; so with this your Majesty will be able to make a just reflection yourself with the right information. The winds, therefore, which most effectively make themselves felt are the Scirocco (SE) and the Tramontana (N), differing not only in their location but also in their effect. The Scirocco hinders the natural course of these waters, causing a great rising; the Tramontana, on the other hand, assists Nature's inclination to make these waters run towards the South. To a greater and lesser degree they manifest themselves in different Seasons: while the Tramontana is felt principally in the Spring and more frequently in the Autumn, the Scirocco is felt, along with other winds, for the rest of the year. The winds are strongest in the afternoon, as I have noticed in other seas, yielding strength as day does to night.

The nature of these winds is contrary not only in their situation: while the Scirocco carries its heat with it, bringing languidness of body to us and serenity to the Skies, the Tramontana brings cold and obscurity, encumbering the Air with clouds and mists. The Barometer is subject to similar and diverse nature, regulating its motion with theirs, and upsetting the order which it maintains in our own Italy, where it is useful for telling the quality of the future day, which is judged to be serene when the Mercury is high, cloudy and disturbed when it is low; but in this Canal, insofar as the Scirocco brings news of the following day it makes the Mercury fall to a low level regulated by the force with which it makes itself felt; by contrast the Tramontana is usually the reason for the disturbed weather, and makes the Mercury rise because of its force; therefore it is necessary to revise the Italian way of foretelling the weather, since the winds have different effects to those in Italy. Therefore, I take the opportunity to tell Your Majesty about the almost imperceptible ebb and flow of the Bosphorus, which is deprived of that fixed order observed in other seas, which in consequence lack many things compared to this, since the continuous motion of the Current does not permit with its great force any other movement. Anyway, that small difference, which one observes, is distinctly explained here to Your Majesty, according to the observations made over several months, together with the winds, which were blowing on those days, and which in a narrow channel like ours make a perceptible difference, as I have already remarked, either in too much restraining or in hastening the waters from their natural course.

But with regard to this insofar as is possible for you to profit by my observations, I have diligently tried to learn, not only the hour, but also the point at which the rising stops and the descent begins, a point called the "del viuo dell'Acque" (the life force of the waters), and likewise the magnitude of these. In this, as Your Majesty's lively mind will discern, I found some sort of law but with no great order, never having been able to establish it precisely. Anyway I found out that near midday it began to abate, and continued until the third hour of the night, and I was able to observe it for two nights, during which time the gates of Galata were kept open for the Public Holiday in honor of the birth of the Sultan's little daughter. After a brief stay of the waters the decrease began again, which I had never been able to see before due to the difficulties involved in spending a night on those shores. With the rising of the Sun the water level was almost always higher than at sunset of the night before and it is probable that the rising which began during the night continues, or rather, grows, until about noon.

From this one could form the hypothesis that for 16 hours a day it rises and it goes down for 8, according to the observations of the 2nd and 3rd of March, but, as I have said, I have never been able to check, either by any sort of enquiry or bribe, and if this should continue, as is likely, on into the summertime, the point of greatest decrease would come to be during the day-time, something which some other interested person will be able to ascertain.

The highest tide observed without any change of wind was in fact close to midday on the day of the Spring Equinox, being of 5 inches, from which one could go on to talk about and compare observations of other seas, subject to greater variations in time, and with a motion precisely subordinate to that of the Moon; however, I will take this up again more leisurely at a later time so that I can talk more maturely about the reasons for such movement.

The greatest decrease is 8¼ inches, as you can see from the Table, in which it is necessary to take the number 24, fixed by me as the supposed level of the waters, in which it will be necessary to subtract the greater or lesser, to obtain the right measurement of the rise and fall.

I shall not refer Your Majesty again to this movement if it is not really the ebb and flow, unknown to many so far because of its small magnitude, which may be attributed, as I said from the beginning, to the movement of the Currents, which with their strength keep down these movements, which appear tenuously and only to the diligent observer. And so I pass onto the other part of that motion which shows itself not visually but by its effects which clearly compensate for that lack.

I mentioned to your Majesty at the beginning of my discourse that my intention in embarking on these travels was not only to contemplate and observe these natural motions which are purely visual, but also to investigate those which are more concealed but just as marvelous to the human mind, and encourage one to discover, for the public good, the reasons. Emboldened by the observations of the surface motion of our channel, happily I was able to prepare myself as best I could, for greater and nobler research, more fruitful since it was more difficult and new, and consequently worthy of Your Majesty's spirited genius.

This is the movement which I call the undercurrent or "Sottana" Current, contrary to the surface motion, or, as I called it before, the Superior, on which I was encouraged to speculate not only by the idea given me by the knowledge of its internal parts, but also by the accounts of many Turkish fishermen, and even more by the suggestions of Sir John Finch, Ambassador to the Porte of His Majesty the King of England and a man well-versed in natural studies, to whom it was introduced by one of his sea captains, who, however, did not arrive at a clear solution, perhaps because of lack of time. Leave aside your incredulity, as I was able to persuade Bigni, the French Jesuit Father, to do; an exemplary man in looking after the poor slaves, and unceasing in the work of the Universal History of Languages, in which he is deeply knowledgeable; I got ready for the experiment in his presence and witnessed by him, and I first tried to find out how the Turks understood this effect. I found that it was by means of the fishing-nets, which in the manner of all fishermen they threw out at random once the boat was stopped. They noticed that down to a certain depth they went along with the Superior Current, but then gave way to the opposite direction with which movement they came up again behind the boat, stretched out to their full length; as well as various other experiments. I also began to learn about it by means of a certain Instrument, made of rope with some pieces of white-painted cork attached, which at considerable depth one could see which way they turned; but more simply, after having stopped the boat well, I used a simple rope with a lead weight attached; this showed me, by its different curvature in which direction it was pushed, never being stationary in the line perpendicular to the center, and as well as this curvature I felt the corresponding tug in my hand. This change of curvature from South to North was sometimes affected with great speed, and maybe within the depth of 8, sometimes 10 and 12 Turkish paces, that is as wide as the width of both arms of a true man: which can be more clearly understood by means of Figure II; supposing that AB is the outline of a section of the channel and CD a hypothetical perpendicular to the center, which begins from the hand which holds it up. There is no doubt that if the water were still, and not moved for any reason, the rope CF with its lead would fall along the perpendicular CD, and it comes out as it does because of the motion impressed on it by the surface current towards A and it is quite certain, that if the same pressure continued down to the bottom, the rope CF would not have turned from the direction of B with so much force and therefore one must certainly conclude that if the effect of the undercurrent is in opposition to that of the Superior current, it must be that the cause

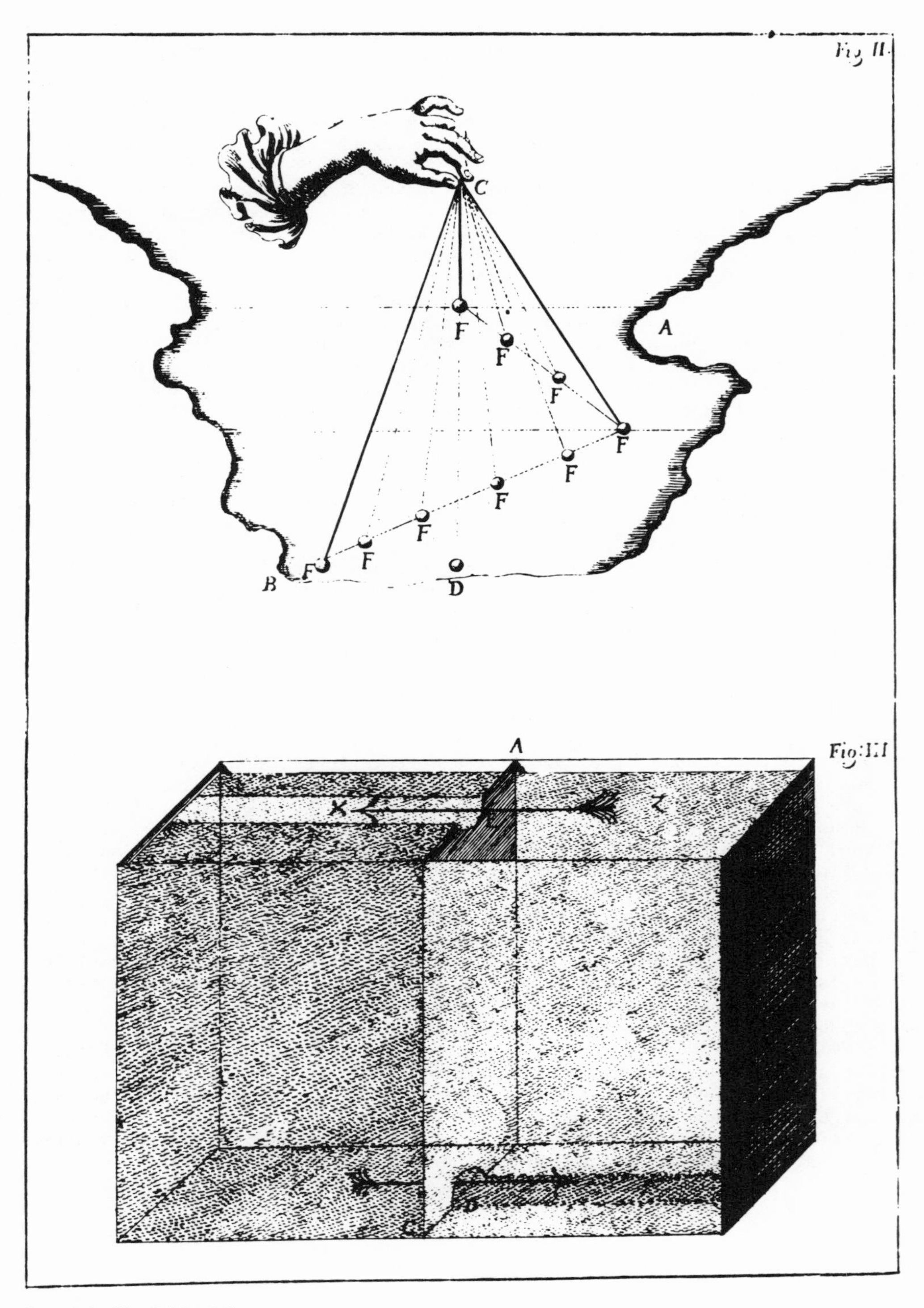
Fig II.
C
A
F
F
F
F
F
F
F
F
F
B
D
Fig: III
A
X
Z

also is the opposite: the cause is the Superficial current towards the South which moves the line CF from the perpendicular: from what is known it is necessary to conclude that there must be in the above-mentioned depths a current which goes from A to B at different speeds, which are themselves governed also by the reasons which govern the Superior, that the sections are more or less wide, having observed that, out of the Canal, that is, at the beginning of the Sea of Marmora, it is imperceptible, as is, in fact, the Superior, and deeper than it is in the Channel.

In order to know, if in this one finds changes at other times, I needed to use the report of those same fishermen, who all assured me that at the end of the Autumn it was almost imperceptible, an effect which I found curious, but it is based on their reports and not known by me personally for lack of time. With regard to the Undercurrent, I explored by every possible means all the effects which might have proved things to be different, like the movement of the boat, and others, and therefore with that security given me by the narrated experiments, I show to Your Majesty this effect, not only new to the Republic of Letters, but also very difficult to explain. I will venture therefore with experiments to touch on something of the reasons for these, submitting myself to the decided judgment of Your Majesty.

So, in the Channel, as I have demonstrated, there are two currents, one opposed to the other, and one above the other (leaving the lateral currents contrary to the Superior which I have explained well enough). The reason, to my mind, may be founded on the principle that the heaviest expels the lighter, therefore my Subject having two densities, one lighter than the other, as Your Majesty will see from the Saltiness. The water of the undercurrent is heavier than that of the Superior by 10 grains and I applied the Subject to the principle presupposed by the gravity and lightness and applied myself to the following experiment, which clearly shows the two contrary currents. I did it in the presence of Signor Luc'Antonio Porzio, a person well-known to Your Majesty not only because he serves You but also because of his literary abilities. One takes a Tank as seen in Figure III, divided in two equal parts XZ crossed by AC which in the lower part has the hole D and in the upper part the hole E. The part X, after having closed up D, is filled up with salt water of the same weight, as the water from the undercurrent. The part Z likewise is filled up with water equal to that of the Black Sea, which I dyed; and then one opens the hole D and immediately one will see the water X pass into Z and that of Z into X, via E, where in fact the two meet, and that movement continues until there is a sufficient deposit of water to make these two waters homogeneous; that cannot happen in the Black Sea since the cause does not stop, which prevents it happening on account of the current of rivers and the undercurrent which both come from perennial sources, one from springs and the other from seas. If one wanted to do similar experiments more continuously and with greater clarity one could bend tubes of twisted glass and insert them into tanks of the above-mentioned waters. So, converting this experiment which itself has all the conditions which I observed in the Thracian Bosphorus, one must infer that the surface current is made up largely by expulsion received from the density of the water brought by the undercurrent, which one can believe works so rigorously meeting, as I have said, the fresh waters of the rivers, and the same current one can say without doubt, from the union of other seas diffuses itself in the Black Sea and therefore with this principle it will not be difficult to persuade oneself that there is no need for that supposed slope which I mentioned at the beginning, and to explain to others who remain amazed, how it is possible for there to be such a large discharge of waters hitherto unobserved and a great addition to the other seas. However the observed motions make this clear and give reason for thinking of more profound investigation on the universal system of the movement of waters, which needs more corresponding observations and more leisure to discuss them in depth, and I am sure that Your Majesty will already have realized from a similar attempt that my weakness is not equal to Your shrewd intelligence. You will forgive me and supply me with better reasons in which I will glory, as taught by Your Majesty, who, to the amazement of the World, knows and judges similar material so well.

From the observations of the movements of the Canal I pass to those experiments which I made to demonstrate the nature of these waters, in the saltiness and color. They are different not only because of the water movement but also because of their constitution, making a comparison principally by means of their weight which gives me reason to speak of the salinity in general, in imitation of the celebrated Robert Boyle, with the following basis.

The color of these waters has a greater reputation than it deserves since everyone calls the major sea by the name of "Black," a color which appears in the expanse of the major sea more than in the channel. I do not know whether it is because of the depth of sand, different to that of other seas and similar to that of rivers, intermixed with pieces of crystal, or else for rarely having the Horizon clear of dense clouds. It is certain that its aspect is gloomy compared with that of other seas, and the water taken up from its bed, and put in glass tubes does not vary in color from that of fresh water, except for that tenuous and almost imperceptible cloudiness which one observes universally in all marine waters, and therefore one must attribute the effect to a reason which is in the sea itself and which to my belief is in the depth of it. During rainy periods the current B distinguishes itself from the Maestra not only by its contrary motion but also by its turbid color, which is brought to it by the above-mentioned rivers of fresh water.

The other part which composes the nature of these waters is the salt taste, common to all seas, in which they are more or less rich, which experiments have shown me by means of the weight raised by the Hydrostatic Balance according to the teaching given me by Signor Doctor Montanari, celebrated Mathematician and my revered Master, who from early days began to demonstrate to me the principles of similar studies, and amongst others, the basics of using that Instrument, and the perfection to which his noble intellect had reduced it, and I still conserve one of his Letters in which he speaks of it at length.

I began therefore in the Archipelago facing Smyrna to measure the waters, and in the places noted in...Table [2].

With the afore-mentioned Balance hermetically sealed, with lead inside weighing 1776 grains of the Venetian Mint I knew the decreasing order of weights (and I put them in the Table without the weight of the Balance itself) following that which one meets in the Black Sea; and the difference between the waters of the harbor at Smyrna and that of the Black Sea amounts to the sum of 29¼ grs., and are even lighter as related by Sailors who use it for cooking: this is the effect of rivers, which cooperate with this to make it happen on their part, as I have explained for the currents.

I was equally curious in speculating on the deep interior, not without some profit; with the use of a tube, closed with a valve, which I was then able to open by way of a rope, even under the water, I found that those waters weighed 10 grains more than the surface waters in the same Canal, an experiment which

Table 2.
Table of various weights of marine waters from the middle of the Archipelago to the Black Sea, and some other places in the Adriatic.

Place	Month	Height
At the Foggie in front of Smyrna Harbor	8 September	85 1/2 Grains
Lemnos 80 miles from the Dardanelles	19 September	81 1/2 grs.
Sestos and Abydos at the harbor mouths	18 October	71 1/2 grs.
Heraclea in the Gulf of Marmora, where I actually observed the source of bitumen	23 October	69 grs.
Toppana in the Canal of Constantinople where the ships drop anchor	22 February	64 1/2 grs.
The New Castles in the same Canal	2 April	64 1/2 grs.
Terzana or Arsenal, where the current of fresh water begins	10 April	46 1/2 grs.
Caragaz, place where there is the Court of King's Horsemen, in the fresh water	10 April	58 1/3 grs.
In the above place, during rain	-----	54 1/20 grs.
Between the two lights at the mouth of the Black Sea	4 August	56 1/4 grs.
At Spalatro in Dalmatia	2 October	91 3/4 grs.
Pelegrino Reef opposite Parenzo in Istria	16 October	86 grs.
On the beach at Venice	-----	75 1/2 grs.
The two following weights show the difference between surface water and deep water		
At Bisectas where there is the King's new Court, the surface water	28 June	61 3/4 grs.
The same place, in deep water	28 June	71 3/4 grs.

I have already used in demonstrating the currents; being complementary, in my view to the afore-mentioned effect.

Boyle says that those of the Ocean are equal in their weight not only in the upper part but also in the lower; speaking of Scaliger who says that he has found about the Kingdom of Candia that the upper was heavier than the lower. However it is true that in the Ocean and in the Aegean it is not possible to make a comparison with the Bosphorus, lacking the currents described, which necessarily must bring waters of different natures into that order though they have different origins.

The low density of these waters, decreasing as one draws nearer the center of the Black Sea, one must attribute to the concourse of the river waters which sweeten that brine brought by the undercurrent, and maybe from other sources, which also explains the diminution of weight, since this depends on the concourse of salt or other materials which I will examine.

4

INVESTIGATION

OF THE

CURRENTS

OF

THE ATLANTIC OCEAN,

AND OF THOSE

WHICH PREVAIL BETWEEN THE

INDIAN OCEAN AND THE ATLANTIC.

BY THE LATE

MAJOR JAMES RENNELL, F.R.S.

OF LONDON AND EDINBURGH, MEMBER OF THE ROYAL INSTITUTE OF FRANCE, OF THE IMPERIAL ACADEMY AT ST. PETERSBURG, AND OF THE ROYAL SOCIETY OF GÖTTINGEN; AND FORMERLY SURVEYOR-GENERAL OF BENGAL.

THY way is in the sea, and THY path in the great waters, and THY footsteps are not known.
PSALM lxxvii. 19.

LONDON:

PUBLISHED, FOR LADY RODD,

BY J. G. & F. RIVINGTON,

ST. PAUL'S CHURCH YARD, AND WATERLOO PLACE, PALL MALL.

1832.

Reprinted from pp. 140–151 of *Investigation of the Currents of the Atlantic Ocean, and of Those which Prevail between the Indian Ocean and the Atlantic,* James Rennell, J. G. & F. Rivington, London, 1832, 359 pp.

ON THE LEVELS OF THE CARIBBEAN AND MEXICAN SEAS, AND THE ORIGIN OF THE FLORIDA OR GULF-STREAM

James Rennell

These SEAS, which receive the continuation of the *Equatorial* current and its adjuncts, and send forth that of FLORIDA, or the GULF-STREAM, may be regarded as a kind of intermediate lakes between them; and the latter of these seas, as the immediate and elevated reservoir of the Gulf-stream. The supposed cause of this elevation, above the general level of the Atlantic Ocean, has perplexed many persons, in their attempts to account for it; and the difficulty has been increased by their supposing that a very great elevation was necessary, in order to account for the velocity of the stream. But perhaps there is no necessity for supposing any great degree of elevation; nor of any cause beyond what is in constant operation elsewhere, although on a less extensive scale. For, the operation of a constant wind, which forces the surface of a vast expanse of water into a more confined space, may raise the level of the included water to such a height, as to occasion a great velocity of stream at the place of its discharge; more particularly when the opening is *comparatively* narrow; as in the case of the Strait of Florida.

In the GENERAL OBSERVATIONS at the commencement of this work, the Author has ventured to offer this opinion, as the cause of the elevation of these seas: and has suggested that it is the effect of the north-eastern (or rather, in that quarter, *eastern*) trade-wind, combined with the position and direction of the coasts of Brasil and Guyana, between Cape St. Roque and Trinidad: the wind forcing the surface of that part of the Atlantic Ocean towards the West-Indian Seas and the coast of America, by interposing itself, obliquely, to the course of the volume of water in motion, reducing the space so operated on, by the wind, to one-half of its original breadth, by the time that the passing waters arrive at the Caribbean Sea. So that the impelled mass of water, being constantly narrowed in its progress by the lateral pressure of the coast of South America, must, of necessity, be forced up to a *higher* and *higher* level, as it advances; like the wave of tide in the Bay of Fundy, the Bristol Channel, and the Gulf of Cambay. For the volume of water cannot *retreat* when pressed laterally; because, besides the effect of its own *momentum*, it is impelled forwards by the accession of fresh supplies, constantly brought on by the same power which originated the movement.

The *inset* of the continuation of the *Equatorial* stream, into the *Caribbean* and *Mexican* Seas, has often been considered as the cause of the elevation of the Gulf of Mexico; but, as this current is known to originate in the Atlantic Ocean, which ought, of course, to be lower than the outfall of the Gulf of Mexico; and as no current *can* raise a head of water, *higher than itself*, this idea cannot be entertained, any more than *that* which accounts for it, from the accession of river water; which is much too inconsiderable, in respect of the volume of water discharged. The current of *inset*, into the Caribbean Sea, may perhaps be regarded rather as an EFFECT of *that* which produces the elevation, than the CAUSE of it.

The current above named *Equatorial*, is so called in conformity to popular opinion; but, although a part of it may be such, the great body of current which enters the Caribbean Sea is doubtless raised by the operation of the N.E. trade-wind, on that part of the Atlantic lying generally to the westward of Cape St. Roque. The vastness of its body proves that it cannot be the Equatorial current alone, because that current, (which is formed by the S.E. trade-wind,) divides itself into three parts; of which, as we have shown, only one passes along the northern coast of Brasil. If it is to be considered as the continuation of the Equatorial current, it is rather that current *renewed*, like that of Lagullas, by the south-east trade-wind.

Although the effect of the trade-wind in impelling the surface, or *drift current*, may be, in the first instance, nearly equal throughout the whole *zone* of the trade-wind, yet the parts nearest to the South-American shore, must necessarily be raised higher than the parts more remote; because of the immediate operation of the lateral resistance produced by that shore; and which effect seems to be clearly shown by the increased velocity of the currents through those channels of the Caribbean chain, situated between the islands Trinidad and Dominica; for the currents there have double the force of those to the northward of Dominica.

For this fact, we have more especially the testimony of Captain Deacon, of H.M. Ship *Niobe*, who had the best opportunities of knowing, from his having been a considerable time employed among those islands. The Spanish work, the *Derrotero de las Antillas*, confirms the fact also, by numerous examples. The Baron Humboldt in the line towards Trinidad, in July, remarks, that the N.W. current increased in velocity, as they advanced towards the coast of South-America. And Captain Rodd, of his H.M. Ship Warrior [1], in May, makes the same remark, in his way from England to Barbadoes. In effect, the existence of a general movement of the surface of the western part of the Atlantic, within the tropic, towards the West-Indies, is well established; not only by common report, and the drift of floating substances towards it, but also by the testimony of the best journals.

In effect, the most elevated part of the great body of water raised by the joint operation of the trade-wind and the coast of America, appears to pass between the Gulf of Paria and Dominica; to the northward of which latter, and between it and Porto-Rico, the currents are weak; as being probably little more than the simple effect of the trade-

[1] Now Rear-Admiral Sir John Tremayne Rodd, K.C.B.—Ed.

wind; (in other words the *drift current*;) but the other, as being the combined effect of both.

Currents are also proved to have set into the Caribbean Sea from the Atlantic Ocean, through the Mona and windward passages[1]; so that there is an *inset* into that sea from *every* passage, save on the west, the Channel of Yucatan, through which its waters communicate with the Sea of Mexico.

But the body of the *drift current* to the northward of the greater West-India Islands (Cuba, Hayti, and Porto-Rico[2],) meeting an opposition at

[1] A deficiency of observations leaves the Author in doubt, whether these two currents are constant, or only occasional. From what has been said it must be inferred, that the *Caribbean* Sea is lower in the quarter towards *St. Domingo* and *Porto-Rico*, than on the side towards the continent; as also, that the Atlantic, to the northward of those islands must be higher than its *general* level, from the accumulation produced by the *drift* current of the N.E. trade-wind, and the opposition of the *Bahama* Islands to it. It appears probable that the direction of the currents through the above straits is regulated by the comparative levels of the two seas, with which they communicate. It might be expected that the current through the *windward* passage ran most commonly into the *Caribbean* Sea.

[2] The Spaniards apply the name of *Antillas* to the whole chain of islands between the Gulf of *Paria* and *Florida*; distinguishing the three greater islands on the north, as the *Great Antillas*; and the chain of smaller islands on the east, (to which we exclusively apply the term *Antillas*) by the *Lesser Antillas*. This appears to be an useful mode of classing them: being at once *comprehensive* and *distinct*.

In some old English maps, the term *Little Antillas* will be found applied to the *Caribbee* Islands.

the Bahamas, collects and forms a stream which is naturally turned to the south and east, and running along the north side of that chain of islands, has carried ships so far out of their course, that commanders who, after passing the Strait of Florida into the Atlantic, intended to go through the *Windward Passage*, have found themselves near the Mona Passage, or at the Virgin Islands.

It may, also, be observed that, most commonly, the drift current, formed by the northern part of the zone of trade-wind in this quarter of the Atlantic, is weak; owing to the irregularity and unsteadiness of the wind beyond the parallel of 25°.[1]

It is certain that no *positive proof* can be adduced that the elevated levels of the Caribbean and Mexican Seas are occasioned by the impulse of the trade-wind: yet it fairly may be presumed that such is the fact:—in the first place, because such effects are well known to be produced in other situations; and, also, because it has been known

[1] The frequent unsteadiness of the N.E. trade-wind is so well known, that ships bound to the West Indies, &c. seldom venture to run down their westing much to the northward of 20°. Many instances of great irregularity might be quoted; but an instance occurred in March, 1819, a month in which ships usually carry the N.E. trade nearly to the Equator. The only regular trade-wind, at this time, was between the parallels of 22¾° and 28½°; and again, between 22° and 13°. Through all the rest of the space the winds were generally N.W. or west; and sometimes S.W.; and these continued to 2½° south.

that the changes in the velocities of the stream, through the *Florida Strait*, have accorded with the different state of the winds (in respect of their strength) in the *Atlantic;* and still more that the seasons, which regulate those winds, produce also a correspondent effect on the state of the current.

But, although it may be well understood that certain portions of the ocean have the level of their surfaces occasionally raised by the wind, yet this has not been hitherto *proved* in respect of any great expanses of surface. A very satisfactory proof, however, has lately come to the knowledge of the Author. It happened within that wide recess in the western coast of Africa which terminates in the Bights of Benin and Biafra, of which the form is calculated to favour the operation.

Captain Lawson, who has had opportunities of viewing the coast of *Guinea*, generally, at all seasons, and who has resided occasionally at *Cape Coast Castle*, reports that the level of the sea, in those parts, is higher, by at least six feet perpendicular, in the season of the strong S.W. and southerly winds, which blow obliquely into the Bay of *Benin*, between April and September, (and which is the rainy season,) than during the more serene weather of the opposite season.

The fact is shown by two different circumstances: the one affording a strong presumptive proof; the other, a most positive one. In the former case, that the trunks of trees, which are thrown upon the shore, during the season of the stormy S.W.

winds, are found at the perpendicular height of six to eight feet above the level of the sea, during the other season. And secondly, that during the stormy S.W. season, the tides *ebb* and *flow* regularly, in the several rivers; but, in the other season, the same rivers run *ebb* constantly; the level of the sea being then too low to allow the tide waters to enter the mouths of the rivers.

With respect to the other question, the different effect of the winds, in the Atlantic, at different seasons, on the velocities of the Gulf-stream; it is well known that the seasons produce periodical winds along the whole northern coast of South America; (making a partial monsoon; the great body of the land being all to the southward;) for, when the sun is far advanced into the northern signs, or on its return towards the Equator, the winds blow more directly towards the Caribbean Sea, and are also stronger; and, of course, impel a great volume of water towards it: and it is then, that is, in July, August, and the early part of September, that the Current of Florida is at its *maximum* of velocity. But, when the sun is in the opposite hemisphere, the winds along the coast of the continent, and amongst the larger islands, and the coasts of Florida, have much *northing* in them; and, of course, less water will then be driven towards the West-Indian Seas. And accordingly, it may be seen, by the tables of velocities of the Gulf-stream, (given hereafter,) that the current is commonly less rapid in the season of the northerly winds, which is that season when the sun is far to the southward [1]. Moreover there is, at that season, a *reflux* of current to the *eastward*, along the coasts of New Granada and Venezuela. The Baron de Humboldt writes, (Vol. iii. p. 378, 9. of his Personal Narrative, English translation,) that, in *autumn*, the wind on the coast of Venezuela, &c. changes from *easterly* to *northerly* or north-west. The currents there, which, during the easterly wind, run to the westward, *cease*, on the arrival of the interval of calm or light winds, between the two periodical ones; and the currents begin to run to the eastward some days before the north-west winds come on. This seems to show that it is the pressure of the wind which keeps up the high level; and that, on the cessation of the former wind, and before the arrival of the other, the water seeks to regain its level. These circumstances must doubtless tend to lower the levels of the Caribbean and Mexican Seas, at this season [2].

To this general statement may be added some particular facts, founded on authorities of consider-

[1] M. Monach, *Cap. de Port* at *Cayenne*, also bears testimony to the increased velocity of the N.W. current, along the coasts of *Brasil* and *Guyaña*, between the months of May and October: when he advises the navigators from Europe bound to *Cayenne*, when arrived in latitude 3° N., and longitude 41⅔° from London, to steer S.W.; which he says is necessary, in order to preserve their parallel; *Cayenne* being in about 5° N. In the other season he says, W.S.W. will suffice.

[2] The same happens in the Bay of Bengal: and doubtless from a like cause. The periodical current changes to the opposite quarter very soon after the cessation of the periodical wind, and long before the opposite wind begins to blow.

able weight, derived from circumstances which took place in March, 1815, and have been already alluded to [1].

In that month, the Gulf-stream was remarked in his Majesty's ships *Gorgon* and *Asia* to be exceedingly rapid, *for the season;* being, by the report on board the Gorgon, 5 miles *per* hour in the narrow part of the Strait of Florida, which is equal to the *maximum* rate of the current in August. Now Captain W. King, of his Majesty's ship *Leonidas,* sailed in the same month from Bermudas to the Equator, and to 2° S. in the direction of Cape St. Roque. He found the trade-wind unusually fresh; and, as he advanced farther south, even *boisterous*; so as to render it prudent to keep away from the wind, although it was his chief object to get as much as possible to windward.

It is proper to remark that the rate of velocity of the Gulf-stream, during the month of March, appears to be extremely various and uncertain. There is a second instance of great velocity, in 1795, when it came up to 97 miles: but in a third, in 1819, it fell to less than one-third of the rate which was remarked in 1815.

Since then the variations in the state of the Gulf-stream are found to accord with the different state of the winds, in respect of their direction and force, can it be supposed otherwise than that the wind alone is the agent which produces the Gulf-

[1] Notwithstanding the above general statements, the "*Derrotero de las Antillas*" has westerly currents along the coast of *Venezuela* in *January*. Perhaps an anomaly.

stream, in the first instance. It certainly appears to the Author that this system accounts, in the most simple and natural way, as well for the high level of the Gulf of Mexico, as for the variations that take place in it.

It has been observed before, that the motion of the waters of the Gulf of Mexico agrees best to the idea of its level being kept up by *pressure*, and not by water supplied by currents. Let the proportions of the *influx* with the *efflux* be compared; and it will then be seen how greatly the latter predominates. And of the proportion of water that flows into it, from the Caribbean Sea, a small proportion only runs to the westward: for one part of it, and that the largest, goes immediately to join the current which runs towards the outlet; and even a part of this latter is returned into the Caribbean Sea, round the Capes of St. Antonio and Corrientes. [*See* Chart V. *of the Florida Strait*, &c.]

Nor does the current of influx advance even so far to the westward as *Vera Cruz*. It may be seen by an inspection of the Chart, that, if a curved line be drawn from *Vera Cruz*, towards the quarter of N.E. by N., so as to pass the parallel of $25\frac{1}{2}^{\circ}$, about the meridian of New Orleans, that line would form, generally, the southern boundary of the *current of the outlet* to that point. From thence, the continuation of that line deviates to the east of south; and afterwards rapidly to the S.E., to the entrance of the Florida Strait; where the curved line terminates.

The western shore of the Gulf of Mexico exhibits most indubitable proofs of the passage of a current along it, to the northward, by the long and narrow alluvial islands and mud banks, by which it is bordered, throughout its whole extent: and very similar to those along the south coast of Louisiana and West Florida; along which it is known the continuation of the same current runs. These appearances cannot be mistaken by those who have been in the habit of observing the operation of currents in forming alluvions. And it must be inferred from these appearances, and the circumstances altogether, that the entire surface of that part of the Mexican Sea, or Gulf, within the space between the above said curved line and the western, northern, and eastern, sides of the same sea, is in motion, either circuitously, or *otherwise*, towards the Strait of Florida: and this space may be reckoned equal to three-fifths, or even three-quarters, of the whole expanse of that sea or gulf; and as the volume of water, so discharged, is obviously so vastly greater than that supplied by the current of *inset*, this appears to furnish an additional argument for believing that the level of the Gulf of Mexico can only be regulated, in the first instance, by that of the Caribbean Sea, and remotely by the effect of the trade-wind on the latter.

5

Reprinted from *Edinburgh J Sci.*, ser. 2, 6:341–345 (1832)

ON THE TEMPERATURE AND SALTNESS OF THE WATERS OF THE OCEAN AT DIFFERENT DEPTHS

E. Lenz

M. Lenz, naturalist to the expedition of Kotzebue, made a series of well-conducted experiments on the temperature and saltness of the ocean in different latitudes and at various depths. The instrument he employed for ascertaining the temperatures was an improvement upon that of Hales, being a large cylinder closed at both ends by valves opening inwards, to one of which was attached a thermometer, and surrounded by a highly non-conducting substance.—The results are contained in the following table :—

	Time of observation.	Places. Lenz.	Lat. N.	Long. W.	Depth in toises.*	Temperature. at surface.	at depths indic.
1	1823. Oct. 10,	Atlant. Oc.	7° 21′	21° 59′	539	25°,80 C.	2°,20
2	1824. May 18,	South Sea.	21 14	196 1	140,7	26,40	16,36
3	,, ,,	,,	,,	,,	413,0	"	3,18
4	,, ,,	,,	,,	,,	665,1	"	2,92
5	,, ,,	,,	,,	,,	914,9	"	2,44
6	1825. Feb. 8,	,,	25 6	155 58	167	21,50	14,00
7	Aug. 31,	,,	32 6	136 48	89,8	21,45	13,35
8	,, ,,	,,	,,	"	214,0	,,	6,51
9	,, ,,	,,	,,	"	450,2	,,	3,75
10	,, ,,	,,	,,	'	592,6	,,	2,21
11	1826. Mar. 6,	Atlant. Oc.	32 20	42 30	1014,8	20,86	2,24
12	1825. Aug. 24,	South Sea.	41 12	141 58	205,0	19,20	5,16
13	,, ,,	,,	,,	,,	512,1	,,	2,14
14	1826. Mar. 24,	Atlant. Oc.	45 53	15 17	197,7	14,64	10,36
15	,, ,,	,,	,,	"	396,4	,,	9,95

From this table the following conclusions may be drawn :

1. Between the equator and 45° the temperature of the ocean decreases regularly to the depth of a thousand fathoms, —beyond this no other experiments have been made.

2. The decrease of temperature is at first rapid, it gradually decreases, and becomes at last insensible.

3. The point where the decrease becomes insensible appears to rise with the latitude. At 41° and 31° it is between 200 and 300 fathoms, at 21° it is near 400. To this remark there appears to be a slight exception at 45°.53, when the temperature at 400 fathoms is still at 10° C. but perhaps that ob-

* A toise = 1.066 English fathoms.

servation is modified by the proximity of the land, since it was made in the Atlantic Ocean only 15° W. from Greenwich, and consequently near the coast of Europe, while the others were made in the south sea far from any continent ; but even in this case the point where the decrease of temperature becomes insensible is still evidently near 200 fathoms.

4. The lowest temperature observed is 2.2 C. (36° F.) and it is perhaps that of all the depths at which the decrease is insensible. The locality of that temperature rises with the latitude; and it would be interesting to know at what latitude it reaches the surface.

The results of M. Lenz, in regard to the saltness of the sea, have been deduced from its specific gravity. It had previously been shown by M. Erman that salt water having a specific gravity of 1.027, the mean of that of the sea, diminishes in bulk gradually down to 25° F., and does not reach its maximum density before congelation. M. Erman's experiments on this contraction extended from 59° F. to 25°, M. Lenz extended them up to 86°, and thence deduced a law for reducing the specific gravity at any one temperature to what it would be at any other. The following table exhibits the specific gravity corrected to the temperature of 63.°5 F., distilled water at that temperature being reckoned unity.

No.	Lat. N.	Long. E.	Depth in toises.	Specific Gravity. at surface.	beneath.	Difference.
1	7° 20′	21° 59′	539,0	1,02574	1,02645	—0,00070
2	21 14	196 1	665,1	1,02701	1,02666	+0,00035
,,	,,	,,	929,4	,,	1,02659	+0,00042
3	25 6	156 58	167,0	1,02706	1,02674	+0,00032
4	41 12	141 58	205,0	1,02562	1,02609	—0,00047
,,	,,	,,	512,1	,,	1,02658	—0,00096
5	32 6	136 48	214,0	1,02678	1,02624	+0,00054
,,	,,	,,	450,2	,,	1,02651	+0,00027
,,	,,	,,	592,6	,,	1,02629	+0,00049
6	32 20	42 30	1014,7	1,02825	1,02714	+0,00111
7	45 53	15 17	396,4	1,02738	1,02732	+0,00006

From this table we see that in the experiments No. 1 and 4 the specific gravity of sea water towards the bottom is a little greater than at the surface, but that the contrary holds in Nos. 2, 3, and 5. In experiment 7 the specific gravity of the surface differs so little from that of the bottom that we may consider them as equal. For the first two cases, we may suppose that a rapid evaporation had at that time determined the slight increase of density at the surface, as abundant rains may have diminished it in experiments 2, 3, and 5. It is remarkable that in the same place the specific gravities are almost exactly the same for different depths, if we except that of the surface. No. 6 alone offers a striking exception, giving at the depth of a thousand fathoms a specific gravity much less than at the surface. We cannot suppose this difference to be due to an error of observation, the specific gravity at the bottom being the mean of three observations agreeing with each other, and that of the surface corresponding with the observations of the day before and the day after. The irregularity may perhaps be due to a current of colder and less salt water flowing at the bottom from the pole to the equator,—a point, however, which can be determined only by repeated observations. Leaving out this latter observation, we may conclude, that, *from the equator to 45° N. lat. the water of the sea to the depth of 1000 fathoms possesses the same degree of saltness.*

M. Lenz gives also two tables exhibiting the results of 258 observations made on the saltness of the sea at the surface, 105 of them made in the Atlantic Ocean, between 56°.41 S., and 50.25 N. lat., and 153 in the South Sea and Indian Ocean, between 57°.27′ S. and 56°.22′ N. Lat. From these tables he deduces the following results:—

1. The Atlantic Ocean is salter than the South Sea; and the Indian Ocean, being the transition from the one to the other, is salter towards the Atlantic on the west than towards the South Sea on the east.

2. In each of these great oceans there exists a maximum of saltness towards the north, and another towards the south,—the first is farther from the equator than the second. The minimum between these two points is a few degrees south of the equator in the Atlantic Ocean, and probably also in the South Sea, though Mr Lenz's observations do not extend to latitudes sufficiently low in the South Sea.

3. In the Atlantic Ocean the western portion is more salt than the eastern,—in the South Sea the saltness does not appear to differ with the longitude.

4. The greatest specific gravity is found in the Atlantic at the maximum point above alluded to, at 40° W. long.=1.02856. In the South Sea at 11.9° = 1.028084.

This last is the only one in the South Sea giving a specific gravity reaching 1.028.

5. In going north from the northern maximum, and south from the southern maximum, the specific gravity diminishes constantly as the latitude increases.

Whence then, says M. Lenz, comes these maxima towards the north and south, why is the maximum not rather at the equator? To answer this question, we must inquire what chiefly determines the saltness of the surface. Evaporation exercises the greatest influence, and by this evaporation the occurrence of these maxima may be explained. In fact, evaporation is influenced both by the heat of the sun, and by the more or less rapid motion of the currents of air. The solar heat is greatest on the equator,—but there, on the other hand, the motion of the air is least. It is remarkable that in the Atlantic Ocean the minimum coincides precisely with the locality of almost constant calms. The vapours raised by the heat of the sun remain suspended above the surface of the water, and prevent farther evaporation. The sea loses thus less of its aqueous particles, and it is consequently less salt than at 12° N. and 18° S. Lat. In these regions the trade winds carry off immediately the vapours formed by the solar heat, which is here little less than at the equator, and give place to other vapours which rise immediately. In this way evaporation proceeds, and the saltness increases rapidly. This consideration would explain also the greater saltness of the western part of the Atlantic Ocean, for we know that the more we approach the coasts of Africa, the more frequent and more continued are the calms. In the south sea great calms are not experienced towards the east, and hence the longitude has no influence on the saltness of its waters.

6

Reprinted from *Nature* 4:97-98 (June 8, 1871)

THE GENERAL OCEANIC CIRCULATION

AMONG the results of the *Porcupine* Expeditions of 1869 and 1870, there are perhaps none more important than those relating to the Temperature of the Deep Sea. For it is only to such accurate determinations of ocean temperatures as have now been made for the first time, not only at the surface and the bottom, but also at intermediate depths, that a really scientific theory can be framed of that great Oceanic Circulation, which, while it eludes all ordinary means of direct observation, seems to produce a far more important effect, both on terrestrial climate and on the distribution of the marine fauna, than that of the entire aggregate of the surface-currents which are more patent to sight. The latter usually have winds for their prime motors, and their direction is mainly determined by the configuration of the land; so that their course and action will change with any superficial alteration which either opens out a new passage or blocks up an old one. The former, on the other hand, depending solely on difference of temperature, will (to use Sir J. Herschel's apposite language) have its movements, direction, and channels of concentration mainly determined by the configuration of the sea-bottom; and vast elevations and subsidences may take place in this, without producing any change that is discernible at the surface.

The history of the doctrine of the general oceanic circulation has been recently given in the Anniversary Address of the President of the Geological Society, with a completeness which (so far as we are aware) had never been previously paralleled. But this doctrine has hitherto rested on the very insecure foundation of observations which were alike inadequate and inaccurate; and it has consequently been discredited, both by physicists and by physical geographers. It is now impossible to assign a precise value to the older observations upon deep-sea temperatures. For it was shown by the careful experiments which were made by Mr. Casella two years ago, under the direction of the late Prof. W. A. Miller, Dr. Carpenter, and Captain Davis of the Admiralty, that the pressure of sea-water at great depths on the bulb of the thermometer—a pressure amounting to about a ton per square inch for every 800 fathoms—exerts so great an influence on even the very best instruments of the ordinary construction, as to cause a rise of eight or ten degrees under an amount equivalent to that which would be exerted at from 2,000 to 2,500 fathoms' depth;* and the error of many thermometers under the same pressure was two or three times that amount. There is reason to believe that some of the thermometers formerly employed, especially in the French scientific expeditions, were protected against that influence; but no such protection appears to have been applied to the thermometers supplied to Sir James Ross's Antarctic Expedition; and the observations by which he supposed himself to have established the existence of a uniform deep-sea temperature of about 39°, now seem to have been altogether fallacious. So again, Captain Spratt's observations in the Mediterranean, though made with great care, were seriously vitiated by this source of error.

It appears from Mr. Prestwich's exhaustive summary, that as long ago as 1812 Humboldt had maintained that such a low temperature exists at great depths in tropical seas, as can only be accounted for by the hypothesis of under currents from the Poles to the Equator. And this view was adopted by D'Aubuisson, Lenz, and Pouillet; the latter of whom considered it certain "that there is generally an upper current carrying the warm tropical waters towards the Polar seas, and an under current carrying the cold waters of the Arctic regions from the Poles to the Equator." Our Arctic navigators had met with temperatures in the Polar seas as low as 29° at 1,000 fathoms; and these observations have been more recently confirmed by those of M. Charles Martins and others in the neighbourhood of Spitzbergen. Several instances are recorded, on the other hand, in which temperatures of from 38° to 35° were observed at great depths nearly under the Equator; and this alike in the Atlantic, Pacific, and Indian Oceans.

The Temperature-soundings taken in the *Lightning* and *Porcupine* Expeditions, with trustworthy instruments, have shown:—(1) That in the channel of from 600 to 700 fathoms' depth which lies between the North of Scotland, the Orkney and Shetland Islands, and the Faroes, there is an upper stratum of which the temperature is considerably higher than the normal of the latitude; whilst there is stratum occupying the lower half of this channel, of which the temperature ranges as low as from 32° to 29°·5; and a "stratum of intermixture" lying between these two, in which the temperature rapidly falls—as much as 15° in 100 fathoms. (2.) That off the coast of Portugal, beneath the surface-stratum, which (like that of the Mediterranean) is super-heated during the summer by direct solar radiation, there is a nearly uniform temperature down to about 800 fathoms; but that there is a "stratum of intermixture" about 200 fathoms thick, in which the thermometer sinks 9°; and that below 1,000 fathoms the temperature ranges from 39° down to about

* Mr. Prestwich cites Dr. Carpenter as estimating the error from pressure "at 2° or 3° or even more." The error is said by Dr. Carpenter to have been from 2° to 3° on the depths of from 500 to 700 fathoms first explored; but would have been from 8° to 10° at the depths subsequently reached.

$36^{\circ}\cdot5$. (3.) That in the Mediterranean the temperature beneath the super-heated surface-stratum is uniform to any depth; being at 1,500 or 1,700 fathoms whatever it is at 100 fathoms, namely from 56° to 54°, according to the locality. To these may be added (4) the observations recently made by Commander Chimmo, with the like trustworthy thermometers, which, in lat. 3° 18½′ S., and long. 95° 39′ E., gave $35^{\circ}\cdot2$ as the bottom temperature at 1,806 fathoms and $33^{\circ}\cdot6$ at 2,306 fathoms. These seem to be the lowest temperatures yet observed in any part of the deep ocean basins outside the Polar area.

It is clear, therefore, that very strong evidence now exists, that instead of a uniform deep-sea temperature of 39°, which, on the authority of Sir James Ross, by whom the doctrine was first promulgated, and of Sir J. Herschel, by whom it was accepted and fathered, had come to be generally accepted in this country at the time when the recent deep-sea explorations commenced, not only is the temperature of the deeper parts of the Arctic basin below the freezing-point of fresh water, but the temperature of the deepest parts of the great oceanic basins, even under the Equator, is not far above that point. And it seems impossible to account for the latter of these facts in any other mode, than by assuming that Polar water is continually finding its way from the depths of the Polar basins along the floor of the great oceanic areas, so as to reach or even to cross the Equator. And as no such deep efflux could continue to take place without a corresponding in-draught to replace it, a general circulation must be assumed to take place between the Polar and Equatorial areas, as was long since predicated by Pouillet.

Such a vertical circulation, it was affirmed by Prof. Buff, would be necessarily caused by the opposition of temperature between the Equatorial and the Polar seas; and this view was adopted by Dr. Carpenter, in his *Porcupine* Report of 1869, as harmonising with the temperature-phenomena which had been determined in the expedition of that year. It has been since contested, however, not only by Mr. Croll and Dr. Petermann, but also by Dr. Carpenter's colleague, Prof. Wyville Thomson, all of whom agree in regarding the amelioration of the temperature of the Arctic Sea as entirely due to an extension of the Gulf Stream, the underflow of Polar water being merely its complement. And the authority of Sir John Herschel was invoked against the idea that any general oceanic circulation could be maintained by difference of temperature alone; though his statements, when carefully examined, only go to prove that no such difference could produce *sensible currents*.

Such was the state of the question when the *Porcupine* Expedition of last year concluded its work; and the results obtained, whilst confirmatory of previous observations, suggested to Dr. Carpenter a definite Physical Theory, which now comes before us with the express approval of the great philosopher who had been said to be opposed to it.

Having ascertained, as our readers have learned from his report, the existence of an outward under-current in the Strait of Gibraltar, which carries back into the Atlantic the water of the Mediterranean that has undergone concentration by the excess of evaporation in its basin, Dr. Carpenter applied himself to the consideration of the forces by which the superficial in-current and the deep out-current are sustained; and came to the conclusion that, as had been previously urged by Captain Maury, a *vera causa* for both is to be found in excess of evaporation, which at the same time lowers the level and increases the density of the Mediterranean column as compared with a corresponding column of Atlantic water. This conclusion, when scientifically worked out, was found to be applicable, *mutatis mutandis*, to the converse case of the Baltic Sound; in which, as was long ago experimentally shown (with a result that has recently been confirmed by Dr. Forchhammer), a deep current of salt water flows inwards from the North Sea, whilst a strong current of brackish water sets outwards from the Baltic, the amount of fresh water that drains into which is greatly in excess of the evaporation from its surface.

Comparing, then, the Polar and Equatorial areas, it is shown by Dr. Carpenter that there will not only be a continual tendency in the former to a lowering of level and increase of density, which will place it in the same relation to the latter as the Mediterranean bears to the Atlantic; but that the influence of Polar cold will be to produce a *continual descent* of the water within its area; thus constituting the *primum mobile* of the General Oceanic Circulation, of which no adequate account had previously been given. This conclusion, as our readers will have seen, has been most explicitly accepted by Sir John Herschel.

Our limits do not admit of our following Dr. Carpenter through his discussion of the relative shares of the Gulf Stream and of the General Oceanic Circulation in that amelioration of the temperature of the Polar area, of which the industry of Dr. Petermann has collected a vast body of indisputable evidence; and for this discussion we would refer such of our readers as are specially interested in the question to the last part of the "Proceedings of the Royal Geographical Society." But as Dr. Carpenter has now shown a capacity to deal not merely with Physiological but with Physical questions, in a manner which has obtained the approval of some of the ablest physicists of our time, we hope that he will not again be accused (as he was by some of those who opposed his views on their first promulgation) of venturing beyond his depth when he began to reason on these subjects, and of advancing doctrines which his own observations refuted. The exclusive doctrine of the thermal action of the Gulf Stream advocated by Mr. Croll, rests, as Dr. Carpenter has shown, upon so insecure a basis, that a very large body of careful observations must be collected before any reliable data can be obtained as to the heat it actually carries forth from the Gulf of Mexico. And how much of this heat is dissipated by evaporation, as well as by radiation, before one-half of the Stream reaches the banks of Newfoundland (the other half having turned round the Azores to re-enter the Equatorial current), is a question which there are as yet no adequate data for determining. On the other hand, in his conclusion that a great body of Ocean water slowly moving northwards, so as to carry with it a considerable excess of temperature even to the depth of 500 or 600 fathoms, must exert a much greater heating power than the thinned-out edge of the Gulf Stream, Dr. Carpenter seems to us to have both scientific probability and common sense on his side.

7

Reprinted from *Nature* 9:387–388 (Mar. 19, 1874)

A NEW THERMOMETER

OUR readers will doubtless recollect a recent discussion in our pages relative to the priority of the invention of protected bulbs for deep-sea thermometers. The discussion has done something more than establish priority of invention, it has been the means of producing what, we believe, will prove to be a new and valuable meteorological instrument, for we have before us a paper by Messrs. Negretti and Zambra, communicated to the Royal Society by Dr. Carpenter at their last meeting, describing a new thermometer of such novel construction that it cannot fail to interest all scientific persons, meteorologists especially. We regret our inability, owing to want of space, to reproduce the paper in its entirety. The following are the main points of this communication.

In Prof. Wyville Thomson's "Depths of the Sea," p. 299, occurs the following passage:—"I ought to mention that in taking the bottom temperature with the Six's thermometer the instrument simply indicates the lowest temperature to which it has been subjected; so that if the bottom water were warmer than any other stratum through which the thermometer had passed, the observations would be erroneous."

Undoubtedly no other result could be obtained with the thermometers now in use, for unfortunately the only thermometer available for the purpose of registering temperature and bringing those indications to the surface, is that which is commonly known as the Six's thermometer—an instrument acting by means of alcohol and mercury, and having movable indices with delicate springs of human hair tied to them. This form of instrument registers both maximum and minimum temperatures, and as an ordinary out-door thermometer it is very useful; but it is unsatisfactory for scientific purposes, and for the object for which it is now used (viz. the determination of deep-sea temperatures) it leaves much to be desired. Thus the alcohol and mercury are liable to get mixed in travelling, or even by merely holding the instrument in a horizontal position; the indices also are liable either to slip if too free, or to stick if too tight. A sudden jerk or concussion will also cause the instrument to give erroneous readings by lowering the indices if the blow be downwards, or by raising them if the blow be upwards. It was on reading the passage in the book above referred to that it became a matter of serious consideration with Messrs. Negretti and Zambra, whether a thermometer could be constructed which could not possibly be put out of order in travelling, or by incautious handling, and which should be above suspicion and perfectly trustworthy in its indications. This was no very easy task. But the instrument submitted to the Fellows of the Royal Society seems to fulfil the above onerous conditions, being constructed on a plan different from that of any other self-registering thermometer; and containing, as it does, nothing but mercury, neither alcohol, air, nor indices. Its construction is most novel, and may be said to overthrow our previous ideas of handling delicate instruments, inasmuch as its indications are only given by upsetting the instrument. Having said this much, it will not be very difficult to guess the action of the thermometer; for it is by upsetting or throwing out the mercury from the indicating column into a reservoir at a particular moment and in a particular spot, that we obtain a correct reading of the temperature at that moment and in that spot.

The thermometer in shape is like a syphon with parallel legs, all in one piece, and having a continuous communication, as in the annexed figure. The scale of the thermometer is pivoted on a centre, and being attached in a perpendicular position to a simple apparatus (which will be presently described), is lowered to any depth that may be desired. In its descent the thermometer acts as an ordinary instrument, the mercury rising or falling according to the temperature of the stratum through which it passes; but so soon as the descent ceases, and a reverse motion is given to the line, so as to pull the thermometer to the surface, the instrument turns once on its centre, first bulb uppermost, and afterwards bulb downwards. This causes the mercury, which was in the left-hand column, first to pass into the dilated siphon bend at the top, and thence into the right-hand tube, where it remains, indicating on a graduated scale the exact temperature at the time it was turned over. The woodcut shows the position of the mercury *after* the instrument has been thus turned on its centre. A is the bulb; B the outer coating or protecting cylinder; C is the space of rarefied air, which is reduced if the outer casing be compressed; D is a small glass plug on the principle of Negretti and Zambra's Patent Maximum Thermometer, which cuts off, in the moment of turning, the mercury in the column from that of the bulb in the tube, thereby ensuring that none but the mercury in the tube can be transferred into the indicating column; E is an enlargement made in the bend so as to enable the mercury to pass quickly from one tube to another in revolving; and F is the indicating tube, or thermometer proper. In its action, as soon as the thermometer is put in motion, and immediately the tube has acquired a slightly oblique position, the mercury breaks off at the point D, runs into the curved and enlarged portion E, and eventually falls into the tube F, when this tube resumes its original perpendicular position.

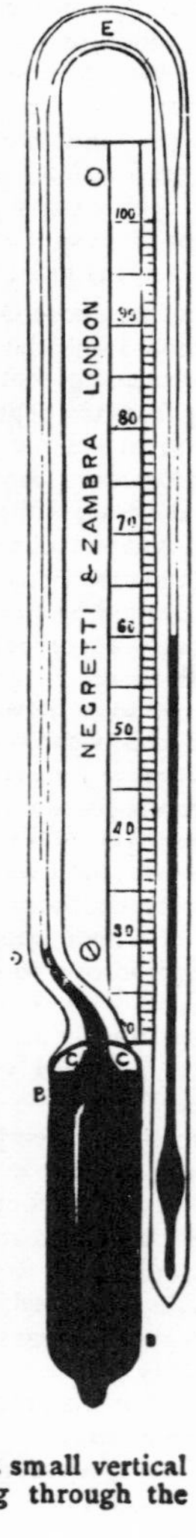

The contrivance for turning the thermometer over may be described as a short length of wood or metal having attached to it a small rudder or fan; this fan is placed on a pivot in connection with a second; on the centre of this is fixed the thermometer. The fan or rudder points upwards in its descent through the water, and necessarily reverses its position in ascending. This simple motion or half turn of the rudder gives a whole turn to the thermometer, and has been found very effective.

Various other methods may be used for turning the thermometer, such as a simple pulley with a weight which might be released on touching the bottom, or a small vertical propeller which would revolve in passing through the water.

[*Editor's Note:* Material has been omitted at this point.]

8

THE ORIGIN OF THE DEEP WATERS OF THE ATLANTIC

Georg Wüst

Institute of Oceanography, Berlin

This article was translated by the Institute of Oceanographic Sciences, from "Der Ursprung der Atlantischen Tiefenwässer," in Sonderabdruck aus dem Jubiläums-Sonderband 1928 der Zeitschrift der Gesellschaft für Erdkunde zu Berlin, *1928, pp. 506–534*

[*Translator's Note:* Significance of "longitudinal." In this paper the adjective "longitudinal" means "along the length of the Atlantic." It does not mean parallel to the circles of latitude. The circulations and sections referred to are, in fact, more nearly meridional.]

The problem of the deep-level circulation of the Atlantic formed, according to the plans of Alfred Merz, the principal task of the German Atlantic Expedition. He left behind him in 1925 a precious bequest to the Expedition in his final conception of the interchange of water in the depths of the oceans, thereby providing the Expedition with a working hypothesis which has proved very productive. Merz was in a position to formulate as the final goal of the Expedition one which would correspond to the systematic lay-out of the stations for which he had provided in the plan of the Expedition. That goal was the numerical evaluation of the deep-level circulation over the whole area of the Ocean. The degree of approximation with which the solution of this difficult task will be obtained cannot as yet be decided at the beginning of the processing of the great amount of observational material which, in a manner faithful to Merz's plans, has been obtained from 310 stations along 14 profiles. That is a question reserved for the full report of the Expedition. It is, on the other hand, now possible to answer some of the questions raised by Merz (1925, pp. 8-10) which refer to the north-south components of motion and whose primary concern is the origin of the individual members of the longitudinal circulation of the Atlantic. We will seek in the following to obtain quantitative insight into the structure of the deep-level waters of the Atlantic and its relation to the large-scale convergences of the surface currents on the basis of new longitudinal sections of temperature and salinity.

ASPECTS OF THE ARRANGEMENT OF THE SECTIONS

A. Merz arrived at the new ideas on the circulation chiefly on the basis of the longitudinal section at 30°W longitude. This, because of the sparsity of observational material for the central parts of the Ocean, was the only one which could be framed. This section lies in the Western Trough in the South Atlantic and, on the contrary, in the Eastern Trough in the North Atlantic. When preparing for the Expedition Merz realized the significance for the circulation of the effects of relief and took careful account of this factor in

[*Editor's Note:* The map (Figure 32) and Plates 33, 34, and 35 have been omitted because of limitations of space.]

planning the lay-out of the cross-sections. As regards the question of the longitudinal circulation he contemplated making a special processing of several longitudinal sections which would cover the troughs on both sides of the Mid-Atlantic Ridge as well as the Ridge itself. At the stations falling in these longitudinal sections the physical, chemical, and biological facts were to be determined as thoroughly and at as narrow intervals as possible down to the bottom. This intention was put into effect throughout the Expedition and already during the voyage a start was made on the combined oceanographic, chemical, and biological processing of both the principal sections along both troughs. The first results on the difference between the circulations of the West and East Atlantic derived from this work are published in the Reports of the Expedition (Expeditionsberichte). The full understanding of the phenomena, especially as regards the source areas of these longitudinal circulations, can, however, only be obtained by fitting these partial sections into the complete system which comprehends both halves of the Ocean between 80°N and 80°S in the same way. An attempt has been made to do this in both the longitudinal sections considered which are at one time based on the stations already used in the Expedition reports and at another time on the usually accessible reliable material of earlier expeditions.

The new longitudinal sections pay as much attention as possible to the morphological conditions. The Mid-Atlantic Ridge, whose S-shaped form is a reflection of the Ocean border, divides the Ocean into two elongated deep-sea troughs. The troughs are, in turn, separated into individual basins more than 5000 m deep by rises and ridges. As the Central Ridge rises, like a closed mountain wall, to depths of 2500 to 3000 m so beneath those depths, apart from a small passage at the Romanche Trench near the equator, the separation of the two systems of deep-sea basins from one another is complete. It is only in the high southern latitudes to the south of 55°S, where the spine of the Atlantic Ocean bends round to the Indian Ocean, that there exists no dividing wall between the west and east Atlantic depths.

Our longitudinal profiles seek out, in a frequently winding course, the greatest depths of the basins and the deepest depressions in the transverse barriers, and provide in the relief of the sea-bottom a direct representation of the continuity of the great depths and of the potentialities of the predominantly meridional exchange of water between them. Some individual small deviations from the primary course have had to be made, especially in the North American Basin, to allow for the location of the scanty number of observation stations available on it. The construction of the sea-bottom relief in the sections and in the general map attached to them is based in the South Atlantic Ocean and in the North Atlantic between 0° and 20°N on the Expedition's preliminary bathymetric charts (cp Defant, 1927), and for the rest of the area on the representation given by M. Groll (1912). The longitudinal sections are not projected on to a mean meridian but the individual stations are entered at the true intervals corresponding to distance in sea-miles. On this account the degrees of latitude in our sections are located at varying intervals. The further apart they are so will the zonal component of the representation naturally be stronger and the closer together they are so will the meridional component be greater. Whatever may be the prejudicial effect of this on the comparability of the two sections one can, on the other hand, hope to come closer to a knowledge of the true propagation of the water masses because this, especially in the deeper localities, is strongly influenced by the morphology.

THE LONGITUDINAL MORPHOLOGICAL CONFIGURATION OF THE TWO TROUGHS

With regard to the longitudinal morphological configuration, the sections, taken together with the general map[1] which shows merely the solid masses of the continental slopes, the central ridge, and the cross-rises, produce a set of characteristics which are important for more than the question of the longitudinal circulation. One must, in this connection, always bear in mind the great increase in the vertical scale of the sections (ratio length : height = 1 : 926).

Both profiles begin in the south at the edge of the Antarctic Continent. The sea-bottom falls relatively quickly down to depths of more than 5000 m in the South Polar Basin. The western section then goes round the South Antilles Arc, passes in the South Sandwich Trench the deep northerly outlet of the South Polar Basin which lies as a preliminary deep area outside the South Antilles Arc, and crosses diagonally the great depths of the Argentine Basin. The sea-bottom here rises slowly to the Rio Grande Rise which probably nearer to the Continent contains an opening about 4300 m deep so that the deep waters of the Argentine Basin above these depths on the Rise are connected with the deep waters of the Brazilian Basin. The sea-bottom sinks again here to depths of more than 5500 m till a small wave in the bottom reveals to us the proximity of the island of Trinidad the base of which we graze. After crossing the Brazilian Basin our western profile turns, following the general structure of the Ocean, to the west and rises to the Para Rise. This is a gentle but extensive dome in the equatorial region. Beyond this the sea-bottom sinks continuously to the great depths of the North American Basin which, south of Bermuda, reach 6300 m. The absence of modern stations compels the use of this wide deflection up to the base of the Bermudas on whose foot there fall some of the "Bache" stations. Between 33° and 40°N, therefore, our profile runs in a predominantly zonal direction, to turn northwards again at the top of the Newfoundland Bank. Here the North American Basin has its northern limit. The rise in the sea-bottom is associated here with a marked narrowing of the West Atlantic Trough, and at about 50°N we cross a rise which has hitherto been absent from the bathymetric charts of Groll (1912) and Schott (1926b), but which was already shown in the Carte Generale Bathymetrique des Oceans (Monaco, 1906?, Edition without contours on land) as probably existing on the basis of soundings. We will see in addition that bottom temperatures show that there is no doubt about the existence of this rise which we intend to name the "Newfoundland Rise." It forms a submarine connection between the Newfoundland Bank and the Mid-Atlantic Ridge. The northern end of the Labrador Trough is formed by the Davis Strait which coincides with a rise going up to 650 m and which leads across to the approximately 2000 m deep Baffin Bay.

In following the east section from south to north we first of all, on leaving the South Polar Basin, cross the Atlantic-Indian Transverse Ridge which rises in the plane of our section to depths of 3000 m. Further east, though, this continuation of the Atlantic Ridge which leads across into the Indian Ocean seems to occupy deeper depressions going down to about 4000 m (cp. Wüst, II, Bericht 1926, p. 248). To the north there follows the Cape Basin with

1. In the terminology of the features of the sea-bed we follow the names laid down by A. Supan in 1903 (A. Supan. Terminology of the most important submarine bottom features. Petermanns Mitteilungen 1903, p. 151), and thus frequently deviate from the terms chosen for the Expedition reports which mainly follow those used by Schott (1926b).

depths of more than 5000 m. The Walvis Ridge, which in our section consists of two elevations towering up respectively to 2000 and 2500 m, separates the Cape Basin from the Congo Basin. This last also has recently been shown to be a basin which is closed all round below 4200 m and which falls in the central part to 5800 m. In the north it is the Guinea Ridge which forms this barrier and which probably also has a somewhat greater depth of sill (about 4200 m) further to the south-west. To the north of this ridge whose existence has recently been established by the *Meteor* there lies the Guinea Basin which is a relatively shallow depression whose depth scarcely exceeds 5000 m. The Sierra Leone Rise (about 4300 m depth on the top of the sill) is the next transverse ridge before the Cape Verde Basin which reaches markedly greater depths (over 6000 m). The shallow dome between the Azores and Madeira which we term the Azores Rise cuts off the Spanish Basin (5800 m). The Iceland Rise provides the natural conclusion to the eastern deep-sea trough.

The two elongated, more or less parallel, depressions of the Ocean exhibit in their longitudinal profiles a number of conformities which are of outstanding significance for the development of the general deep-sea circulation of the Atlantic. Both sea-bottom profiles possess, on the one hand, the pronounced cutting-off of the Arctic Basin from the Atlantic depths in 60° to 65°N latitude and, on the other hand, the broad opening into the Antarctic Basin. In both troughs at about 30°S there is a remarkable morphological disturbance in the shape of the Rio Grande Rise-Walvis Ridge system which turns out to be the greatest zonal transverse ridge of the Ocean. A second, less marked, transverse rise spanning both basins occurs in the equatorial region. This constriction coincides with a pronounced contraction in the width of the Ocean so that here the vertical cross-section available for the exchange of water between the two hemispheres is reduced on two accounts. Finally, there is a striking degree of conformity in both bottom profiles in regard to the greatest depths of the basins (about 6000 m), and, if we imagine the deep sea-bottom between 70°S and 50°N to be smoothed out, also in regard to the medium depths (about 5000 m).

In contrast to these conformities there exists in the detailed structure of the two troughs a series of differences which produce remarkable differences between the West-Atlantic and East-Atlantic circulations. The East-Atlantic trough is markedly more strongly organized into separate parts. Not only is the number of ridges and rises in it greater, but they also rise up to greater heights. Whereas in the West-Atlantic trough passages at least 4300 m in depth are everywhere available to the lower arms of the currents in the East-Atlantic trough the Walvis Ridge completely cuts off the deep waters of the Antarctic below about 3000 m. In addition the meridional exchange of the sea-bottom water will obviously be restrained here by the remaining ridges and rises.

THE SOURCE-MATERIAL

In order to judge the soundness of our new results it is necessary to be acquainted with the original data from which we construct the sections of temperature and salinity.

In the South Atlantic there are 23 *Meteor* stations (13 in the east and 10 in the west section) which have already been used in preparing the longitudinal sections in the Expedition reports. These values,[2] which are based on double readings of temperature and numerous titrations of the chlorine content, constitute a reliable set of data. Also, in conformity with the systematic character of the Expedition, they cover uniformly all depths and all geographic latitudes. Though they go only down to 2000 m depth it was possible for the Weddell Sea to use the careful determinations in six series due to Brennecke (1921).

In the North Atlantic up to 20°N a sure basis was provided by nine additional *Meteor* stations. In higher latitudes, especially in the western section, many gaps were found in the material. As yet there are no modern determinations of salinity and temperature at all from the great abysses of the North American Basin. It is here that the most painful gap in our present knowledge is to be found. It is however, to be hoped that the stations of the *Dana* Expedition will soon produce the first modern deep-reaching serial observations from these unexplored depths. At present here in the west section between 15°N and 35°N it is only for the three *Challenger* stations (stations numbers 20, 29, and 62) that reliable temperature values to 3000 m and from the bottom are available. Further the American surveying ship *Bache* has made a number of modern determinations of temperature and salinity down to 1800 m in the neighborhood of the Bermudas (Report 1917). This material is sufficient to construct the temperature section. For the construction of the salinity section we are left with an indirect procedure which uses the relation between temperature and salinity given by Helland-Hansen (1918).[3]

Further north in the west section the *Michael Sars* Expedition of 1910 (Murray and Hjort 1912) obtained excellent observational material of which, unfortunately, only extracts have so far been published.[4]

For the south Labrador Basin we have available one station of the *Scotia* Expedition (Bruce 1906) which gives observations down to only 1000 m. Further north the stations of the *Ingolf* Expedition form the basis of our representation. The measurements of temperature and, most remarkably, also those of chlorine content of this by now long-past Expedition thoroughly measure up to

2. These data still lack the reduction to true depth on the basis of the simultaneous readings of the two types of reversing thermometer, one unprotected and the other protected against pressure. However, this correction would not on the scale of our representation produce any appreciable change in the outline.
3. We determine by a graphical method the relation t/S for all the stations falling in the Sargasso-Sea area (*Meteor* 287, *Deutschland* 46, *Margrethe*, *Bache* 10179 and 10181, *Michael Sars* 63) and obtain a set of curves from which a systematic change with latitude in this relation can be deduced. We then interpolate the curves for the latitudes of the *Challenger* stations (20, 69, and 62) and derive from them the salinities corresponding to the deep-water temperatures. Because of the sparsity of observational material a certain insecurity attaches to this procedure and we cannot exclude the possibility that here our salinity section will be altered by the results from new stations.
4. We quote from the publication of Murray and Hjort (1912) the values for station 63 published on page 246, and derive from the Newfoundland Bank-Iceland cross-section reproduced on page 250 on a, to be sure, very small scale the intersections of the isohalines and isotherms for station 82 which falls in our longitudinal section.

modern requirements as a comparison with the modern Norwegian stations in the northern part of the Labrador Basin (published in Bulletin Hydrographique 1915-1923 du Conseil perm. international pour l'exploration de la mer) shows. These stations at which observations were made by the Norwegians (1924) within the framework of international marine research provide the first modern determinations of temperature and salinity from the great depths of the southern part of Baffin Bay. (In our sections use is made of the station of 13 August 1924 at 67°19'N, 58°52'W at 1335 m depth.) In the northern part of Baffin Bay **the measurements of the Nordenskjöld Expedition of 1883 (Hamburg 1884) give us a sure basis.**[5]

In the eastern section two stations of the *Deutschland* Expedition (Nos. 36 and 38) link up with the *Meteor* stations, while in the further sector between 30° and 80°N our construction is based predominantly on the works of the Norwegian scientists Helland-Hansen and Nansen (1909 and 1926) who have extended our knowledge of the northeast Atlantic Ocean and the Norwegian Sea in an admirable way. This Norwegian material is supplemented by observations at some Danish stations [*Thor* (Nielsen 1907), *Margrethe* (Jacobsen 1916), *Dana* (Helland-Hansen and Nansen 1926)] and by those made at two stations by the cruiser *Berlin* on the suggestion of A. Merz (Möller 1926a).

For depths below 1000 m we notice that the deep-level material is distributed relatively uniformly and that it is only in the western section north of 15°N that large gaps are to be found. This time, in opposition to earlier sections, only corresponding values of temperature and salinity are used, so that the run of the isotherms is strictly comparable with that of the isohalines. On the other hand in consequence of seasonal differences a certain degree of inhomogeneity of the material continues to exist. In high latitudes our sections reproduce the summer situation but in middle and lower latitudes the observations derive from differing seasons. Many unevennesses in our sections are doubtless due to these sources of error.

THE DEVELOPMENT AND PROPAGATION OF THE DEEP-LEVEL ATLANTIC WATERS

Our two new sections confirm in their main features the picture we had gained earlier for 30°W and show the already known basic facts of the temperature and salinity structure. The agreement of the sections is surprising if we consider that the section at 30°W in the South Atlantic was constructed from very fragmentary material, and that especially in the crucial region between 30°S and 55°S the construction still remains quite hypothetical, above all in regard to the upwelling of North Atlantic Water first recognized by Merz and the entry of this Water into the Weddell Sea. The *Meteor* Expedition has thus brought the final proof of the correctness of these new ideas. In particular the new sections because of their arrangement to take account of the morphology lead to a series of new insights, and already partially answer some of the questions thrown out by Merz in his last paper (1925) about the origin and disappearance of the different sorts of deep-sea water.

5. The temperatures of this Expedition were taken from Hamberg's paper (1884). The salinities, however, as the original transactions were not accessible, were taken from Krümmel's Handbook (1907, p. 348) which gives the values already reduced by means of Knudsen's Tables.

In earlier times the symmetry in the structure of the Ocean on both sides of the equator was exaggerated and it was certainly, as is well known, this exaggeration which offered the basis for the idea of a symmetrical circulation. This earlier, exaggerated, degree of symmetry, which was first reduced to its true measure by Merz, is reduced still further by our new sections and now continues to exist for only a shallow surface layer. In the depths of the Ocean an astonishing degree of asymmetry prevails. Two gigantic primary centers of action lie opposite one another. These centers are the cold low-salinity South Polar area and the warm high-salinity North Atlantic Ocean. The substantially smaller spaces of the North Polar Basins and the Mediterranean are barred off and therefore only come into question as secondary areas of action. The effects of the Indian and Pacific Oceans also seem to be of minor importance. These centers of action are the areas of development of the deep-level waters which spread out from them in the various layers of the Ocean. In the depths of the Atlantic Ocean we have to do with two principal types of water, "Antarctic Water" and "North Atlantic Water" which respectively correspond to the two principal source regions.[6]

If one tries to characterize the water types more closely by temperature and salinity and, in doing so, leaves out of consideration the uppermost roughly 200 m thick surface layer which is directly subjected to climatic effects, a further subdivision corresponding to the four layers in the structure of the Ocean is found to be necessary. If we use the source region as a title, then on the basis of our sections and imitating the well-known designations of the deep-water currents, we arrive at the classification [in Table 1, p. 72].

[*Translator's Note:* I have translated literally the German term Unterwasser applied in this paper to the Type I Water Mass. Its properties, location in depth, and source regions appear to correspond roughly to those of the mass termed "Central Water" in *Oceanography for Geographers,* ch. 3. (King 1962) following "The Oceans" (Sverdrup and others) but I thought it best to translate the word literally as "Central Water" may imply the use of information not available in 1928.]

The extraordinarily sharp discontinuity layer in the temperature at about 100 m depth forms the lower boundary of the vertical turbulent exchange of the tropical water formed at the sea-surface. It coincides with the lower boundary of those "tropical undercurrents" which are so beautifully recognizable in the salinity and which carry water of high salinity below the low-salinity equatorial waters.

Below this discontinuity layer there spreads out cooler less-saline water of subtropical origin which we term Subtropical Under Water. Its lower boundary is roughly determined by the 10° isotherm and the 35.5 ‰ isohaline. We know that this water type has its greatest vertical thickness in two regions located to some extent symmetrically on both sides of the equator at about

---

6. The separation into the two types "polar water" and "tropics water" suggested by E. v. Drygalski (1926) has the great advantage of simplicity of the sort which perhaps attaches to the meteorological distinction between polar air and tropical air. It does not on this account recommend itself for the Atlantic Ocean, because the principal mass of "tropical water" does not originate from tropical latitudes and because, further, it does not bring out the asymmetry with respect to the equator of the two source regions.

Table 1. Classification of the deep-level (below 200 m) waters of the Atlantic

Type of Water	Approximate limits of Salinity $^{o}/oo$	Approximate limits of Temperature C^{o}	Approximate position in depth, m	Remarks
I. Subtropical Under Water	36.7-35.5	20-10	200-700	Two source regions.
II. Subpolar Intermediate Water				
a. Subantarctic	34.9-34.0	10-4	about 1000	
b. Subarctic	34.9-34.5	10-4	about 1000	
III. North Atlantic Deep Water	35.5-35.0	10-4	1000-2000	In the inflow into the Mediterranean in front of the Straits of Gibraltar S $^{o}/oo$ up to about 36.5 and t^{o} rising to 11°.
IV. Bottom Water				As Deep-Level Water also in the South Atlantic Ocean.
a. North Atlantic	35.0-34.85	4-2	2000-bottom	
b. Antarctic	34.8-34.65	< 2	1000-bottom	
c. Arctic	34.92-34.88	< 2	500-bottom	

30° latitude. In the North Atlantic it can be found down to depths of as much as 1000 m while in the South Atlantic it extends down to only about 500 m. On the other hand in a broad equatorial zone between 15°N and 15°S this Subtropical Under Water has a vertical thickness of only about 300 m. It now turns out that these accumulations of relatively warm and highly saline water in deep water in both hemispheres do not coincide with the locations of maximum salinity at the sea-surface which are found at about 25°N and 18°S. Still less do these accumulations coincide with the areas of greatest evaporation which, indeed, according to earlier investigations (Wüst 1920) are displaced still further towards the equator. Thus the idea, repeatedly put forward, that there is a direct connection between the two phenomena is found to be untenable. It is not by convection that has been strengthened by high evaporation and salt enrichment that heat is transported into the depths, but as has before now been stated, especially by Sandström (1918), it is the dynamical processes which play the main part in the production of these accumulations of warm and highly saline water in deep water. The map of sea surface currents produced by Hans H. F. Meyer (1923) which, indeed, replaced the system of current-free centers of symmetrical circular streamlines by regions stirred by currents with convergence lines makes these dynamical processes clearly recognizable. In both the South Atlantic and North Atlantic a marked convergence line runs at about 30° latitude in a winding course over almost the whole breadth of the Ocean. At this line the water masses of the West-wind zone meet those of the Equatorial Currents producing components of motion directed vertically downwards. We term it, for brevity, the "Subtropical Convergence."[7] Our two longitudinal profiles cut these subtropical convergences in different geographical latitudes; the points of intersection coincide fairly closely with the position of the strongest warm-water accumulations in our sections, the comparison [in Table 2, p. 74] proves.

This agreement, especially also as regards the difference of latitude between the west and east sections, makes the causal connection between the two phenomena clearly recognizable. From these convergence areas the Subtropical Under Water spreads out in various directions, including equatorwards, and, in doing so, is more and more subject to mixing with water masses of different origin.

The Subtropical Under Water accumulates most strongly in the northern North Atlantic Ocean as here the Subpolar Intermediate Water sets a limit to its development to a far smaller extent than in the south. The forces arising at the Subtropical Convergences squeeze the warm and highly saline water into the depths. Corresponding to the stronger development of the convergence area in the Northern Hemisphere the accumulation of warm and highly saline waters extends down to greater depths in the north than in the south. This phenomenon is doubtless connected with the passage of South-Atlantic water into the Northern Hemisphere which takes place continuously at the surface along the northern branch of the South Equatorial Current. However other factors, with which we will deal later, also come into this matter.

It is remarkable that the convergence phenomenon starting from the surface

7. We have not counted the convergence located further south between the Falkland and Brazil currents as the southern subtropical convergence; it is in certain respects comparable to the "Cold Wall" in the North Atlantic and it is therefore better to consider it to be a secondary convergence which is ascribable to the southern polar front.

Table 2. Relation between the position of the Subtropical Convergence and the Subtropical Warm-Water Accumulation in both longitudinal sections

Geographical position of	West section	East section
the Northern Subtropical Convergence (according to Meyer) . . .	28°-32°N	29°N
the Northern Subtropical Warm-Water Accumulation	31°N	29°N
the Southern Subtropical Convergence (according to Meyer) . . .	34°S	29°S
the Southern Subtropical Warm-Water Accumulation	34°S	29°-30°S

at 30°S almost coincides with the already mentioned morphological disturbance constituted by the two transverse barriers of the Walvis Ridge and Rio Grande Rise. This greatest of the zonal elevation has, as we will see later, a profound effect on the propagation of the water types. Thus the area around 30°S is, for two reasons, found to be a zone of disturbance in which through the appearance of vertical components in the flow and morphological effects the surface currents and deep-water movements are hindered and diverted.

The next layer of the Ocean is filled with the Subpolar Intermediate Water which betrays its presence in almost all latitudes between 40°S and 60°N by a pronounced intermediate salinity minimum at about 1000 m. Specifically, this Intermediate Water derives from two source regions, a subantarctic one and a subarctic one.

It is formed in enormous amounts at the edge of the ice-covered Antarctic. W. Brennecke (1921) and E. v. Drygalski (1926), in particular, have devoted themselves to the problem of its origin. The German Atlantic Expedition was able to make a thorough study of the processes in this area, which it crossed several times along transverse and longitudinal sections, and to report the most important results whilst still at sea (cp II.Report, 1926, pp. 245 ff.).

Under the influence of the polar climate and the melting of the ice, cold and low-salinity water masses are formed at the surface in the Antarctic and flow away to the north. In the southern summer these water masses are covered by a warmer but equally low-salinity surface layer so that we can encounter a cold Antarctic undercurrent in the temperature stratification north of 65°S. This can be traced at the 100 to 200 m level in the western section as far as 53°S and in the eastern section even to 49°S. Because of heavy atmospheric precipitation the Antarctic water masses advancing northwards between 50° and 53°S undergo a further reduction in salinity at the surface. After this, between 45° and 50°S they sink down to the true Intermediate Water and, in doing so, undergo mixing, if only to a small extent, with warmer water masses. The Intermediate Water thus attains its final form outside the Antarctic and its source area lies at the surface between 40° and 50°S. Now we can answer the question put forward by Merz (1925, p. 9), as to whether in relation to this branch of the circulation the preferable adjective is "Subantarctic" as chosen by Brennecke or "Antarctic" as formerly used by us, in favor of Brennecke's term.

There is no doubt that the large-scale subsidence in question of the Antarctic water masses to the Intermediate Water is connected with a remarkable convergence line which divides the West Wind Drift in the central part of the Oceans, which has hitherto been interpreted in the current charts as a uniform whole, into two ocean current areas. Already in 1923 Meinardus had recognized the bifurcation of the West Wind Drift from the temperature distribution at the sea-surface and even conjectured the connection between this phenomenon and the subsidence of the cold Antarctic water to the "Subantarctic deep-level current" as he terms the Intermediate Water (1923, p. 540). Later Schott (1926a) named this boundary line the "Meinardus Line" and interpreted it as a closed convergence line of the current set for the whole southern water ring. For the meridian of 60°E he represents it schematically by current arrows placed perpendicularly one against the other, and locates it at 47°S on the basis of a longitudinal salinity section which points to the occurrence of subsidence at that latitude. M. Willimzik has sought with success in a paper completed in 1924 (published in 1927) to represent the south Indian convergence line from the certainly sparse observations made by expeditions of differences between dead reckonings and observed positions, and has established its close connection with the deep-level circulation and the pack-ice boundary. His treatment unquestionably shows that the meridian of 60°E is not crossed by a convergence line in the current set between 40° and 50°S. The observations of currents and temperatures of the *Valdivia* Expedition make it very probable, when they are taken together with the run of the pack-ice boundary, which exactly between 50° and 60° longitude shows a pronounced protrusion towards the south, that in the region of these meridians a convergence of currents exists in higher southerly latitudes and also that here the subsidence of the cold water masses ($\lessapprox 2^{\circ}$) already takes place more to the south. Moreover one must rid one's mind of the idea of any systematic rule that the subsidence of the polar water masses is recognizable everywhere from the presence of a closed convergence line of the current set at the surface. This will only be the case here and there. In other areas the convergence line is absent, and the subsidence to be seen in the longitudinal sections can, when there is no inward flow from the side, be associated with a dynamic convergence, that is to say with a decrease in the speed of a current which as a rule flows in approximately the same direction. Similar conditions are indeed known to us from the analogous conditions in the atmosphere, e.g., from the polar front which is indeed also shown only locally by a convergence in wind direction. In general the processes which take place in the ocean and the atmosphere when cold polar masses encounter the warmer masses of temperate latitudes exhibit close agreement. Following a suggestion by Prof. Defant[8] we therefore adopt the suggestive meteorological term for the oceanographic process and term this convergence line of the West Wind Drift the oceanic "Polar Front." In the North Atlantic it is known to us from the descriptions of currents, namely as the "Cold Wall" and as the eastern boundary of the East Greenland Current. In the South Atlantic it has hitherto been absent even in the treatment of the speeds of flow of the currents made by Hans H. F. Meyer (1923) from

8. A. Defant: The systematic investigation of the ocean. This special volume, p. 459. Further, in *Dynamic Oceanography*. In preparation with J. Springer, Berlin.

the hydrodynamical viewpoint, which, otherwise, brings together all the important surface current convergences in a representation made in accordance with the modern point of view. Dr. Hans H. F. Meyer now offers for this region a modified representation of the currents, which is, indeed, still hypothetical but, nevertheless, probably correctly reproduces the fundamental principles (cp the comments of Dr. Hans H. F. Meyer in the footnote[9]). This chart makes it probable that in the central parts of the Ocean between about 40°W and 10°E

---

9. Comments on the map of surface currents: The . . . current chart is a reproduction of the chart of surface currents of the Atlantic Ocean published in *Veroffentlichungen des Instituts für Meereskunde,* N. F. Reihe A, Heft 11, Berlin 1923, in which the discernible important convergence lines are made clear by a plotted line. In addition, using information provided by G. Wüst the principal source regions of the Antarctic Bottom Water have been entered at the positions conjectured by him, just as there have been entered the positions of the principal source regions of the Arctic Bottom Water in approximately the position reported by Nansen (F. Nansen, The bottom water and the cooling of the sea, *Internat. Rev. Gesamten Hydrobiologie Hydrographie,* Bd. 5, Heft 1, 1912, pp. 15-24).

A part that is altered is the representation of some of the Antarctic surface currents in order to take account of knowledge derived from, to the most part, the recent treatment of the deep-level circulation of the Atlantic and Indian Oceans. This knowledge enables us to conjecture with high probability the existence of a convergence line at about 50°S. In regard to this it was possible to assume the existence north of this convergence line between 40°W and 10°E of a purely westerly current whose existence was verified by the few drift observations available there. To the south of this convergence line and especially between 30°W and 10°E the drift observations made by the *Meteor* on profile V (Jan./Feb. 1926) indicate a more northerly set of the current than was shown on the 1923 chart. The *Meteor* drift observations made in October on profile III (48½°S) also show a northerly component. Unfortunately the drift observations made by the *Carnegie* in January 1916 *(Researches of Terrestrial Magnetism,* Vol. 5, Washington 1917, p. 148) could not be taken into consideration as, probably because of special weather conditions, they show currents which with few exceptions are directed to the south or west. These observations are directly opposite to the overwhelming majority of all other drift observations and, also, cannot be reconciled with our other conceptions of the water circulation.

It was possible to leave the representations of the currents in the Weddell Sea unchanged. The same applied to the entry of a southerly current at 60°S and 40° to 50°E whose existence is made probable by the *Valdivia* drift observations, the recession of the pack-ice boundary, and the thermal favorableness of Kerguelen, and is supported by the investigation of M. Willimzik (1927, p. 26). The question as to the extent in latitude of the west current on the edge of the Antarctic must still remain undecided.

The more northerly current set shown between 40°W and 20°W now also provides a better explanation for the well-known cold tongue of Bouvet Island (G. Schott, Oceanography and maritime meteorology, *Wiss. Ergebnisse der "Valdivia"-Expedition,* Bd. 1, Jena, 1902, p. 133), which exists because this island lies in the area of current streams which come by a direct route from the edge of the Antarctic Continent. Because of the sparse amount of data . . . blank spaces on the arrow shaft[s] indicate that they refer to a "conjecture based on scattered observations" (Hans H. F. Meyer).

the southern polar front is developed as a convergence line in the current sets and runs here in approximately the position in latitude deduced by Meinardus from the temperature differences. As in the Indian Ocean it shows a marked parallelism with the pack-ice boundary. In the border areas of the Ocean it is not perceptible in the current sets.

Each of our two longitudinal sections crosses the southern polar front at points which from Meyer's chart lie at about 50°S for the west section and about 48°S for the east section. The salinity sections give the same geographical latitudes for the subsidence of the water masses and, indeed, even the small 2° difference in latitude between the east and west sections is common to both phenomena.

Before the southern polar front, consequently, there forms the Subantarctic Intermediate Water. From here it subsides in its core-region to about 1100 m in the west and 900 m in the east. It reaches this deepest position at about 35°S and then advances predominantly horizontally towards the north and only rises again quite gradually up to 700 m in the equatorial region. This advance can be followed in a particularly beautiful fashion in the salinity sections but does not appear clearly in the temperature sections. The principal reason for this is to be found, as we have already shown earlier (Merz and Wüst, 1922, p. 21), in the conditions at the surface of the subsidence area. Between 45° and 50°S the salinity is made uniformly low by precipitation while the temperature even from 45°S rapidly increases towards the equator. The vertical structure of the Antarctic Intermediate Water during its horizontal advance can be conceived as the image of the horizontal distribution of temperature and salinity at the surface of the subsidence area, specifically in such a way that the upper parts of the Intermediate Water correspond to the northern parts in the source area. And as, moreover, the underlying North Atlantic Deep Water has approximately the same temperature but markedly higher salinity, the advance of the Intermediate Water is marked by a pronounced salinity inversion, whereas in the temperature field its advance is indicated by only a weak inversion or by uniformity of temperature (homothermy). This is connected with the further phenomenon that the salinity minimum lies about 300 m higher than the temperature minimum; the latter is associated with the boundary with the Deep Water while the former marks the core-layer of the Intermediate Water itself.

The heat and salinity contents of our Intermediate Water are changed during the course of its progressive advance by turbulent mixing[10] in the boundary layers. However in the core-regions this effect can only be extraordinarily small as even under the equator at about 700 m depth the observed salinity is only 34.5 °/oo. On the far side of the equator mixing with the adjacent water masses of higher salinity seems to take place more intensely. The salinity rises more quickly in the direction of propagation of the water while the temperature even further on increases only gradually. The Antarctic Intermediate Water consequently becomes heavier and up to 20°N subsides again in its core-region to a depth of 1000 m. In this latitude there now sets in a strong mixing process with the warm and highly saline water squeezed downwards at the Northern Subtropical Convergence. Here also there begins the subsidence of the mixed water

10. L. Möller (1926) has drawn some important conclusions on the mixing processes associated with the formulation and propagation of the Intermediate Current and the Deep Current from the relation between temperature and salinity which had been determined for the 30°W section. These conclusions have in the main been confirmed by our sections.

down to the next layer which is that of the North Atlantic Deep Water. The last traces of Antarctic effects can be followed up in the intermediate salinity minimum up to 30°N in the west section and 31°N in the east section.

Also at the Northern Polar Front cold waters of low salinity subside and form the Subarctic Intermediate Water. However corresponding to the small catchment area of polar melt-water the Subpolar Intermediate Water is only produced on a very small scale in the North Atlantic. The Northern Polar Front encloses only a small area indeed of Arctic Water Masses. This Front borders the Labrador Current on the American side. In the middle of the Ocean it becomes broken and then south and east of Greenland it forms the outer boundary of the East Greenland current. It is not crossed by our east section and only just touched at 60°N in the west section. Thus we see that the formation of this Intermediate Water and its subsidence from the surface are still only just indicated in the west section at the southern edge of Baffin Bay where its principal source-region lies. In the east section, however, the connection of the Intermediate Water with the surface is absent. Both sections, however, possess the intermediate salinity minimum at about 1000 m. In the west this is substantially more strongly marked and can be followed in our section up to about 40°N. In the east, however, it can be recognized at this level only around 50°N since further south the salinity inversion is obviously affected by the addition of Mediterranean water. The damming-up of the warm and highly saline water masses of the Atlantic Current in front of the Iceland Rise forces the intermediate salinity minimum down to greater depths (1500-2000 m) in the east section and simulates an ascent of the Arctic Intermediate Water towards the south. Corresponding to the location of the section on the Polar Front we are concerned here, however, with a sideways influx of Intermediate Water from the west. The small depth of sill of the barrier surrounding the east Arctic Ocean scarcely permits the formation in the east section of a branch of the Subarctic Intermediate current flowing from the north.

While the Subantarctic Intermediate Water can be followed in the middle level of 1000 m over 70 degrees of latitude (i.e., almost 8000 km), the influence of the Subarctic Branch scarcely extends over 20 degrees. Already at 50°N the salinity of this latter water has risen to 34.9 °/oo and between 50° and 40° it is subjected to strong mixing with the warm and highly saline Under Water of the Northern Subtropical Convergence. The mixture of water produced out of both of them forms in this way the second root of the North Atlantic Deep Water.

In the temperature stratification the Subarctic Intermediate Water cannot be detected south of 50°N. This is the reason why its propagation in the northern North Atlantic Ocean was first brought to our notice by the modern salinity determinations. A. Merz (1925, p. 12) and L. Möller (1926a) were probably the first to recognize the last traces of its propagation as far as the Azores Plateau in their discussion of the observations of the cruiser *Berlin*, while Helland-Hansen and Nansen (1926) have recently pursued the study of this phenomenon in greater detail.

The formation of the North Atlantic Deep Water which fills up the next layer of the Ocean is obviously related to three processes, viz, the subsidence of relatively warmer and more saline water masses at the Northern Subtropical Convergence, the mixing of the latter with offshoots of the Subpolar Intermediate Water (Antarctic and Arctic Intermediate Water), and the influx from the side of Mediterranean Water through the Straits of Gibraltar which can be clearly recognized in the east section between 30° and 40°N by an isolated core-region of high salinity at a depth of 1000 m. The Deep Water is thus supplied from four sources and we agree with L. Möller's assertion (1926, p. 44) that mixing processes play a decisive part in the source-region of the North Atlantic Deep Water. It is developed most strongly in the east in consequence of the higher proportion of Mediterranean Water. It is in this way that within the sphere of the North Atlantic Convergence Area there forms an accumulation of warm and highly saline water which extends down to a depth of about 2000 m in the west and 2500 m in the east. This water is precisely the North Atlantic Deep Water which is propagated from here in the levels between 1200 and 2000 m in depth. This propagation clearly takes place along the most diverse directions and not, as had hitherto to be supposed from the information of the 30° west section, only with a southerly component. Our two salinity sections show without doubt that it also moves towards the north. In other words, from a center of action located below the Northern Subtropical Convergence the North Atlantic Deep Water spreads out more or less on all sides.

It pushes most strongly towards the Southern Hemisphere. We see from the elongated tongue of the area of over 35 °/oo salinity that its strongest movement occurs at a depth of about 1600 m; it can be followed at this concentration up to 10°N in the west section but, on the other hand, in the east section only up to 14°N. It can be identified from temperatures exceeding 4° between depths of 1000 m and 2000 m even as far as 10°S in the west section but, on the other hand, only up to 6°S in the east section. This Deep Water is thus the principal source of the portions situated above 2000 m of that highly saline deep-water layer which we encounter in the South Atlantic Ocean between depths of 1200 and 4000 m.

There is no doubt that the North Atlantic Water advances from its place of formation towards the north, i.e., polewards in the North Atlantic, but it does so in a substantially weaker fashion than in the southward movement described above. It can be followed in a tongue of highly saline water, which in the east Atlantic contains a strong admixture of Mediterranean Water, at depths between 1000 and 2000 m up to north of 50°N. At about 51° to 52°N it sinks to the bottom and is subjected to mixing with the Arctic Bottom Water formed further to the north. This provides the explanation of the striking phenomenon that the highest bottom-level salinities of more than 34.9 °/oo occur at about 50°N in the North Atlantic while to the north and south of there the salinities have the otherwise typical value for the bottom of 34.88 to 34.90 °/oo.

The bottom layers of the North Atlantic Ocean below about 2000 m depth are thus filled up with a Bottom Water which we must conceive as having been formed by the mixing of Arctic Bottom Water (in Nansen's sense) formed by cooling at the surface with the northern branch of the North Atlantic Deep Water. This North Atlantic Bottom Water now advances towards the south and in both sections flows over into the Southern Hemisphere. It forms the principal mass of that highly saline deep-level layer which hitherto has been thought of as a unit under the name "North Atlantic Deep Current." Between 1200 and 2000 m this layer is chiefly fed from the North Atlantic Deep Water. Between 2000 and 4000 m, on the other hand, it is fed from the North Atlantic Bottom Water, which contains an admixture of water of Arctic origin, and therefore exhibits lower temperatures (below 4°) and higher oxygen contents. Also, in the East Atlantic Trough it forms the principal mass of the Bottom Water as far as the Walvis Ridge. In the West Atlantic Trough, on the other hand, it slides away in about 10° to 20° north latitude above the Antarctic Bottom Water and here it is, in consequence, evolved into a "Deep Water." There is a most remarkable difference between the East and West Troughs in the extent to which this North

Atlantic Bottom Water spreads into the Southern Hemisphere. In the West Trough it can be traced from the salt concentration of more than 34.9 °/oo and temperature of over 3° up to almost 40°S, while in the East Trough its advance is already over at 20°S. Here it sends off merely a weak offshoot over the Walvis Ridge while in the West it slides away over the Rio Grande Rise. This difference in spread of North Atlantic Water explains why in the East Trough the temperature inversion is so much more weakly developed and indeed in the peripheral areas is in many places entirely absent.

Then both sections have in common between 40° and 50°S that ascent of the offshoots of the North Atlantic Water from around a depth of 3000 m up to the layer between 500 and 1000 m whose existence was first recognized by Merz. We find the last traces of this water of North Atlantic origin in the somewhat warmer and highly saline intermediate layer in the Southern Ocean though even here it shows a preference for the western half. While Merz was able to recognize the last traces of this water in Brennecke's observations as far south as 65°S, in the east, according to the *Meteor* observations, its extreme boundary is to be set at 56°S.

Polar Deep Water is formed on a most imposing scale in the Antarctic. Here, as Antarctic Bottom Water, it fills up at depths below 1500 m the great expanses of the South Polar Basin. It is characterized by ice-cold temperatures (from 0° to -0.5°) and an almost uniform salinity (from 34.64 to 34.67 °/oo). It is, however, separated from the cold low-salinity surface layers by a warmer and somewhat more saline intermediate layer which is obviously fed from two sources. One of these supplies of water comes from the north along the offshoots of the North Atlantic Deep Water and the other comes from the Indian Ocean along a coastal current (cp II Bericht 1926, p. 246). How are we now to understand the method of formation of the Bottom Water which in our sections seems everywhere to be cut off from the surface by a somewhat warmer and more saline intermediate layer? Probably one cannot, as Brennecke (1921) wanted to do, derive the enormous water masses of the Antarctic deep-sea regions exclusively from the shelf-water which sinks into the depths in winter on the Antarctic continental slope. Brennecke was, indeed, able to show that the separating intermediate layer was absent at times in the shelf and coastal waters of the Weddell Sea and that the prerequisites for such an advective process are satisfied here. On the other hand, the attempt at an explanation by Drygalski (1926) is also, as already Schott (1926a) emphasized, entirely unsatisfactory. According to Drygalski cooling at the surface plays no part in the formation of the Bottom Water since this Water nowhere at all comes into contact with the atmosphere. On the other hand it seems to Schott (1926, p. 428) "not yet settled that the mode of origin of the Bottom Water of the Southern Hemisphere is fundamentally and exclusively something different from that of the Northern Hemisphere," as Nansen had explained it. We will now show in the following text that in fact, as Schott conjectures, Nansen's trains of thought are also valid for the Antarctic. Drygalski thought of "the Bottom Water as Tropical Water which at the edge of the continental shelf rolls and mixes with the Polar Water flowing there in the opposite direction, and is in this way cooled, made heavier, and then sinks to the bottom, while its salinity remains in general unaltered and it contains only island-like inclusions of water masses of different salinity" (Drygalski 1926b). Apart from the fact that the Antarctic Bottom Water only just contains traces of North Atlantic Water (and it is to be noted, moreover, in this connection that this latter water type originates predominantly from higher latitudes), and it is therefore difficult to imagine it as being a "Tropical Water," this explanation is not in harmony, at least as far as the Atlantic Ocean is concerned, with a remarkable feature of the stratification in the Southern Ocean which, because of sparsity of observations, is only hinted at in the west section, whereas in the east section it is proved to exist in a more detailed way through the *Meteor* stations. This remarkable feature is that the water columns which are coolest and have least salinity are not to be found near the edge of the continental shelf, where Brennecke and Drygalski suppose them to form, but rather at the northern boundary of the true Antarctic at about 58°S in the east section and 64°S in the west one. It is here, as the run of the zero degree isotherm shows in both sections, that the ice-cold deep-level water approaches most closely to the surface. This considerable up-welling of the deep-level water by 1000 to 2000 m occurs just where the intermediate layers in the Atlantic and Indian Oceans are most weakly developed and are consequently separated by a somewhat cooler and less saline water column. If we recall that our sections describe the summer situation it is natural to suppose that the polar Bottom Water in the south is formed in a similar way to the one Nansen (1912) demonstrated for the north, namely, that it forms at the edge of the ice-covered zones in autumn and winter as a result of intense cooling and through freezing of the surface layers. In the Southern Ocean this process is also associated with dynamical conditions, namely those which exist at the convergence areas of the cyclonic Antarctic vortices whose water masses exhibit a small horizontal circulation. The principal sources of origin of the Antarctic Deep Water are thus to be sought in those centers of the Antarctic current vortices which are presumed, from Meyer's chart, to exist outside the pack-ice at about 65°S, 30°W in the Weddell Sea and 60°S, 30°E in the East Atlantic South Polar Basin. The conspicuous agreement in the geographical position of both phenomena, which is shown on the one hand in the chart of the surface currents and on the other in the two sections, speaks for such a connection with the currents. Also the latitudinal difference between the east and west sections in the occurrence of the phenomena is common to both treatments of the matter.

In these regions convection extends steadily deeper in the course of the autumn and winter and finally also involves the water masses of the intermediate layer. The subsiding water particles are replaced by warmer and more saline particles which are continually fed into the intermediate layer from the side. These warmer and more saline particles ascend in the intermediate layer up to the surface film where they give up their heat content, but approximately preserve their salinity. Finally at times the remaining small density difference between the surface and the deep-levels is entirely overcome and the convection extends over the whole water column.

Unfortunately we have no observations for the winter half-year from the crucial areas. Brennecke's important serial measurements made during the winter drift of the *Deutschland* in the ice lie outside these central areas and more on the western edge of the Weddell Sea. Nevertheless, the comparison, set out in Table 3, of the observations made in summer and spring in about 64°S (at the end of the wintertime drift in the ice) at two of the *Deutschland* stations, still makes it possible to recognize the effect of the convection which occurs in winter [see Table 3, p. 82].

If the convection had entirely stopped in summer owing to the melting of the ice, then obviously in autumn and winter an equalization of all the factors (temperature, salinity, and density) had taken place from the surface down to the intermediate layer. This refers to a convective equalization which at the end of the winter in this peripheral region even extends down to a depth of 200 m.

Table 3. Temperature, Salinity, and Density Stratifications of the Northern Weddell Sea

Depth	6 January 1912			27 November 1912		
	Deutschland Station 116 63°47S 28°9W			Deutschland Station 166 63°44S 35°50W		
m	t°	S °/oo	σ_t	t°	S °/oo	σ_t
0	-1.64	34.06	27.43	-1.72	34.49	27.77
50	-1.77	34.40	27.70	-1.76	34.47	27.76
100	-1.69	34.44	27.74	-1.83	34.49	27.77
150	-0.44	34.58	27.81	-1.81	34.48	27.77
200	0.19	34.66	27.84	-1.60	34.51	27.79
300	0.37	34.67	27.84	0.10	34.65	27.83
400	0.44	34.69	27.84	0.38	34.69	27.85
600	0.38	34.67	27.84	0.35	34.69	27.85
800	0.29	--	--	0.28	34.69	27.85
1000	0.22	34.69	27.86	0.21	34.69	27.86
1500	0.00	34.68	27.86	-0.01	34.69	27.87
2000	-0.19	--	--	-0.17	34.67	27.87

To sum up, we have arrived at the following ideas which, indeed, still remain to be confirmed at some time by systematic observations in summer and winter:[11] the Antarctic Deep Water is mainly formed by mixing of the cold and low-salinity water masses of the surface layer with the more saline and warmer masses of the intermediate layer. It takes its low temperatures from the surface and owes its high salinity to the extreme offshoots of the North Atlantic and Indian Ocean Water in the intermediate layer. This formation takes place mainly in autumn and winter in specific areas of cyclonic movement located outside the pack-ice. With these convective processes (in Nansen's sense) which proceed down from the surface there are associated, on the edge of Antarctica proper, advective processes (in Brennecke's sense) but these latter probably play a markedly less important part than the former.

The Antarctic Deep Water advances towards the equator. At about 50°S along that important hydrographical dividing wall which is almost vertical in our sections it is subjected to strong mixing with the offshoots of the North Atlantic Water. It then presses on as Antarctic Bottom Water with somewhat higher temperatures but with little change in salinity into the Argentine Basin. It is then obstructed by the Rio Grande Rise and flowing through the Rio Grande Trench fills up the depths of the Brazilian Basin below 4000 m. It sends out an offshoot over the Para Rise into the North Atlantic but this soon vanishes on account of increasing mixing with the North Atlantic Bottom Water which pushes against it and also thrusts it upwards. The last traces of water of Antarctic origin are to be found in salinity at about 10°N and in temperature at about 17°N. The effect of relief is revealed beautifully in this propagation. Corresponding to the streamlines the isotherms and isohalines rise above the upward

11. This would be a worthwhile task for an expedition which carries out oceanographic researches throughout a number of years in these areas as the *Discovery* Expedition is doing at the present time.

projections on the sea-bottom and sink again in the troughs. This parallelism between the relief of the bottom and the stratification can often be recognized even up to more than 1000 m above the bottom.

The bottom relief has a decisive effect on the propagation of the Antarctic Bottom Water in the east section. Thus already on reaching the Atlantic-Indian Ocean Transverse Ridge only the upper portions are allowed to pass into the Cape Basin and the Walvis Ridge brings this movement to a complete end. Beyond this latter Ridge the bottom layers of the East Atlantic Trough are filled up with almost homogeneous water masses which come, essentially, from the North Atlantic Ocean. It is only in the Guinea Basin, as Böhnecke (1927) has shown, that traces of Antarctic Bottom Water occur; they arrive there through a narrow depression in the Mid-Atlantic Ridge near the Romanche Deep. They are so trivial that they do not appear in our representation.

While the polar Bottom Water forms on a huge scale in the Antarctic the same process takes place, on the other hand, to only a negligible extent in the Arctic. Because of the predominance of land masses the polar catchment area is only a small one here and, moreover, these limited sea spaces are almost completely barred off from both Atlantic troughs by rises. True Arctic Bottom Water is probably only formed, in the way so beautifully described by Nansen (1912), in autumn and winter in two areas of cyclonic movement. One of these is the Labrador Basin to the south of Greenland and the other the northern ocean to the north (Greenland Sea) and south (Norwegian Sea) of Jan Mayen. Neither of these two processes can be recognized in our sections because these represent the summer situation. In its salinity (34.88 to 34.95 $^{o}/oo$) the Arctic Bottom Water exhibits essentially stronger North Atlantic influences than the Antarctic one does. It fills up the bottom layers of the Labrador Basin down to about 40°N but is absent further south where bottom temperatures exceeding 2° were measured. This fact, together with the results of the sounding series, leads to the conclusion that the Labrador Trough is shut off at about 50°N by a rise which we will term the Newfoundland Rise. It bars off the cold Arctic Bottom Water below about 3000 m depth on the south so that the Arctic Bottom Water current in the West Atlantic Trough here comes to an end. Its upper layers engage in turbulent mixing with the North Atlantic Bottom Water which overlies them and presses against them, and thus form by mixing with this the North Atlantic Bottom Water which, with its temperatures of 2° to 3° and salinities of 34.85 to 34.90 $^{o}/oo$ fills up the North American Basin.

Conditions in the west section are quite similar. We are, in fact, confronted here with ice-cold water in considerable thickness, yet its source-region, the Greenland/Norwegian Sea area, is substantially more strongly barred off from the Atlantic Trough as the Iceland Rise whose saddle is only about 600 m deep stops all exchange of water at lower depths. The cold deep-level water of the Greenland and Norwegian Seas flows on only a small scale over this Rise into the East Atlantic Trough where, however, it quickly mixes with the here more strongly advancing North Atlantic Deep Water to form together with the latter a Bottom Water of similar properties to that of the West Atlantic Trough.

CONCLUSIONS ON THE NATURE OF THE LONGITUDINAL
CIRCULATION IN BOTH TROUGHS

We have hitherto treated the thermal and haline structure of the Atlantic Ocean along two planes of section rather descriptively from the uniform viewpoint of the dispersal of different water masses defined by means of their temperature and salinity and, in doing so, not assumed the existence of deep-level currents in the sense of surface currents.

Undoubtedly movement occurs in all layers of the Ocean but the sections, in the first place, only enable us to advance conjectures about the nature of the movement. Actually what we recognize in the vertical distribution of temperature and salinity is the spatial propagation of different sorts of water along definite levels corresponding to their respective specific gravities from definite centers of action which are associated with areas of convergence at the surface. In other words the deep-level circulation presents itself in such a treatment as the spatial filling up of thick ocean-wide layers with definite types of water which from the action centers mentioned spread out in more or less all directions. It is natural, therefore, to apply the convection concept also to the horizontal deep-level water movements and, thus, to consider these as a kind of horizontal convection. One immediately recognizes that such a treatment cannot provide an answer to the question of the true nature of the circulation processes which forms the core of the problem. For what we see in the thermal and haline structure is, so to say, the end effect of these deep-level water movements on the stratification; the detailed movements remain mysterious. In which regions in depth are we concerned with quite gradual convective processes? Where are there systems of complicated, yet measurable, currents which are characteristic of the individual levels similar to those of the surface, and which procure through endlessly tangled tracks the complete filling of the individual layers with definite nearly homogeneous types of water? These questions can only be finally answered by the mathematical handling of the problem, that dynamical treatment of the whole of the observational material which Merz had in mind as the final goal. That this will bring us a good part of the way forwards is, according to our first researches (Merz and Wüst 1923; Wüst 1924) and the results of the works of Helland-Hansen and Nansen (1926), to be hoped for. In fact, we believe that the dynamical treatment will bring still further important information if we drop the assumption made by Helland-Hansen and Nansen more on practical grounds that the speeds everywhere decrease with depth and that the water movement in 2000 m depth which was normally the greatest depth of observation from the *Michael Sars* and *Armauer Hansen* can be neglected, and, instead, start from the actual nil layers deduced from the vertical structure.

Happily we have available, as our new treatment of the salinity section shows, in the Atlantic Ocean one, we might say universal, null-surface at a depth of between 1000 and 1500 m. This surface is the boundary layer between the Polar Intermediate Water and the North Atlantic Deep Water and can be considered as a relatively quiet separation layer most easily as a result of the opposed directions of propagation of the two types of water. It is only in the region of the northern Subtropical Convergence that it is interrupted. Thus there exists a prospect that we will be able to progress in the dynamical treatment from relative to absolute values of the speeds by taking this exactly ascertainable null-layer as a basis and that finally we will be able to obtain, naturally in the form of approximations, a quantitative spatial representation of the circulation processes. The deep-level current measurements made at the, certainly less numerous, anchored stations will provide a desirable check on the calculation.

In the meantime we are thus more or less left to making conjectures about the true nature of the deep-level water movements (whether flow or convection) and above all about their tracks. For we must always bear in mind that our longitudinal sections only allow us to ascertain the component of the motion which lies in the plane of the section and this can depart from the true motion by up to 90°. With this reservation we turn back to the already introduced designations of the principal members of the currents of the deep-level circulation (Intermediate Current, Deep-Level Current, Bottom Current). The Subantarctic Intermediate Current corresponds to the propagation of the Subantarctic Intermediate Water. It originates at the Southern Polar Front and shows in its further course the effects of the diverting force of the Earth's rotation. The intermediate layer is tilted from west to east and, corresponding to this effect, dips to the left from the current direction. Also it is most strongly developed in the western half. All these indications show that at least as far as 20°S we are dealing with a measurable current in this branch of the circulation.

The propagation of the Antarctic Bottom Water corresponds, especially in the West Trough, to the Antarctic Bottom Current. Its propagation bears a close relation to the relief to which the isohalines and isotherms cling like streamlines. It is bounded at the top by a small, but for these great depths unusually sharp, layer of discontinuity in temperature and salinity. These facts again allow us to conjecture the existence of a current even if it is a very slow one.

The position is different with the North Atlantic Deep Current and the North Atlantic Bottom Current which, according to the statements we have made above, are by no means so homogeneously developed as had previously been assumed. If one understands by North Atlantic Deep Current the propagation of the North Atlantic Deep Water formed in the northern Subtropical Convergence Area then its lower boundary is, corresponding to the position of the 4° isotherm, to be set at a depth of about 2000 m. Accordingly, opposite to our previous ideas, its axis lies immediately below the Intermediate Current at a depth of 1600 m. It advances in the North Atlantic Ocean, however, not only towards the equator but also towards the Pole from the northern Subtropical Convergence. The deeper and substantially thicker layers of the intermediate South Atlantic salinity maximum below 2000 m belong essentially to the North Atlantic Bottom Current which in the West Atlantic Trough thrusts itself in between the North Atlantic Deep Current and the Antarctic Bottom Current. Its water is formed under quite different conditions, namely in 40° to 60°N by the mixing of the North Atlantic Deep Water with Arctic Bottom Water. It is now in this way that the high oxygen contents of the layers between 2000 and 4000 m in the South Atlantic Ocean, which are by no means to be derived from the Sargasso Sea, find, as already Wattenberg has conjectured, their interpretation (cp Wattenberg, III Bericht 1926, p. 140). In the East Trough the North Atlantic Bottom Current between 2000 m and the bottom can be traced as far as the Walvis Ridge.

Neither of these current members--the North Atlantic Deep Current and Bottom Current--gives any sign in our sections of being influenced by the deflecting force of the Earth's rotation. Indeed, in the South Atlantic Ocean, both of them in opposition to the effect of this force, are most strongly marked in the west which is to the right of the direction of propagation. Obviously in their case we have to do with a quite gradual somewhat convective type of propagation which provides the compensation for the Antarctic current members and spreads out over a huge cross-section more than 1000 m thick. This idea that the North Atlantic Deep and Bottom Currents form a kind of compensatory current for the Antarctic members should make it understandable why the West Trough is the preferred one in their propagation. In the East Atlantic Trough just one Antarctic current member--the East Atlantic Bottom Current--is missing and therefore here the substitution by North Atlantic water is that much weaker. The principal mass of the latter is drawn over to the west. Hence, probably, one should not, when dealing with these current members, speak of a current flow in the proper sense. This applies especially to the East Trough of the South Atlantic where the cross-ridges obstructively oppose them.

By comparison with the Antarctic current members the Arctic ones--Subarctic Intermediate Current and Arctic Bottom Current--are very weakly developed in our sections. These, too, give preference to the West Trough over the East Trough.

SUMMARY

The treatment of the two longitudinal sections has, besides many new individual features, brought confirmation of the fundamental aspects of the ideas obtained by Merz on the nature of the longitudinal circulation of the Atlantic. It is only in the North Atlantic, above all to the north of 30°N, where our earlier longitudinal section "did not allow us to draw a, in some degree, trustworthy conclusion" (Merz 1925, p. 8), that we have arrived at a new understanding.

Further, our new sections, taking account as they do of the morphology, have now shown us, as it has already been possible to point out in the Expeditions report for the South Atlantic Ocean, that there exist considerable differences between the meridional components of the West and East Atlantic, and that these differences are mostly in the sense that the exchange of water masses proceeds in a more lively fashion in the West than in the East Trough. These deviations can essentially be referred to the effects of the morphological differences between the two Troughs. In the lower layers of the Ocean (below 3000 m) they are of such a decisive nature that, corresponding to the morphological dichotomy, one must distinguish two longitudinal circulation systems--those of the West and East Atlantic respectively.

Our sections have, in conclusion, permitted us, to differentiate the individual deep-level water masses from one another according to temperature and salinity and to follow up their formation and distribution more closely. We have found the roots of their formation in the great surface current convergences and in the centers of cyclonic motion. Located at the boundary of the ice these latter are the principal source regions of the Bottom Water and this applies not only in the north, where Nansen studied the factors of its formation, but probably also in the Southern Ocean where similar processes seem likely to take place on a substantially greater scale. The formation of the Subpolar Intermediate Currents (Subarctic and Subantarctic) is linked with the subsidence of subpolar water masses at the southern and the northern Polar Front as well as at the convergence line between the west wind drift (or the North Atlantic Current) and the Polar Current. The subtropical accumulations of warm water at deep levels are related in both hemispheres to the extensive convergence areas at the surface which we have termed subtropical convergences.

In the North Atlantic Ocean and in the Sargasso Sea these dynamical processes are so deep-reaching that they participate in the formation of the North Atlantic Deep Current.

The treatments, made from a dynamical point of view, of the deep-level circulation and the surface currents lead accordingly to a harmonious overall picture of the surface currents and the movements of the deep-level water as being the members of a huge cycle essentially closed in the convergence and divergence regions of the Ocean.

BIBLIOGRAPHY

1927 Böhnecke, G.: Bottom water temperatures in the Romanche Deep. The German Atlantic Expedition on the survey and research ship *Meteor* 1925 to 1927. Zeitschr. Gesell. f. Erdkde. Berlin. 1927.

1921 Brennecke, W.: The oceanographic work of the German Antarctic Expedition 1911 to 1912. Aus dem Archiv der Seewarte, XXXIX, No. 1. Hamburg 1921.

1906 Bruce, W.: Report on the scientific results of S.Y. *Scotia* during the years 1902 to 1904. Edinburgh 1906.

1926 Bulletin Hydrographique, English supplementary appendix of surface observations for the period 1915 to 1923. Conseil permanent international pour l'exploration de la mer. Kopenhagen. 1926.

1927 Defant, A.: Report on the results of the *Meteor* Expedition. Z. f. Geophysik 3, pp. 340 to 350. Brunswick 1927.

1926a Drygalski, E. v.: The Ocean and Antarctica. Oceanographic researches and results of the German South Polar Expedition of 1901 to 1903. (Vol. VII of results: Oceanography. Deutsche Südpolar Expedition.) Berlin 1926.

1926b Drygalski, E. v.: Remarks on the deep-level currents of the oceans and their relations with Antarctica. Sitzungsber. Bayer. Akad. d. Wiss., Math-nat. Abt. Munich 1926.

1912 Groll, M.: Isobaths of the oceans. Veröff. d. Inst. f. Meereskunde, N. F., Reihe A, Heft 2. Berlin 1912.

1884 Hamberg, A.: Hydrographical Observations of the Nordenskiöld Expedition to Greenland 1883. Proc. R. Soc. 1884, p. 569.

1909 Helland-Hansen, B., and F. Nansen: The Norwegian Sea. Kristiania 1909.

1926 Helland-Hansen, B., and F. Nansen: The Eastern North Atlantic. Geofysiske Publikasjoner. Vol. IV, Nr. 2. Oslo 1926.

1916 Jacobsen, J. P.: Contributions to the hydrography of the Atlantic. Researches from the M/S *Margrethe* 1913. Meddelser fra Kamn. f. Havundersøgelser. Serie Hydrografi. Bind II. Copenhagen 1916.

1899 Knudsen, M.: Hydrography. The Danish *Ingolf* Expedition. Vol. 1, No. 2. Copenhagen 1899.

1907 Krümmel, O.: Handbook of Oceanography. Vol. 1. Stuttgart 1907.

1923 Meinardus, W.: Meteorological results of the German South Polar Expedition 1901 to 1903 (Deutsche Südpol-Expedition 1901 bis 1903, III Bd. Meteorologie, I Bd. I. Hälfte). Berlin 1923.

1922 Merz, A.: Temperature stratification in the South Atlantic Ocean according to the *Challenger* and *Gazelle* observations. Zeitschr. Gesell. f. Erdkde. Berlin 1922.

1925 Merz, A.: The German Atlantic Expedition on the survey and research ship *Meteor*. Sitzungsber. Preuss. Akad. d. Wiss., Phys-math kl. XXXI. Berlin 1925.

1922 Merz, A. and G. Wüst: The vertical circulation of the Atlantic. Zeitschr. Gesell. f. Erdkde. Berlin 1922.

1923 Merz, A. and G. Wüst: The vertical circulation of the Atlantic. 3. Contribution. Zeitschr. Gessel. f. Erdkde. Berlin 1923.

1923 Meyer, Hans H. F.: The surface currents of the Atlantic Ocean in February. Veroff. Inst. f. Meereskunde. N. F., Reihe A, Heft 11.

1926a Möller, L.: Hydrographic investigations of the cruiser *Berlin*. Marine Rundschau, Heft 3. Berlin 1926.

1926b Möller, L.: Methodological aspects of the vertical sections along 35.4°S and 30°W in the Atlantic. Veröff. Inst. f. Meereskunde. N. F., Reihe A, Heft 15. Berlin 1926.

1912 Murray, J. and H. Hjort: The depths of the Ocean. London 1912.

1912 Nansen, Fr.: The bottom water and the cooling of the sea. Int. Rev. d. ges. Hydrobiologie u. Hydrographie. Leipzig 1912.

1907 Nielsen, J. N.: Contribution to the hydrography of the north eastern part of the North Atlantic. Meddelelser fra Komm. f. Havundersøgelser. Serie Hydrografi. Bind I. Kopenhagen 1907.

1917 Report of the United States Commissioner of Fisheries for the fiscal year 1915 with appendices. Washington 1917.

1884 Report on the Scientific Results of the Voyage of H.M.S. *Challenger* during the years 1873 to 1876. Physics and Chemistry. Vol. I. London 1884.

1918 Sandström, W. J.: The hydrodynamics of Canadian Atlantic waters (Canadian Fisheries Expedition 1914 to 1915). Ottawa 1918.

1926a Schott, G.: The deep-level water movements of the Indian Ocean. Together with a review of E. V. Drygalski "The ocean and Antarctica." Ann. d. Hydrogr. usw. Berlin 1926.

1926b Schott, G.: Geography of the Atlantic Ocean. 2. Edition. Hamburg 1926.

1926- Wattenberg, H.: Report on the chemical work I-IV. The German Atlantic
1927 Expedition on the survey and research ship *Meteor*. Zeitschr. Gesellsch. f. Erdkde. Berlin 1926 and 1927.

1927 Willimzik, M.: The Antarctic surface currents between 50°E and 110°E. Veröff. Inst. f. Meereskunde. N. F., Reihe A, Heft 17. Berlin 1927.

1920 Wüst, G.: Evaporation over the sea. Veröff. Inst. f. Meereskunde. N. F., Reihe A, Heft 6. Berlin 1920.

1924 Wüst, G.: The Florida-Antilles current. A hydrodynamical investigation. Veröff. Inst. f. Meereskunde, N. F., Reihe A, Heft 12. Berlin 1924.

1926- Wüst, G.: Report on the oceanographic investigations I-IV. The German
1927 Atlantic Expedition on the survey and research ship *Meteor*. Zeitschr. Gesell. f. Erdkde. Berlin 1926 and 1927.

9

THE NORTHERN BOUNDARIES OF ANTARCTIC AND SUBANTARCTIC WATERS AT THE SURFACE OF THE WORLD OCEAN

George E. R. Deacon

This article was translated expressly for this Benchmark volume by George E. R. Deacon, from "Die Nordgrenzen antarktischen und subantarktischen Wassers im Weltmeer," in Annalen der Hydrographie und Maritimen Meteorologie 4:*129–136 (April 1934). Published with the permission of the Discovery Committee.*

The subject of my talk has been chosen as one of special interest. We were ourselves concerned with it mainly because of its close connection with questions of plankton distribution. My conclusions are based on measurements made by the Royal Research Ships *Discovery, William Scoresby,* and *Discovery II* --the track and seasonal distribution of the sections worked by the *Discovery II* (1932-1933) have been described by Professor Schott in the October number, 1933, pages 342-44, of the *Annalen.* The present investigation is based mainly on observations made between 1926 and 1933.

The principal features of the water circulation in the Antarctic seas and neighboring waters can be seen in a meridional section of temperature, like that along 30°W in Figure 1, Plate 14. The thin surface layer at the southern end of the section consists of cold Antarctic surface water, composed mainly of water from melting ice, upwelling deep water, and the high precipitation of the Antarctic. It behaves in many respects as two strata: the upper moving as a wind drift and subject to summer warming, and the lower moving as a colder undercurrent not so much changed by summer conditions. Throughout the layer the water moves east or west according to the prevailing wind, but also to the north, partly because of the effect of the west wind at the surface, and partly because of a general tendency of cold Antarctic water to sink to a lower level towards the equator. It flows in a northerly direction at the surface till it meets lighter subantarctic water below which it sinks and continues to the north at a deeper level. The line of convergence is generally well marked. To identify it clearly we have called it the Antarctic convergence. It was first recognized by Meinardus and given the name Meinardus Line by Schott. Defant and Wüst later called it the Antarctic Polar Front.

Just north of this Antarctic convergence the southern part of the subantarctic zone appears as a region of much deeper vertical mixing, in which the sinking Antarctic water becomes incorporated as subantarctic. In the Atlantic ocean two branches can be distinguished: the secondary temperature minimum recognizable almost as far as the equator in our section seems to be formed mainly from the lower, winter-cooled Antarctic water mentioned above, while the salinity minimum, known since the *Challenger* Expedition, seems to be derived

---

[*Editor's Note:* Figures 1, 2, and 4 have been omitted because of limitations of space.]

largely from the surface stratum of Antarctic water. The mixed waters sink deeper as they flow northward in the Antarctic Intermediate Current.

The meridional section of salinity in 30°W shows that in the northern part of the subantarctic zone the water above the level of minimum salinity has a different origin. Between the depths of 80 m and 200 m a relatively high salinity indicates southward movement. Surface water still moving northward under the influence of the west wind appears to sink and turn back to the south when it meets higher subtropical water, and finally--after further mixing to sink towards the north in the Antarctic intermediate current--above the level of minimum salinity. In addition to the northward movement of Antarctic water at the surface, there is also a northward flow at the bottom which is dealt with later in my talk.

Figure 3 shows the continuity of the Antarctic and subtropical convergences round the southern hemisphere. They are the boundaries between different types of surface water, marked by horizontal changes of temperature and salinity. They do not follow lines of latitude: near Bouvet Island Antarctic water reaches as far as 48°S but to only 62°S in 80°W. South of the Agulhas and Brazil currents, subtropical water reaches as far as 43-48°S but to only 37°S in the Greenwich meridian. These striking variations become understandable when we look at the surface currents.

THE POSITION OF THE ANTARCTIC CONVERGENCE

There seem to be two good reasons for particularly strong northward flow of Antarctic water in the Atlantic ocean. The west wind drift from the Pacific ocean is deflected northward by the Antarctic peninsula, and it is joined by a strong current carrying water to the north and east from the Weddell sea. This current has a strong influence on the hydrological conditions of the Southern ocean: it seems to be the main reason for the northward advance of the Antarctic convergence in the eastern half of the ocean and will repay further study.

In both Atlantic and Indian oceans the prevailing wind south of 66°S is easterly, and the same seems to be true of the Pacific ocean though a few degrees farther south. This produces a coastal current to the west which trends northward when the land or a submarine ridge extends northward. Along the eastern side of the Antarctic peninsula it flows sufficiently far north to come under the influence of the west wind. This turns it back to the east, leaving only a small part to continue westward round the northern end of Graham land into the Bransfield strait.

In the Indian ocean the Antarctic convergence lies between 48°S and 53°S. It follows the current where it bends northward over the ridge between Marion island and the Crozet Islands, and over the ridge south of Kerguelen. Conversely it bends southward over the deep soundings between the ridges. It is instructive to compare the position of the convergence with the northern limit of pack-ice. According to Drygalski this lies in 48°S near Bouvet island and in 62½°S south of Kerguelen. Although the convergence continues to the east near 50°S under the influence of the Antarctic current from the Weddell and Bellingshausen seas, the drift ice is not carried so far, at least in summer. In the Cape Town meridian there is a rise in surface temperature south of the cold tongue of water that extends eastward across the Atlantic ocean from the northern part of the Weddell sea, and it seems quite possible that even in winter, or towards the end of winter there may be open water between the Weddell

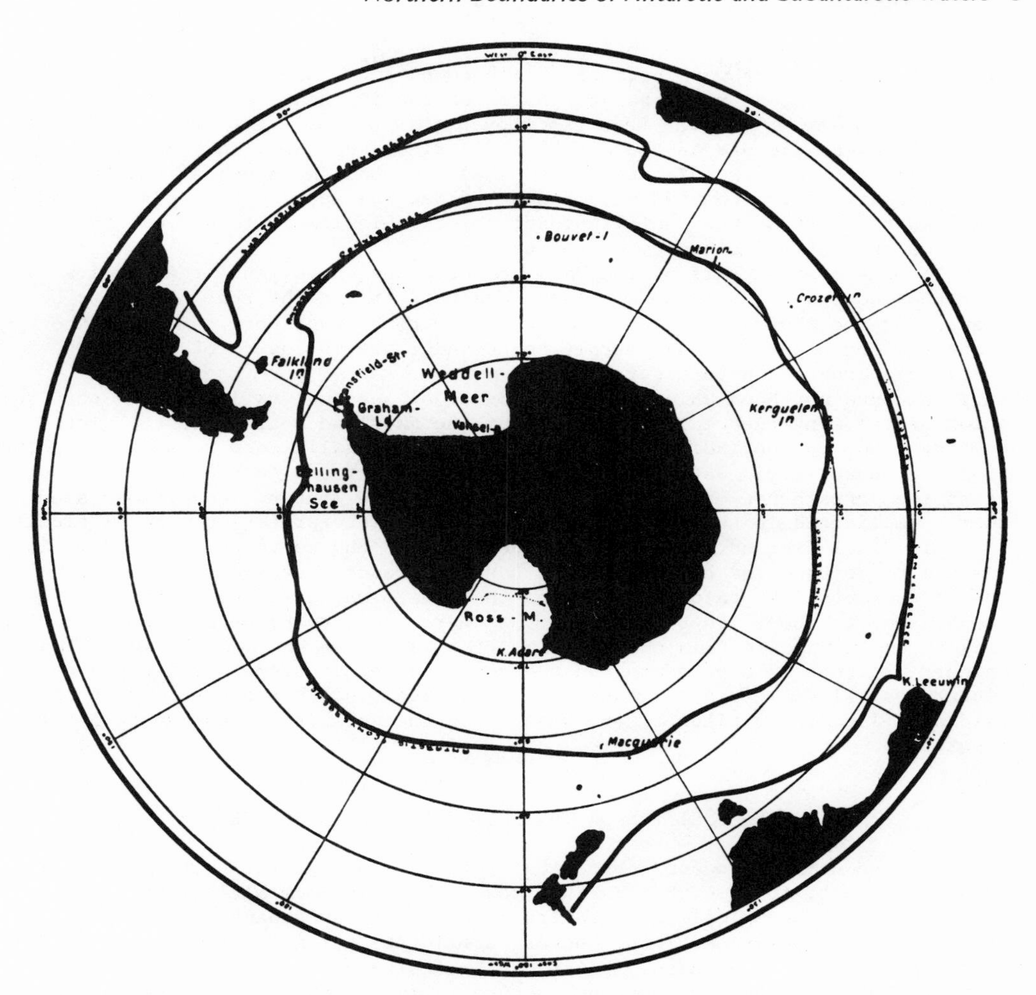

Figure 3. The position of the Antarctic and subtropical convergence.

sea current and the ice that borders the continent.

The southward bend of the convergence in 90°E may be explained by a southward component of the surface current which may itself be related to the prevailing southward trend of cyclones in the region west of Australia as accepted by Meinardus and Griffiths Taylor. This is the explanation suggested by Mecking (*Ann. Hydrog.* 1932, p. 227) for the southward trend of the northern limit of drift ice in the same area. The northward advance in 120°E may be due to a gap we discovered in the zonal ridge that bends eastward from the Kerguelen rise. South of eastern Australia and the Tasman sea our line bends polewards to 55-56°S, and east of New Zealand to 58°S. This retreat must be due to a weakening of the flow of Antarctic water or to a southward trend to the current under the influence of New Zealand.

The strong northward bend to 55°S between 120°W and 150°W in the middle of the Pacific ocean might be attributed to the influence of new Antarctic water streaming out of the Ross sea though this did not seem to be confirmed by our present picture of the plankton distribution. We also found a rise with soundings of 2700 m running to the northeast beyond the deep basin with soundings of 4500 m to 5500 m southwest of New Zealand and then east across the central part of the Pacific ocean; the rise is likely to deflect the surface current and convergence to the north.

In the eastern half of the Pacific ocean the convergence reaches its farthest south--62½°S in 80°W. There can be little doubt that this must be caused by the eastward flow bending south to round Cape Horn. East of the cape the convergence bends sharply north under the influence of the South Antillean arc and surface current.

We have not been able to find evidence of seasonal variations in the position of the convergence in our data from the Falkland sector but the winter observations in the Indian ocean and south of Australia seem to show a tendency for the subantarctic water to thrust southward over the Antarctic water, so that the convergence may be displaced southward in winter. Conditions seem more complicated in the Falkland sector and the interpretation more difficult.

In concluding our survey of the position of the convergence it seems interesting to compare the effects of the Ross and Weddell seas on the Southern ocean currents. Captain Scott showed that there was a strong current to the west along the Ross ice barrier and yet there seems to be no cold current flowing to the northeast from the Ross sea like that from the Weddell sea. The higher latitude of Cape Adare and the coast beyond it seem to allow the cold water that flows west along the barrier and bends north along the coast of Victoria land to continue to the west along the Antarctic coast south of Australia.

THE POSITION OF THE SUBTROPICAL CONVERGENCE

In the Atlantic ocean subtropical water is carried far south in the Brazil current, to 47-48°S. Farther south there is much mixing and sometimes offshoots of subtropical water entirely surrounded by subantarctic water. The most northerly position of the convergence in the Atlantic ocean is 37-38°S near the Greenwich meridian. South of the Agulhas current there is a southward bend to 43-44°S. Farther east across the Indian ocean its latitude is uncertain. The observations of earlier expeditions seem to give no certain information and this part of the ocean is still too little known for such a study. We were able to find the convergence at several places near Cape Leeuwin, finding a southward trend like that south of the Agulhas current only much smaller, probably due to a warm flow from the South Australian bight. We were able to make observations across the west Australian current. At the first station, 185 miles west of Fremantle, the surface temperature was 21.02°C and the salinity 35.82 °/oo. Ten miles farther east the temperature rose sharply to 22°C and at the second station (22.03°C and 35.57 °/oo) there was a strong southward current sufficient to make it difficult to prevent a large wire angle. The temperatures at 200 m were 12.32°C at the first and approximately 195°C at the second, where the warm water extended much deeper and the salinity was lower. At the third station 40 miles from Fremantle the surface temperature was again higher by 0.41°C and the salinity lower by 0.09 °/oo. Entering the Swan river the temperature fell to 19.5°C. These findings are in full agreement with the view expressed by Schott in his recent study of upwelling off the west coast of Australia (*Ann. Hydrog.* 1933, p. 225).[1]

East of Australia we found that the convergence advanced to 46°S under the influence of the warm current along the coast of New South Wales. In the western half of the Tasman sea the subantarctic and subtropical waters seem so well mixed that there is no sharp temperature change. The convergence has been drawn along the 14°C isotherm and 34.9 °/oo isohaline.

In the Pacific ocean we had insufficient fuel to reach the subtropical convergence though we cannot have been far from it. It should be possible to find an approximate position between our observations and those of the *Challenger* and the *Carnegie*.

MOVEMENT OF DEEP WATER IN THE OCEANS

Our measurements have provided some indication of the relative strength of the meridional flow of Antarctic and subantarctic waters in the Antarctic intermediate current in the different oceans. In the South Atlantic ocean we were able to show that the increase of salinity and decrease of oxygen concentration towards the north in the intermediate current is not uniform. At the level of minimum salinity this may give some indication of the rate of flow of the mixed water formed largely from sinking Antarctic water. The salinity and oxygen content at the depth of minimum salinity at stations made in 1931 along 30°W is shown in Figure 4. It shows a general increase in salinity and decrease in oxygen content towards the north and the change is not regular and sometimes faster and sometimes slower. The explanation for this may lie in the seasonal variations in the sinking Antarctic surface water which is likely to be colder with a higher salinity and lower oxygen content in winter, and warmer, less saline, and richer in oxygen in summer. The differences might show up more clearly if more observations were made. Assuming that the separation between the maxima and minima is the progress made in a year gives speeds of about 1½ miles a day in the south and 2½ miles a day farther north in the narrower part of the ocean. Similar observations have been taken on the east side of the Atlantic and some information may be obtained for the other oceans when the meridional sections are plotted. Speeds of the same order of magnitude were calculated by Castens using meridional temperature variations.[2,3]

1. The *Discovery* observations were made on 9-10 May, 1932.
2. Castens. *Ann. Hydrog.* 1931, p. 335 and 1932, p. 39.
3. The possibility of measuring the progress of deep currents composed of water sinking from regions marked with large seasonal variations by locating maxima and minima along the subsequent path of the current was suggested by Brennecke in 1909, Krummel in 1911, and Wattenberg in 1927. Now that we know more about the variability of water movements at all depths in the ocean an argument based on the assumption of a more or less steady northward flow seems most questionable, and, in any case, the observations are so widely spaced that quite different curves could be fitted to them. While the author of the paper, now being reproduced after 40 years, still feels that 1 or 2 miles a day must be about the right speed for residual motion in the deep layers, he believes that the method cannot be justified, at least till more observations are available, and has asked the editor to add this 1978 amendment to the earlier publication.

THE SPREAD OF ANTARCTIC INTERMEDIATE WATER TO THE NORTH

In the western Atlantic ocean the 34.30 ‰ isohaline extends as far as 26°S in the poorly saline intermediate layer below the Brazil current (Figure 2), and to 34°S on the eastern side of the ocean. The origin of the strong flow in the Atlantic ocean is likely to be found in the strong outflow of Antarctic surface water from the Bellingshausen and Weddell seas west and east of Graham land. South of the Agulhas current in the Indian ocean water of the same salinity reaches to only 43°S and in this area the intermediate water has to sink below a very large mass of subtropical water before it can continue its flow to the north. The observations of the German South Polar Expedition (1901-1903) indicate that the 34.30 ‰ isohaline must be farther north in the middle of the Indian ocean, but farther east it retreats to 45°S and to 50°S south of Australia and New Zealand. This might well be expected in this region where northward flow is hindered by land masses. East of New Zealand water with a salinity of 34.30 ‰ reaches 45°S, and in mid-Pacific ocean to about 44°S. Off the west coast of South America there is clearly a stronger northward movement and the 34.30 ‰ isohaline advances to 31°S. As the sub-antarctic water crossing the Pacific ocean approaches the land part it turns to the north while the remainder bends south round Cape Horn. The favorable conditions for upwelling in the region of the Humboldt or Peru current are probably helped by the rich supply of Antarctic intermediate water.

THE ANTARCTIC BOTTOM WATER

Finally I should give a brief note on our preliminary conclusions on the outflow of water from the Antarctic at the bottom of the ocean. Figure 5 is a chart of potential temperature at the deep ocean floor, showing the 0.0°, -0.5° and -0.8°C isotherms. The coldest bottom water occurs in the Atlantic sector. It appears in a narrow tongue with potential temperature less than -0.8°C in the northwest corner of the Weddell sea and extends eastward across the Atlantic ocean. There can be no doubt that this coldest bottom water is formed as first stated by Drygalski and Brennecke, on the Antarctic continental shelf, though I was not able to find any direct evidence of cold water sinking down the continental slope since no observations have been made along the east coast of Graham land. The bottom temperatures (Figure 5) show that cold bottom water flows northward as well as eastward, and Wüst (*Wiss. Ergeb. d. Dysch. Atl. Exp. a. d.* Meteor, VI, 1, 1933) has traced it as far as 40°N in the west Atlantic basin and to 34°N in the eastern basin. On the eastern side of the Atlantic its northward flow is restricted by the Walfish ridge.

May I now call attention to the position of the 0°C potential temperature isotherm. The *Meteor* data show that in the west Atlantic basin it extends northward as far as 5°S and in the eastern basin to 40°S; farther east it recedes towards the Antarctic continent. In the middle of the Pacific ocean the bottom water in 65°S is no colder than that below the equator in the west Atlantic basin. The continuous increase in bottom temperature towards the east seems to show quite clearly that the Weddell sea is the main source of Antarctic bottom water. Other sources, on other parts of the continental shelf must be insignificant in comparison. There is a northward bend of the 0°C isotherm in the eastern part of the Pacific ocean which might be attributed to flow of cold bottom water from the Ross sea, but its general position and the

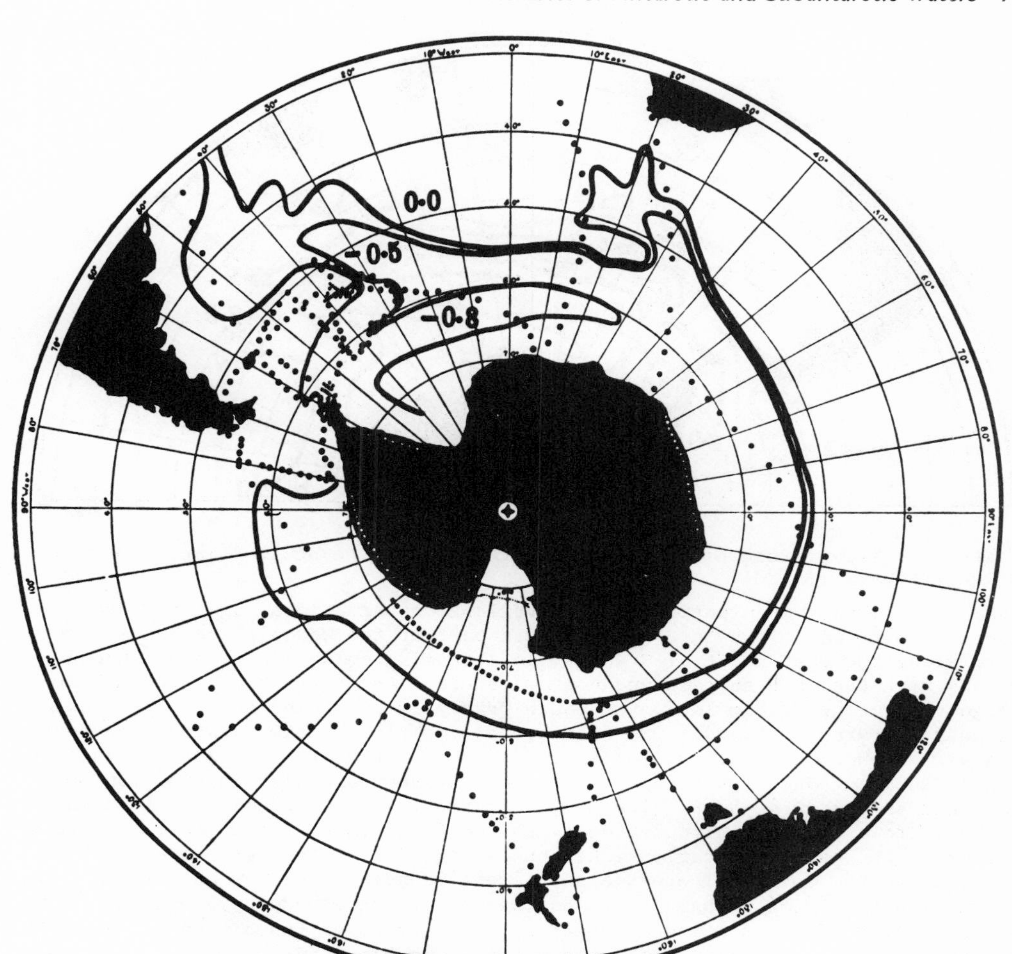

Figure 5. Circumpolar distribution of potential temperature at the sea bottom.

straightness of the isotherms near the Ross sea make this explanation seem unlikely. It seems more likely to be due to a northward spreading of the bottom current as it approaches the rise in the sea floor west of Cape Horn.

It is still rather early to attempt further conclusions about the movements of the bottom water, but an explanation can be suggested for the powerful influence of the Weddell sea. Figure 6 shows the maximum temperature in the warm deep current in the Atlantic Antarctic sector. There is a cold-water region interposed between a warm deep current that flows westward into the Weddell sea along its southern margin and the main part of the warm circumpolar current farther north. The cold-water tongue may not be due only to upwelling in the divergence region between the westerly and easterly flows,

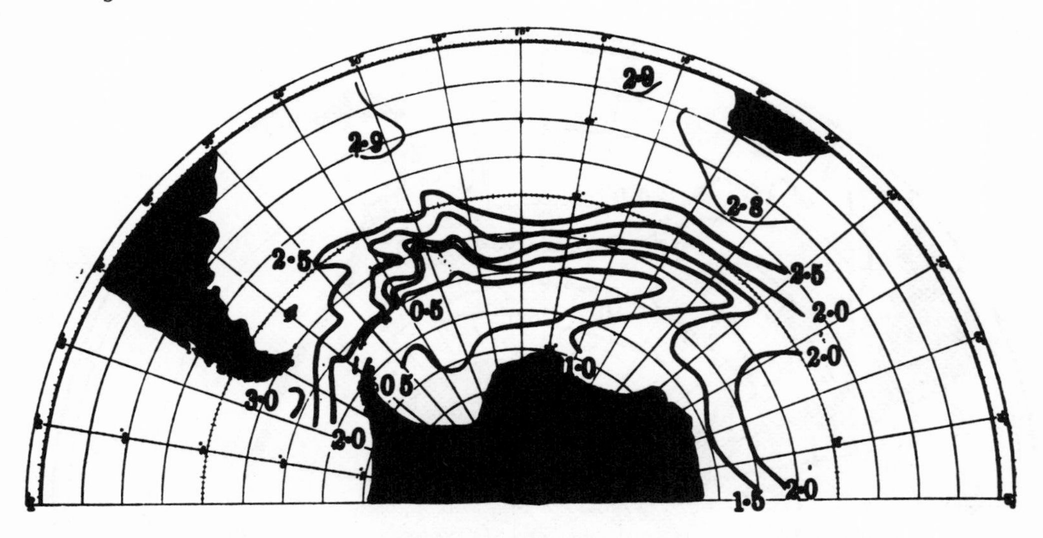

Figure 6. Temperature of the warm deep current in the South Atlantic Ocean.

and a preliminary examination of the available observations suggests that the deep current that flows into the southern part of the Weddell sea bends northward along the east coast of Graham land, becoming colder and colder through mixing with the cold surface and bottom waters. As it reaches the prominent ridge connecting Graham land, the South Orkney islands, and the South Sandwich islands it turns eastward. One branch seems to flow northward along the South Sandwich trench and another, continuing eastward from the cold water tongue across the Atlantic ocean. Its effect on the deep temperature distribution can be followed as far as 40-50°E. The surface current in the Weddell sea is also forced into a cyclonic circulation by the configuration of the land, and the effect of the earth's rotation makes the water sink on the left side of the current, along the coast. There are sufficient soundings to show an extensive continental shelf east of Graham land, and the measurements made by Brennecke in the southern part of the sea and by the Swedish South Polar Expedition (1901-1903), and by the *Discovery* near the northern tip of Graham land, show that the surface water on the shelf is cold and highly saline. The conclusion that Antarctic bottom water is largely formed by the sinking of shelf water in this region seems very reasonable. The sinking shelf water must mix with the already cooled deep water to form the large volume of bottom and deep water that flows north and east from the Weddell sea.

In certain respects conditions similar to those of the Weddell sea must occur in the Ross sea. Our observations there show that in at least part of this shelf sea cold bottom water is formed, though it still seems, as mentioned earlier, that a submarine ridge restricts its outflow into the Pacific ocean. There is another example of such a basin with cold bottom water in the Bransfield strait, and, as shown by Brennecke, near Vahsel bay at the southern extremity of the Weddell sea.

I apologize for the rather uncertain character of many of my conclusions. We hope that further study of the observations will give more information on problems such as the position of the subtropical convergence in the Pacific ocean, the origin and movements of bottom water in the Indian and Pacific oceans and the outflow from the Ross sea, and that such information will appear in the forthcoming *Discovery* reports.

10

Reprinted from *Deep-Sea Research* **3**:74–81 (1955)

A neutral-buoyancy float for measuring deep currents

J. C. SWALLOW

Abstract—Floats designed to stabilize themselves at a given depth, and fitted with means of sending out acoustic signals to indicate their position, have been made at the National Institute of Oceanography. Some current measurements were made with them on a recent cruise of R.R.S. "Discovery II".

INTRODUCTION

A body which is less compressible than sea water will gain buoyancy as it sinks ; if its excess weight at the surface is small, it may at some depth gain enough to become neutrally buoyant, when no further sinking will occur. Following the movement of such a float would give a direct measurement of the current at that depth, free from the uncertainties involved in using a conventional current-meter from an anchored ship. The possibility of using this method for measuring deep drift currents over a long period has been suggested in a recent note by STOMMEL (1955).

THE FLOAT AND ITS SIGNALLING EQUIPMENT

In the present design, tracking the floats is made possible by fitting them with acoustic transmitters, capable of sending out a short pulse every few seconds for two or three days. Besides having a sufficiently low compressibility, the float must provide enough spare buoyancy to carry this transmitter, and must not collapse at the greatest working depth.

Aluminium alloy scaffold tubing (alloy specification HE-10-WP, described in B.S. 1476 (1949)) has the required mechanical properties, and can be made into convenient containers for the electrical circuits and batteries. The compressibility of a long tube, closed at its ends, is (see, for example, NEWMAN and SEARLE, 1948)

$$-\frac{1}{v}\frac{dv}{dp} = \frac{R_1^2/\mu + R_2^2/k}{R_2^2 - R_1^2}$$

where R_1 and R_2 are the internal and external radii, v is the external volume, and μ and k are the rigidity and bulk moduli of the tube material. In Fig. 1 this compressibility is plotted against the ratio wall thickness : mean radius. The buoyancy and collapsing depth are also shown, the latter being calculated from formulae given in B.S. 1500 (1949). It can be seen that, by reducing the thickness-to-radius ratio of the standard tube to 0·16, the buoyancy can be doubled without seriously impairing the low compressibility or restricting the working depth. The outside diameter of the tubes was reduced uniformly, by solution in caustic soda, to make the ratio 0·16 ($\pm$ 0·01), and in this condition 6 m of tubing is needed for each float. For convenience in handling, this is made in two lengths of 3 m, one containing the transmitter circuit and batteries, and the other providing buoyancy.

Fig. 2 is a sketch of the float, with the end plugs shown in detail. The ends of the tubes are bored out to fit individual plugs ; this method of sealing scaffold tube has been tested to 4,500 m depth. The transmitter consists of a nickel scroll resonant at 10 kc/s, wound toroidally and energized by discharging a capacitor through a flash tube. Originally the transmitter was arranged so that, besides sending out a steady series of pulses, it would respond when it received a signal from the ship's echo-sounder. Since no current measurements were made with this arrangement, however, only the simple circuit is shown here.

The mean density of each complete float and transmitter is adjusted to an accurately known value by immersing it in a salt solution of known density and temperature

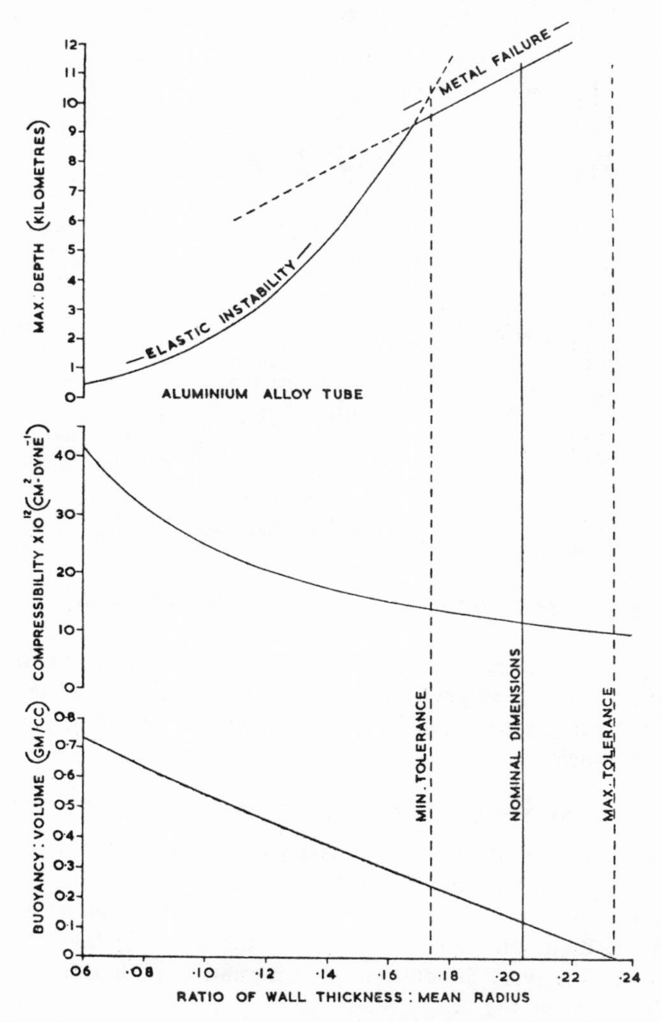

Fig. 1. Compressibility, buoyancy per unit volume, and maximum depth for aluminium alloy tubes.

(in the present case this was 1·0264 at 14·1°C) and adding weights until it is neutrally buoyant. This adjustment can be made to 1 gm. without difficulty, with a float weighing about 10 kg. in air. The density can be altered to any desired value by adding or subtracting weights, in proportion to the total weight of the float. All the extra weights are put inside the buoyancy tube, so that no change in volume has to be allowed for. Before launching any of the floats, temperature and salinity observations are made and the water densities *in situ* calculated from tables (ZUBOV and CZIHIRIN, 1940). The extra weight required to take the float down to any desired depth can then be determined, from the known density at that depth and the calculated compressibility of the float. Allowance has to be made for the tempera-

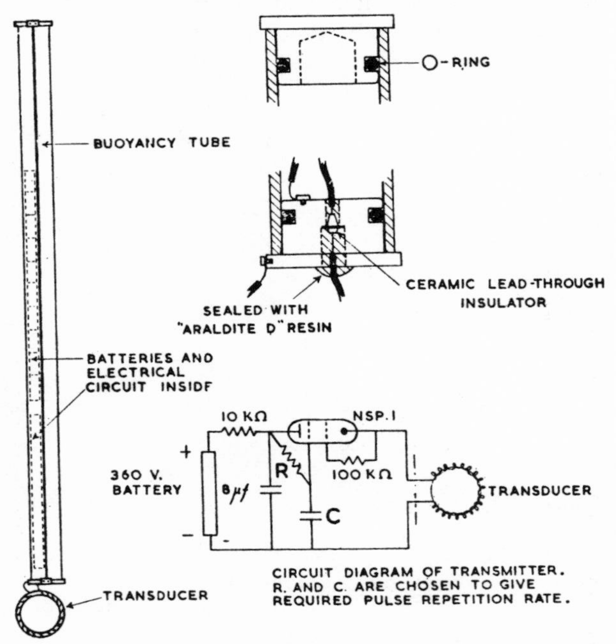

Fig. 2. Sketch of float and end plugs, and circuit diagram of acoustic transmitter.

ture change, but the effect of this is small. For an example of the order of magnitude of these extra weights, in the area where the present measurements were made, a float arranged to stabilize itself at 1,000 m depth had 38 gms. negative buoyancy at the surface.

METHOD OF TRACKING A DRIFTING FLOAT

The original plan was to follow the floats by means of the ship's echo-sounder, observing the responses of the floating transmitter when it was triggered by an outgoing echo-sounder pulse. This would have the advantage of indicating directly the depth of the float, but the narrowness of the beam of the echo-sounder made searching for the float very difficult and an alternative method had to be used. This scheme is illustrated in Fig. 3.

With the ship stopped head-to-wind, two hydrophones are lowered over the side, as far apart as can be conveniently arranged. The signals from them are fed via separate tuned amplifiers to a double-beam oscilloscope, the time-base of which can be triggered from signals applied to either beam. Pulses from the floating transmitter are received at different times at the two hydrophones, and the magnitude and sign of this time-difference can be measured on the oscilloscope.

As the ship's head falls away from the wind direction, the time-difference is observed as a function of the bearing of the line joining the hydrophones. It follows a " figure-eight " polar diagram, with sharp zero-values when the bearing is at right angles to the line from the ship's position to the float. Usually, observations over an arc of about 120° are enough to indicate the bearing from the ship to the float. The process is repeated with the ship in other positions, and the intersections of these bearings locate the float in a horizontal plane. The ship's position is determined by radar range and bearing from an anchored dan buoy, and the movement of the dan buoy itself is checked by sounding over small but recognizable nearby features on the sea bed.

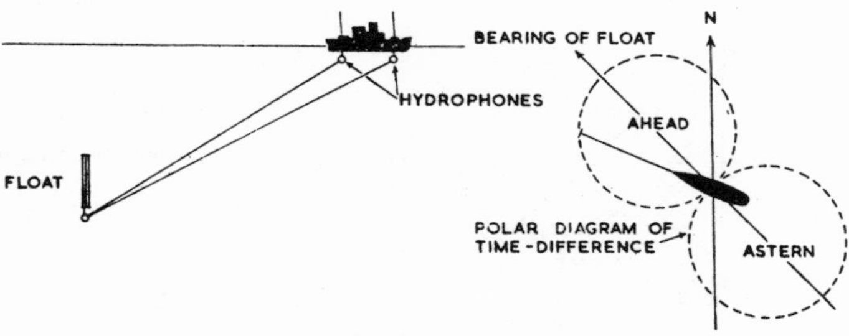

Fig. 3. Method of locating float.

The depth of the float can be estimated each time a bearing of it is taken, from the size of the " figure-eight " pattern obtained when the time-differences are plotted. The ratio of the maximum time-difference observed, (when the ship is heading directly towards the float) to the direct travel-time from one hydrophone to the other, is the cosine of the angle between the horizontal and the ray coming from the float to the ship. The depth of the float can then be found when the horizontal distance between it and the ship is known. The direct travel-time between hydrophones is measured by floating a transmitter on the surface and observing the maximum time-difference at the two hydrophones, as the ship is swung round.

To avoid errors in bearings and in the time-differences, the hydrophones are kept fairly shallow (about 7 m) and are weighted to prevent their cables from straying too far from the ship's side.

OBSERVATIONS

Of the six floats used during the May-June (1955) cruise of " Discovery II," one was lost in trying out the echo-sounder method of tracking, two others disappeared within a few hours of being released, one developed an electrical fault after 8 hours running, and the other two worked satisfactorily. No explanation can be offered for the loss of two of the floats, though it may possibly have been due to undetected flaws in the tubes causing the compressibility to be higher than the calculated value.

The track followed by one of the floats is shown in Fig. 4, together with the positions of the dan buoy used as a reference mark. These are plotted relative to the southern end of a submarine ridge, rising about 200 m above the surrounding plain. The sides of the ridge are quite steep, and the depth to the plain is 5,330 m. Trouble was experienced with the dan buoy dragging its sinker along the sea bed, and the interpolation between successive fixes is uncertain, so that only the average drift can be determined. This is 5·7 cm/sec, in the direction 250° (true). The float was loaded to sink to 900 m, and estimates of its depth ranged from 800 to 1,500 m.

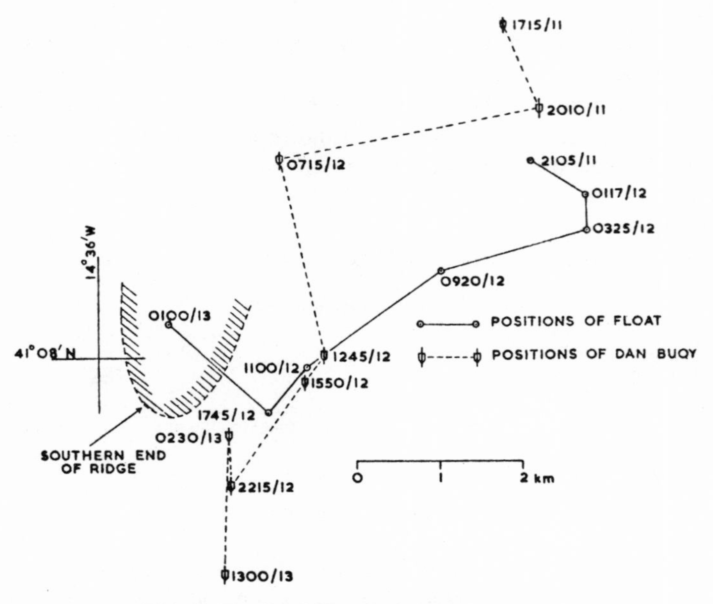

Fig. 4. Track of float, 11th-13th June, 1955.

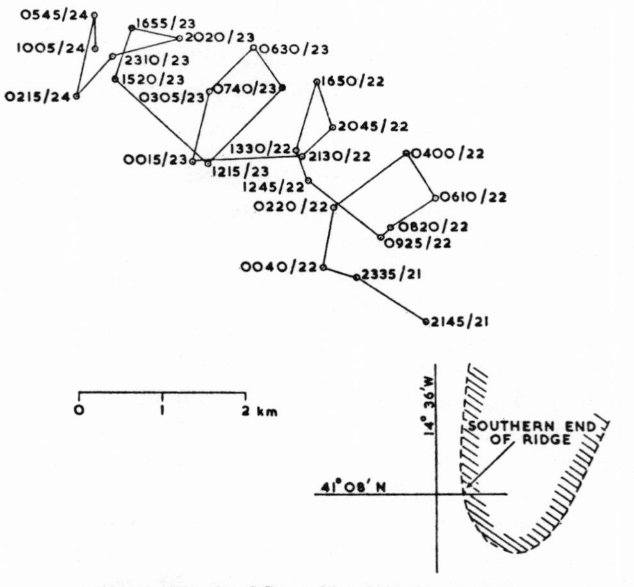

Fig. 5. Track of float, 21st-24th June, 1955.

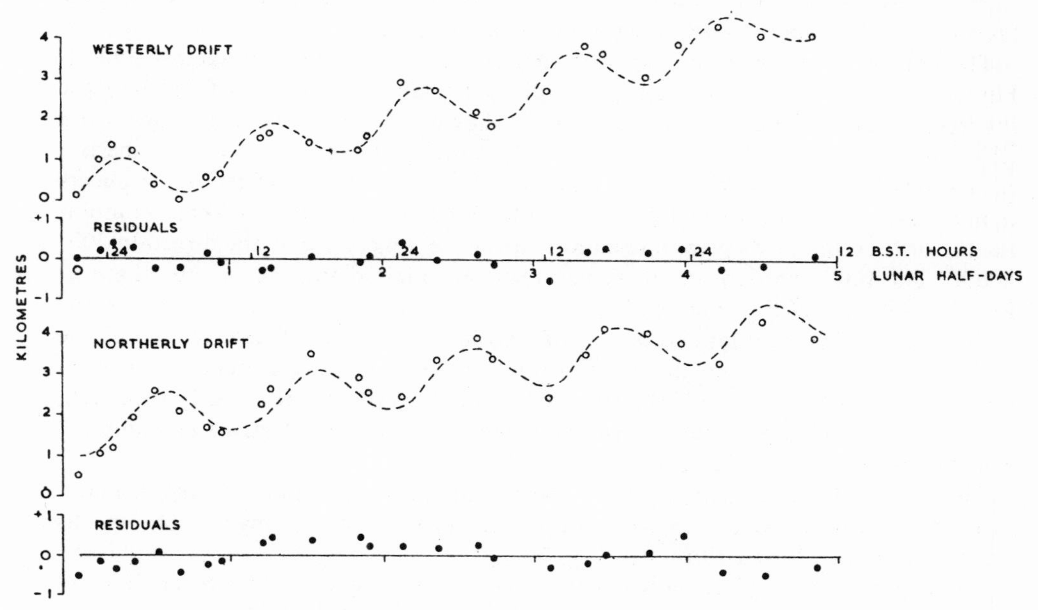

Fig. 6. Northward and westward movements of float plotted against time.

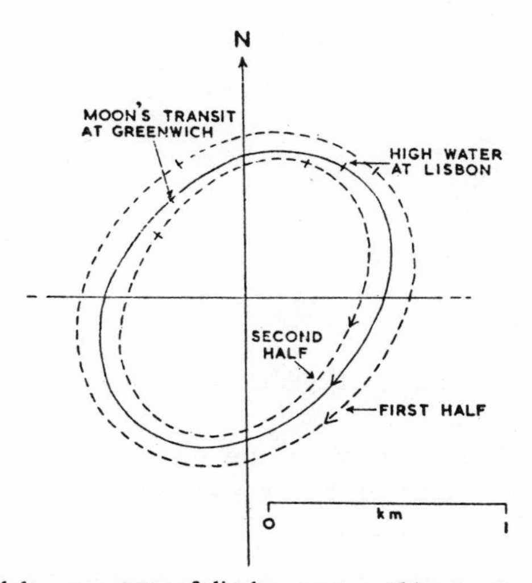

Fig. 7. Tidal components of displacement combined to form an ellipse.

Another float, loaded for 400 m depth, followed the track shown in Fig. 5. More frequent fixes were made, and the dan buoy was more securely anchored, so that tidal oscillations can be seen superimposed on the drift.

The northerly and westerly displacements of the float are plotted against time in Fig. 6. A steady drift plus a lunar semi-diurnal oscillation has been fitted by least squares to each of these, leaving the residuals shown. There is some tendency for the residuals to show an 18-hour fluctuation (approximately the period of inertial oscillations in this latitude) but it is not very significant, since the estimated uncertainty of each observation of the displacement of the float is about $\pm$ 0·2 km. Combining the drift components gives a resultant current of 2·4 cm/sec. in the direction 300° (true). The tidal components of displacement are plotted together as an ellipse in Fig. 7. The ratio of the lengths of the axes is 0·73, which agrees fairly well with the theoretical ratio (for an infinite ocean) of 0·68 in this latitude. The direction of the major axis, and the *cum sole* direction of rotation, are also in agreement with theoretical predictions (BOWDEN, 1954). The tidal current, varying from 7·3 to 10·0 cm/sec., is similar in magnitude to those measured by the " Meteor " and " Armauer Hansen " expeditions (DEFANT, 1932 : EKMAN, 1953).

To estimate the uncertainty of fitting these tidal oscillations, the observed displacements have been divided into two groups and separate fittings made. The results are shown as the dotted ellipses in Fig. 7.

The depth measurements show considerable scatter (Fig. 8), with a mean value of

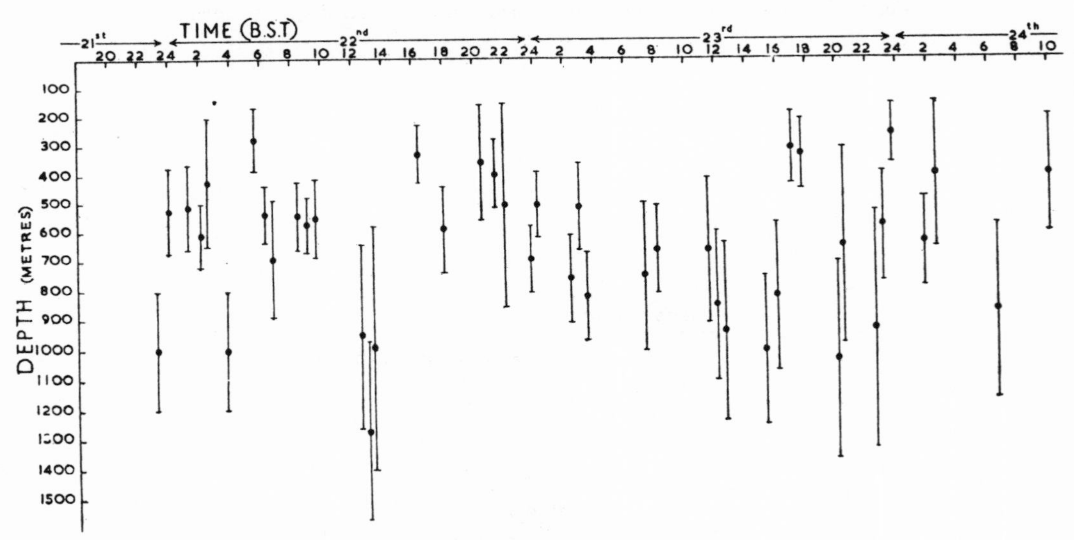

Fig. 8. Depth estimates.

about 600 ($\pm$ 200) m. The observations are too uncertain to show whether any genuine depth oscillations occurred, but they are sufficient to demonstrate that the sinking rate, after the first few hours, is very small.

ACKNOWLEDGEMENTS

The author wishes to express his thanks to many colleagues in the laboratory, and in " Discovery II," who took part in the construction of the apparatus and in the experiments at sea. The assistance of the Chief Scientist and members of the staff of H.M. Underwater Detection Establishment, Portland, in making the acoustic signalling equipment, is gratefully acknowledged.

National Institute of Oceanography, Wormley, Godalming.

REFERENCES

BOWDEN, K. F. (1954), *Deep-Sea Res.* **2**, 33-47.
B. S. 1476 (1949), Wrought Aluminium and Aluminium Alloys. *British Standards Institution.*
B. S. 1500 (1949), Fusion-Welded Pressure Vessels. *British Standards Institution.*
DEFANT, A. (1932), " *Meteor* " *reports.* Bd. VII. 1 Teil.
EKMAN, V. W. (1953), *Geofys. Publik.* **XIX**. (1).
NEWMAN, F. H. and SEARLE, V. H. L. *The General Properties of Matter.* Arnold & Co., London. 4th ed. 1948, p. 122.
STOMMEL, H. (1955), *Deep-Sea Res.* **2**, 284-285.
ZUBOV, N. N. and CZIHIRIN, N. J. *Oceanological Tables.* Moscow, 1940.

11

Reprinted from *J. Geophys. Research* **67**(8):3173-3176 (1962)

Velocity Measurements in the Deep Water of the Western North Atlantic[1]

Summary

J. Crease

National Institute of Oceanography
Wormley, Godalming, Surrey, England

In general our knowledge of the deep circulation of the ocean has been deduced indirectly either by a study of the geographical distribution of water masses represented by different temperature and salinity characteristics or by the use of the geostrophic equation. To obtain the pressure field needed in this equation requires, by present techniques, several days' work. Further, the performance of existing instruments precludes, in the face of extremely small pressure gradients, estimates of all but the larger scales of geostrophic motion. Velocities typically of the order of 1 cm/sec or less below a depth of 2000 meters in open ocean are suggested by the observations.

Direct observations with current meters of various types were reviewed by *Bowden* [1954], and for the whole Atlantic he is able to quote only 20 observations lasting from half a day to 4 days in the period 1910–1938. Again, the difficulties of building robust and sensitive instruments to be used from a buoy or drifting ship have limited the accuracy obtainable.

In recent years neutrally buoyant floats fitted with acoustic transmitters (analogous to constant level balloons) and developed originally by *Swallow* [1955] have yielded information in the Atlantic and Pacific on the amplitude and variability of the flow on scales not previously susceptible of observation. This paper is concerned with some aspects of a joint program of Woods Hole Oceanographic Institution and the National Institute of Oceanography, England, with Dr. J. C. Swallow leading the expedition, to observe with these floats the currents at great depth in a limited area and for a prolonged period.

Most observations were made over a 14-month period in 1959–1960 in a region of abyssal plain about 5000 meters deep and 200 miles west of Bermuda. Individual observations of a float determined its position to somewhat better than 0.2 mile, and the trajectory was generally followed for 4 to 10 days with one or more position fixes a day. To make the data as homogeneous as possible, attention was focused on two depths, 2000 and 4000 meters; even so, the total number of 72 trajectories is far too small to discuss the statistical structure of the velocity field in any detail.

Typical of the trajectories are those shown in Figure 1. They have with one exception only slight curvature, and, for example in the last series in August 1960, the fluctuations about a steady speed are barely detectable above the errors of observation. Speeds are of the order of 10 cm/sec, and direction is extremely variable from one series of trajectories to another. This is contrary to the widely held view that the deep ocean is relatively quiescent with velocities of the order of 1 cm/sec. There is a significant contrast between the variability of trajectories between one series and another and the steadiness of individual floats over a duration not much less than the interval between series. This suggests the possibility that the Lagrangian scale is noticeably larger than the Eulerian scale, although no definite figures can be stated at present. It is of interest that these high velocities in the open ocean are of the same order as those found by *Swallow and Worthington* [1961] below the Gulf Stream.

The last series of trajectories illustrates a fur-

[1] Based on a paper presented at the International Symposium on Fundamental Problems in Turbulence and Their Relation to Geophysics sponsored by the International Union of Geodesy and Geophysics and the International Union of Theoretical and Applied Mechanics, held September 4–9, 1961, in Marseilles, France.

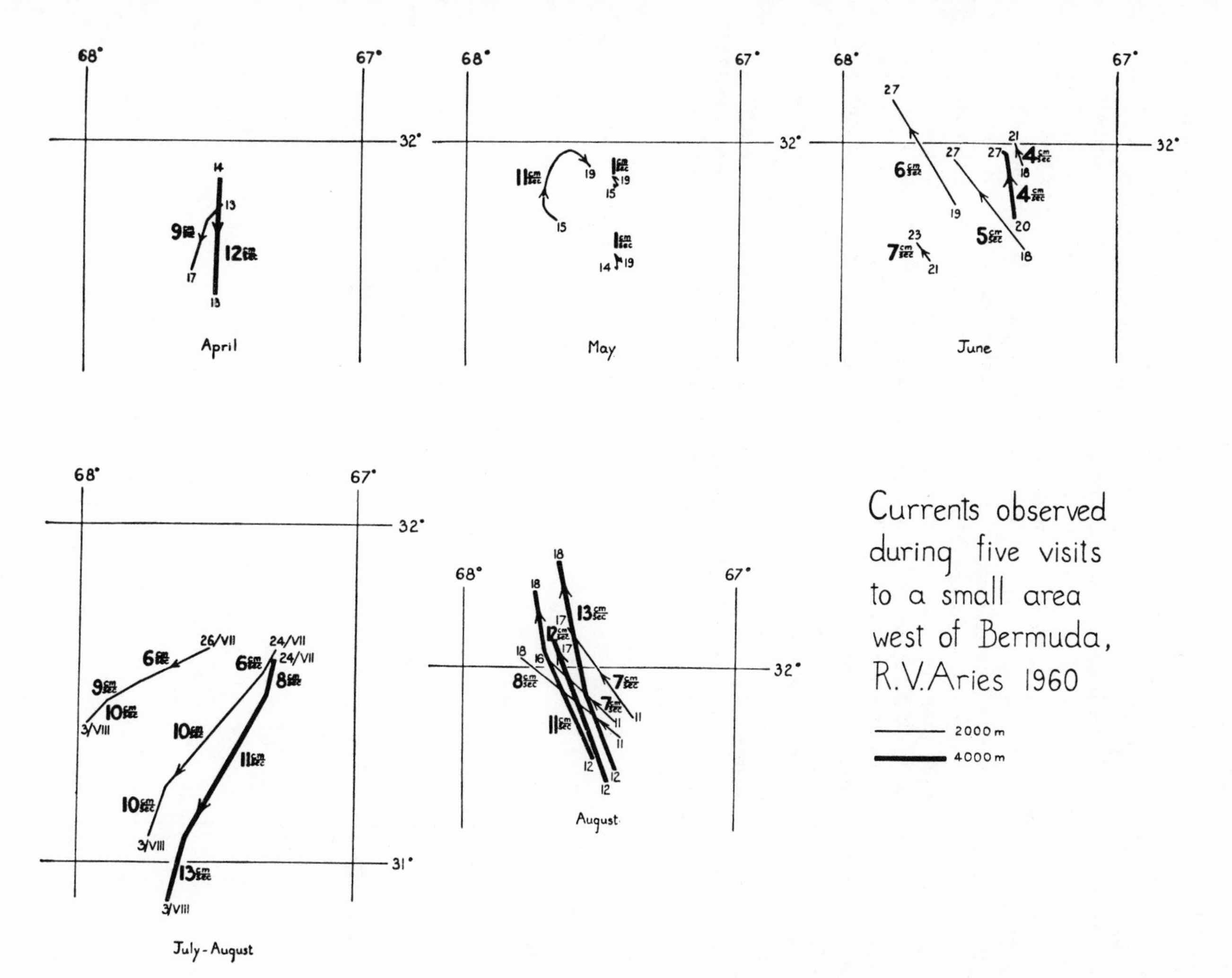

Fig. 1. Trajectories of five series of floats. Figures at ends of trajectory are starting and finishing dates. Figures beside trajectory are average speeds.

ther characteristic of the flow that was invariable over the duration of the measurements. Whenever there were simultaneous measurements on floats at 2000 meters vertically above one at 4000 meters, the deeper float moved in a direction closer to the meridian, and in all but one case its velocity was slightly greater. This preferred direction for the baroclinicity is unexplained.

Some information can be gained on the horizontal scales of the transient motions by considering the differences in velocity between pairs of trajectories at varying separations. In Figure 2 pairs of trajectories have been arranged in groups with separations 0–10, 10–20, 20–30, and 30–50 nautical miles. The mean value $\overline{\Delta V^2}$ for each group is the ordinate, where ΔV is the magnitude of the vector difference in velocity between two trajectories. There is apparently a large increase in the contribution to $\overline{\Delta V^2}$ in scales around 30 nautical miles. The last point on the curve at 40 nautical miles has a $\overline{\Delta V^2}$ equal to the mean square fluctuation in velocity of the individual tracks, suggesting that half the energy is contained in eddies up to 40 nautical miles in extent. This is a surprising result, and without further more extensive investigation it should be viewed with extreme caution in view of the limited data. (There are only six to ten pairs in each group.) There appears to be little in the way of external factors (e.g., depressions in the atmosphere or bottom topography) that would generate motions directly in these scales. Concerning this result it is of interest that an isolated identifiable water mass in the upper 1000 meters of water that drifted through the area and was some 80 nautical miles in extent drifted with a steady velocity for 2 months. This would be consistent with the view that most of the transient energy was on a scale less than the size of the water mass.

In an earlier paragraph the variability in direction from series to series was remarked on, but on even larger time scales the variability is still very noticeable. Over the whole 14-month period the velocity at 2000 meters was $2\frac{1}{2}$ cm/sec

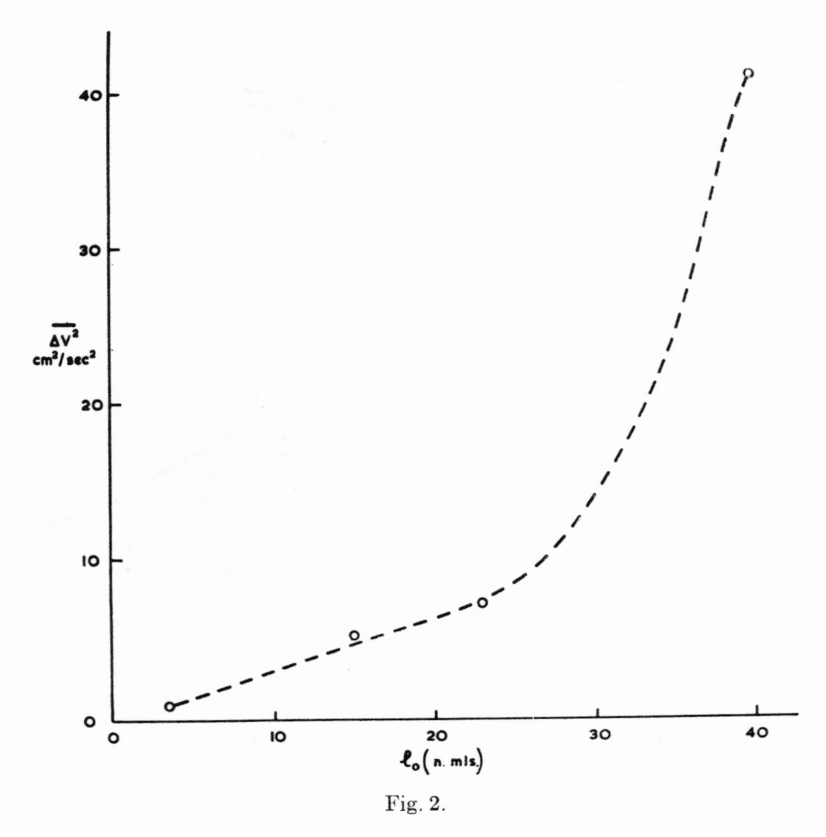

Fig. 2.

due west, but for the first and second halves of the period the directions were 30°–40° south and north of west respectively. It seems doubtful whether the figure of $2\frac{1}{2}$ cm/sec is representative of a long-term mean, and a far longer series of measurements would appear necessary to measure such a mean accurately. This raises the question of just how important the time-independent mean circulation is in the deep water of the open ocean compared with the eddy transport of properties by the transient motions.

References

Bowden, K. F., The direct measurement of subsurface currents in the oceans, *Deep-Sea Research, 2,* 33–47, 1954.

Stommel, H., *The Gulf Stream,* University of California Press, 1958.

Swallow, J. C., A neutrally buoyant float for measuring deep currents, *Deep-Sea Research, 3,* 74–81, 1955.

Swallow, J. C., and L. V. Worthington, An observation of a deep countercurrent in the western North Atlantic, *Deep-Sea Research, 8,* 1–19, 1961.

12

Reprinted from *J. Geophys. Research* **68**(22):6171-6186 (1963)

Large-Scale Air-Sea Interactions over the North Pacific from Summer 1962 through the Subsequent Winter[1]

Jerome Namias

Extended Forecast Branch, U. S. Weather Bureau
Washington, D. C.

Abstract. The extensive and persistent warmth of central and eastern North Pacific waters during the last half of 1962 and early 1963 is described and attributed to prevailing abnormal wind and weather systems. Evidence is presented to indicate that the evolution of the large-scale prevailing atmospheric systems was partly determined by the anomalous water temperatures as well as forcing influences due to a seasonal change in the elevation of the sun. The evidence is partly statistic-synoptic and partly physical, involving computations of sensible and turbulent heat exchange as well as radiation balance. Among the more outstanding weather abnormalities treated are the strange behavior of typhoon Freda and the extremely abnormal winter of 1962–1963 over much of the northern hemisphere.

Introduction. In an earlier report *Namias* [1959] investigated some large-scale interactions between the atmosphere and the North Pacific Ocean which may have been responsible for climatic fluctuations on a time scale of months and seasons and a space scale larger than the ocean itself. During the summer, fall, and winter of 1962–1963 some equally strong climatic anomalies have been observed in sea surface temperature over the North Pacific and in atmospheric circulation over the northern hemisphere. It is the purpose of this report to interrelate some of these abnormalities and to propose a modus operandi whereby lag effects of the order of a season or two can lead to sizeable climatic fluctuations in both sea and air. The feedbacks envisioned are not simple cause and effect relationships but are complexly coupled mechanisms established by the eternal abnormality of large-scale states of both atmosphere and sea. This type of interaction renders futile any attempt to discover an 'ultimate cause' of climatic anomalies in air or sea, because one abnormal state in either medium leads to abnormalities in the other, and the longevity of the disturbed condition differs between atmosphere and ocean.

Generation of warm surface water over the eastern North Pacific during the summer. During much of 1962 in the eastern Pacific near 42.5°N, 173°W, extensive pools of warm surface water departed more than 4.4°C (8°F) from the normal for September. The remarkable extent and growth of this warm pool is indicated in Figures 1 and 2. From the curves shown in Figure 3 it is clear that a rapid rate of warming occurred in August and September, leading to a displacement of the annual maximum from August, when it normally occurs, to September. Once established, the relative warmth continued through the succeeding fall and winter. Notice that before the August-September warming at 40°N, 170°W, a large warm patch of water persisted from May through September in the area roughly bounded by longitudes 165°W and 135°W and latitudes 30°N and 45°N.

In interaction problems involving the sea and air there is no definite point at which to begin to discuss causes; the choice is arbitrary. The warmer than normal seas observed from May to July appear to have been related to a greater than normal frequency of southeasterly components of air flow at sea level, particularly during May and June, which in turn drove more southerly (warmer) surface waters about 45° to the right through the operation of wind stress on the Ekman layer. We shall not document this statement, except to mention that computations of water displacement made from the observed anomalous wind stresses by the method described by *Namias* [1959] support this contention. We shall treat more thoroughly the pronounced September warming farther west which had in fact begun in late August (Figure 2).

[1] Presented at the SCOR Session on Air-Sea Interaction, Halifax, Nova Scotia, April 8, 1963; Symposium on the General Scientific Framework for the Comprehensive Study of the World Ocean.

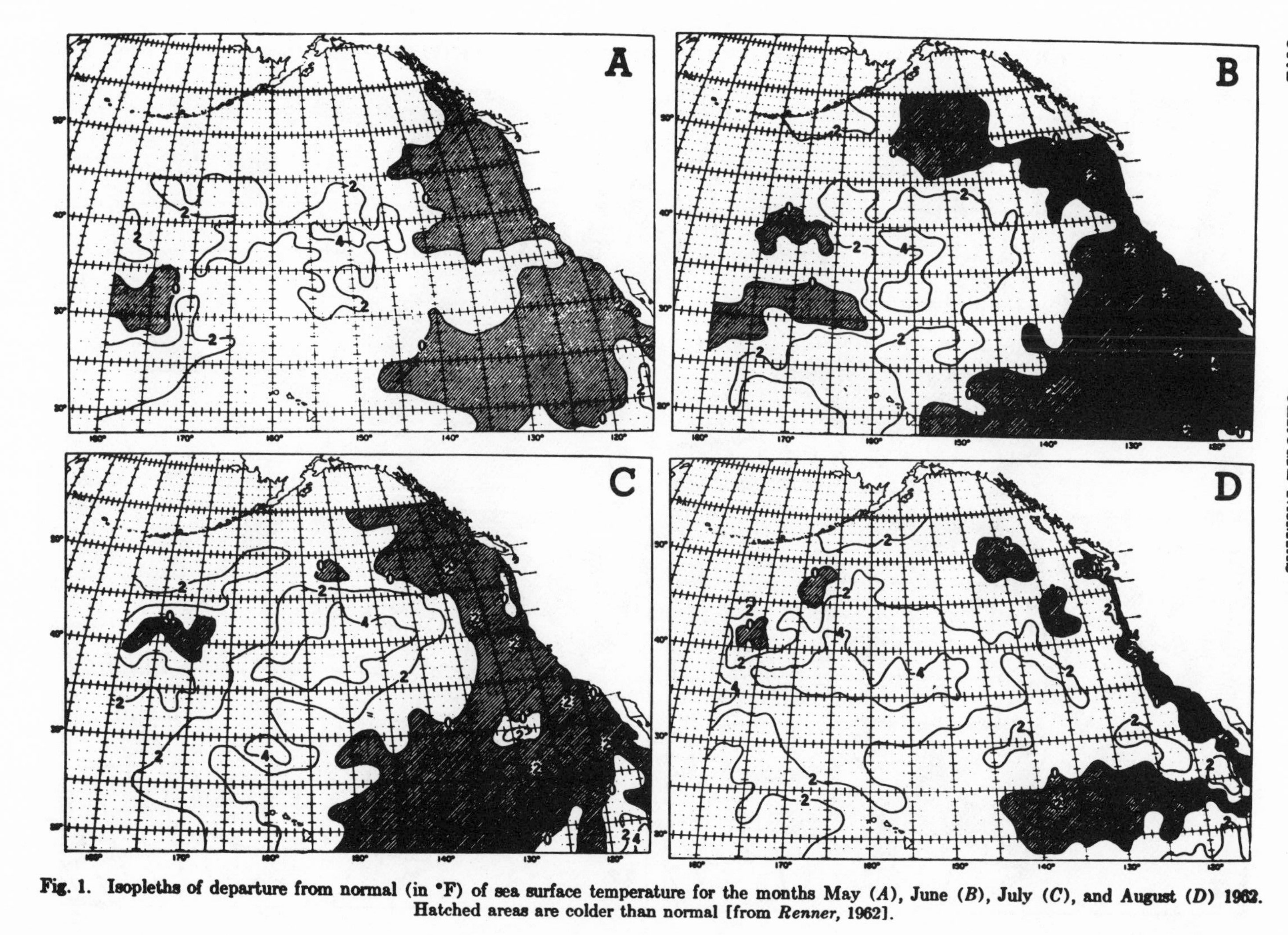

Fig. 1. Isopleths of departure from normal (in °F) of sea surface temperature for the months May (*A*), June (*B*), July (*C*), and August (*D*) 1962. Hatched areas are colder than normal [from *Renner*, 1962].

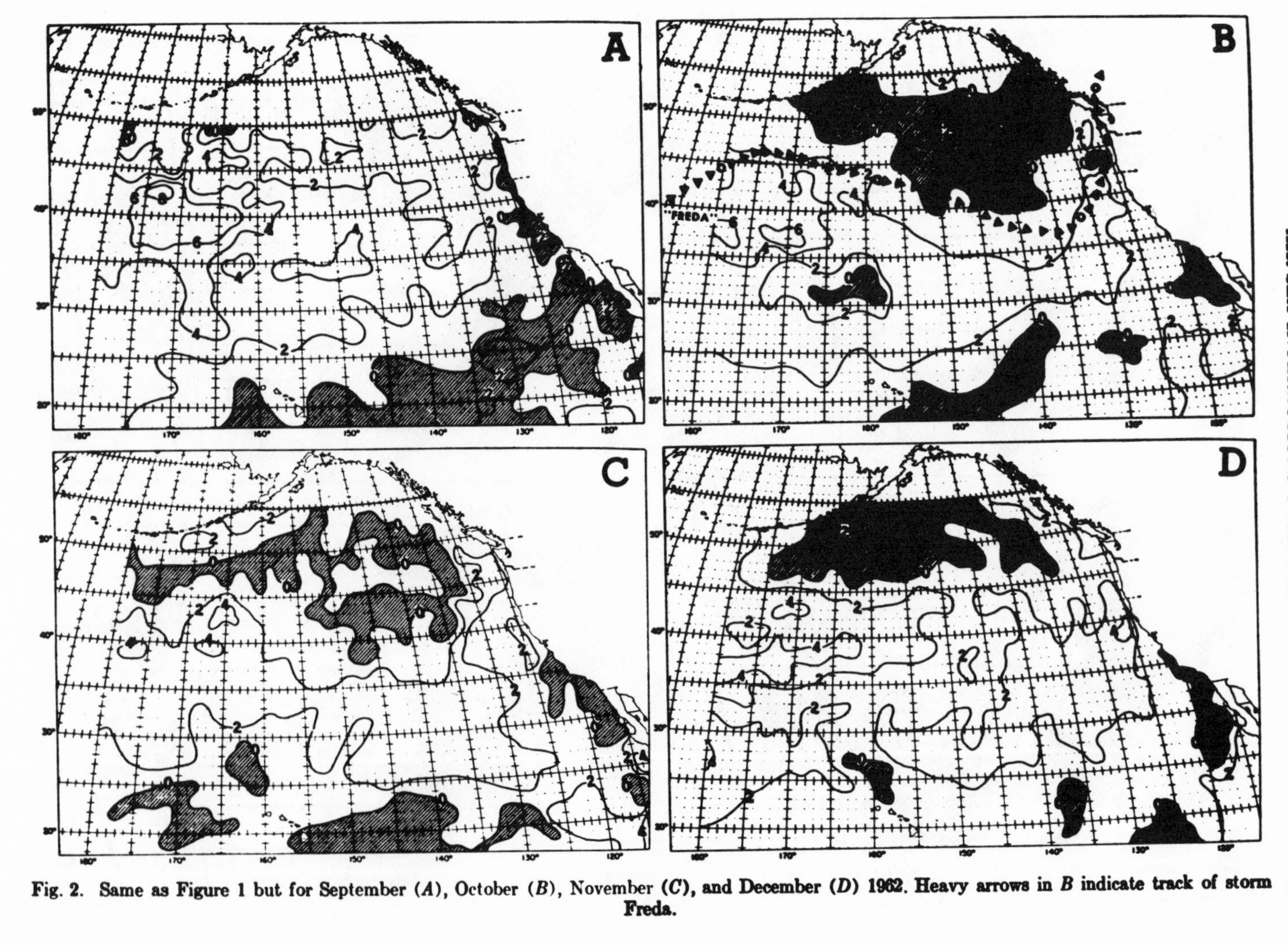

Fig. 2. Same as Figure 1 but for September (*A*), October (*B*), November (*C*), and December (*D*) 1962. Heavy arrows in *B* indicate track of storm Freda.

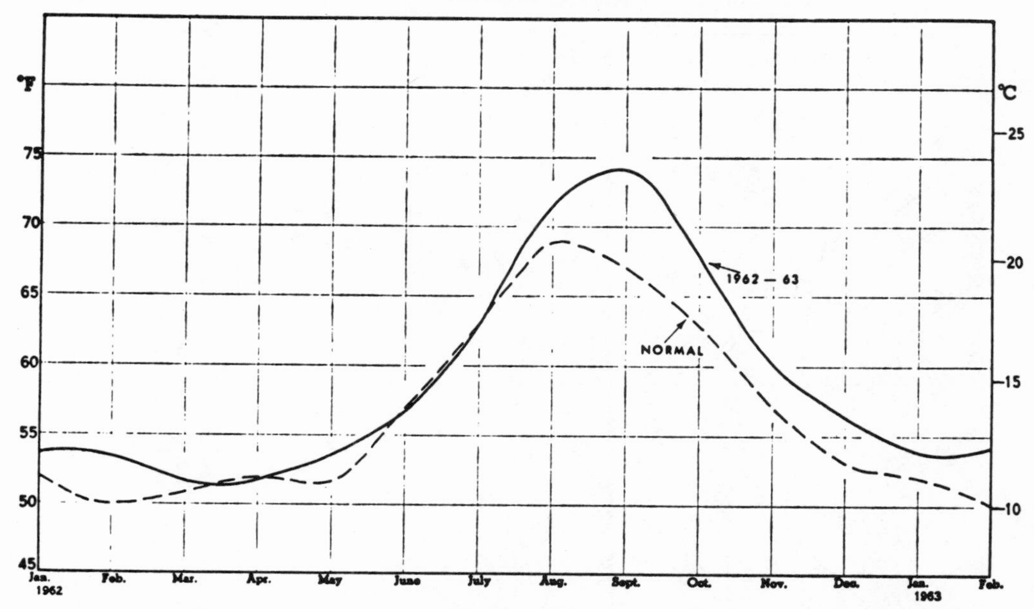

Fig. 3. Observed and normal monthly mean temperature of the sea surface at 40°N, 170°W, for the period January 1962 through February 1963.

We get considerable information on the nature of this warming from the contemporaneous mean atmospheric wind systems overlying the Pacific. These can be obtained from Figures 4*A* and 4*B* by using the geostrophic winds. We grasp the significance of the anomaly isopleths by considering that the total geostrophic wind (given by the isobars) consists of two vector components: normal and anomalous (represented by the broken lines). The geostrophic relationship of the anomalous wind to the isopleths of anomaly is the same as that of the total wind (the resultant wind) to the isobars.

From Figures 4*A* and 4*B* it is clear that during September the North Pacific anticyclone was much stronger than normal from the Aleutians southeastward. In fact, the maximum departure of 6 mb at sea level is slightly more than twice the standard deviation for the Septembers of 60 years at this point. This large area of positive anomaly at the surface, as well as aloft, and the strong anticyclonic curvature of the sea level isobars reflect the relative absence of cyclonic activity in this area and the predominance of anticyclonic systems—a fact which can be noted from the day to day weather maps of the month (not reproduced). Only one cyclonic center (a weak one of only 1012-mb depth) passed within 500 miles of the center of maximum warming.

The peculiar distribution of anomalous wind components (and the associated prevalence of anticyclones) indicates a strong horizontal convergence of surface water when account is taken of the effect of wind stress on the motion of the surface layer. This is shown by the arrows in Figure 4*C* which indicate the water displacement (relative to normal), assuming 45° between anomalous wind components and anomalous surface drift. (If the entire depth of the mixed layer is associated with the transport of the surface water, then the drift is at 90° to the surface wind. This does not materially affect the argument.) The horizontal convergence is especially strong in the area of maximum warming. Besides, some of the horizontally converging surface water (to the south and east of the anticyclonic ridge) is being advected from farther south, and therefore from warmer water sources than normal, as indicated in Figure 4*C*. However, the anomalous components east of the axis of greatest anomalous convergence suggest increased transport from *colder* water sources as may be seen from the northerly component

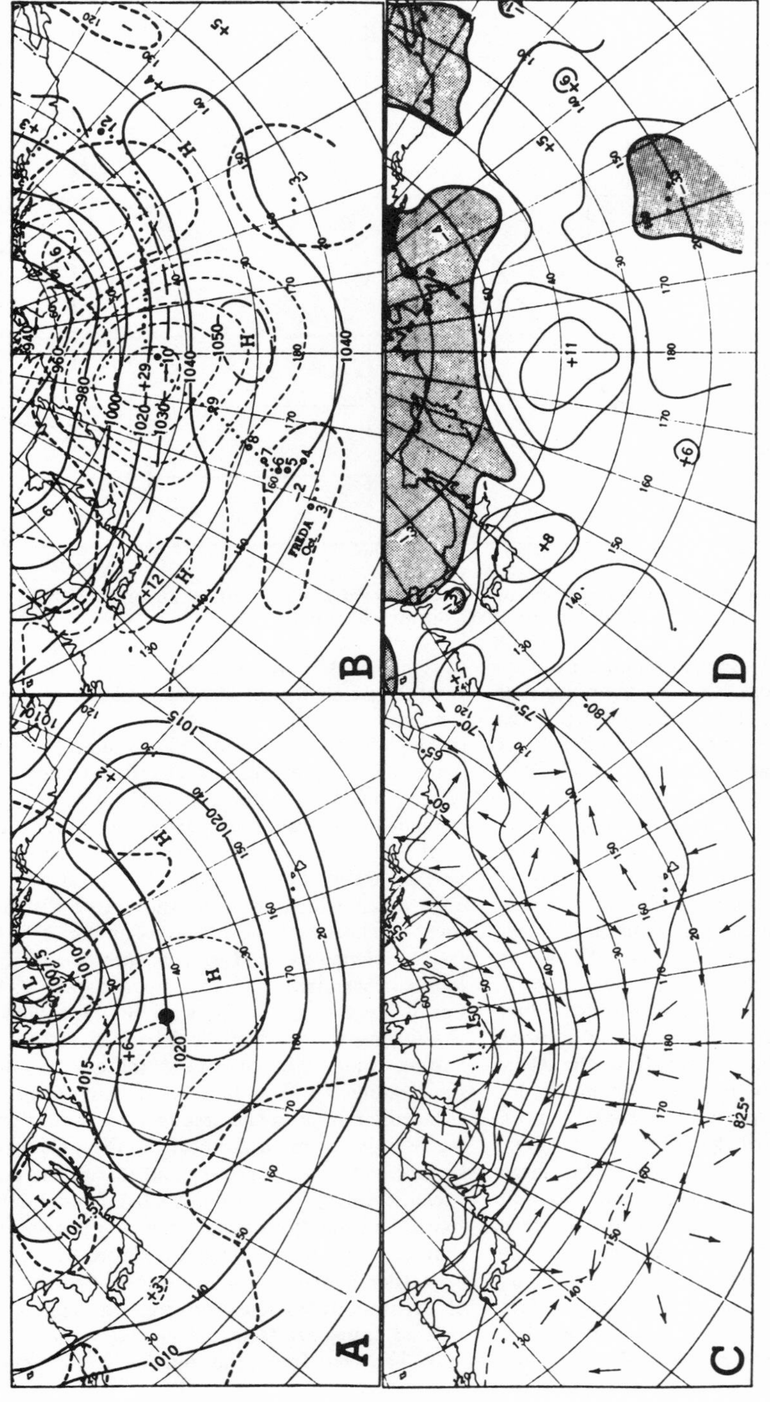

Fig. 4. *A*. Mean sea level isobars (solid) and isopleths of departure from normal (broken) for September 1962, both in millibars. *B*. Mean 700-mb contours (solid) and isopleths of departure from normal (in tens of feet, broken) for September 1962. *C*. Anomalous displacement of surface water (arrows) indicated by isopleths of sea level pressure anomaly (shown in *A*) and normal sea surface isotherms (°F) for September. *D*. Mean 1000- to 700-mb thickness departures from normal (in tens of feet) for September 1962.

arrows superimposed on the normal sea surface isotherms.

The warming thus seems to be explainable in the following manner: The horizontal convergence of surface water led to some downwelling, or certainly an absence of upwelling, relative to normal. The anticyclonic systems and relative absence of cyclones led to frequent periods of atmospheric subsidence with the associated dissipation of middle and high clouds. The effects of subsidence are suggested by the greater than normal thicknesses (synonymous with above-normal temperature) observed in the lower troposphere (Figure 4*D*) and by tabulations of cloud observations from ships in the area. The effects of subsidence and the resulting less than normal vertical temperature gradient are also suggested by the facts that the air-sea temperature difference was about half its normal value and the relative humidity higher than normal. As a result of these factors and of wind speeds slightly less than normal, *Jacobs'* [1951] formulas indicate less evaporation and less sensible heat transfer than normal. These processes—less upwelling, less turbulent heat exchange with the air, and advection of warmer water masses from a more southerly source than normal—would all operate to promote warmer than normal surface waters.

Also, it is possible that the thin cloud and lack of middle and high (or cumuliform) clouds resulted in somewhat greater than normal absorption of solar radiation by the sea surface. However, because of the unsatisfactory nature of the cloud observations, both in quantity and quality, we are not able to demonstrate conclusively the contribution from radiation.

The net result appears to be a pronounced warming of the layer of water down to the thermocline which, according to *Berson* [1962], is at about 30 m in this area in September.

After first treating an especially unusual meteorological event whose cause is probably related to the air-sea abnormalities described above, we shall turn to the problem of why the relative warmth of the surface water persisted into the following winter.

The life history of typhoon Freda. During the first week of October, a typhoon was observed in the area around 20°N, 160°W, and was subsequently tracked along the uncommon path indicated by the dotted lines in Figure 4*B*. This storm achieved notoriety because of its rapid intensification as it plunged southward toward the west coast of the United States producing exceptionally heavy rains and destructive winds (up to 120-mph gusts). *Decker et al.* [1962], in describing this storm, wrote that 'it took the greatest toll in death and destruction of any wind storm in the history of the Pacific Northwest.'

Let us now examine the relationship of this unusually intense storm to the large-scale time-averaged state described earlier. In the first place the area in which the storm formed was already marked as a favorable breeding ground by the form of the quasi-stationary (mean) circulation established in the *preceding* month (September), as can be inferred from Figures 4*A* and 4*B*. These figures not only show a lower than normal surface pressure and upper-level height distribution in the area where Freda was subsequently born but suggest a high probability of storm formation by virtue of the fact that the subtropical ridge line was far north of normal and the mid-tropospheric subtropical easterlies were correspondingly displaced and increased in strength. These factors are well recognized as favorable conditions for the creation of easterly waves and hurricanes [*Riehl,* 1954].

Another related inference may be made from the change in curvature of the mean sea level isobars and 700-mb contours from strong anticyclonic around 180°W to practically straight in the area stretching some 20° of longitude westward. Since the monthly mean can be considered to be a steady state, this change in curvature implied a strong horizontal convergence to offset the anticyclonic trajectories (and streamlines) required if the absolute vorticity of the northward curving stream were conserved. The manifestations of this lowest tropospheric convergence are the numerous easterly waves and disturbances generated in the area. The aberrant circulation pattern thus made it probable that the tropical systems would develop in the area of anomalously low pressure, and Freda happened to be one of these systems at the beginning of October.

After several days' hesitation in the zone of generation, Freda was picked up by an individual mid-tropospheric wave trough moving eastward from the Asiatic coast and was then carried northeastward by the upper-level pre-

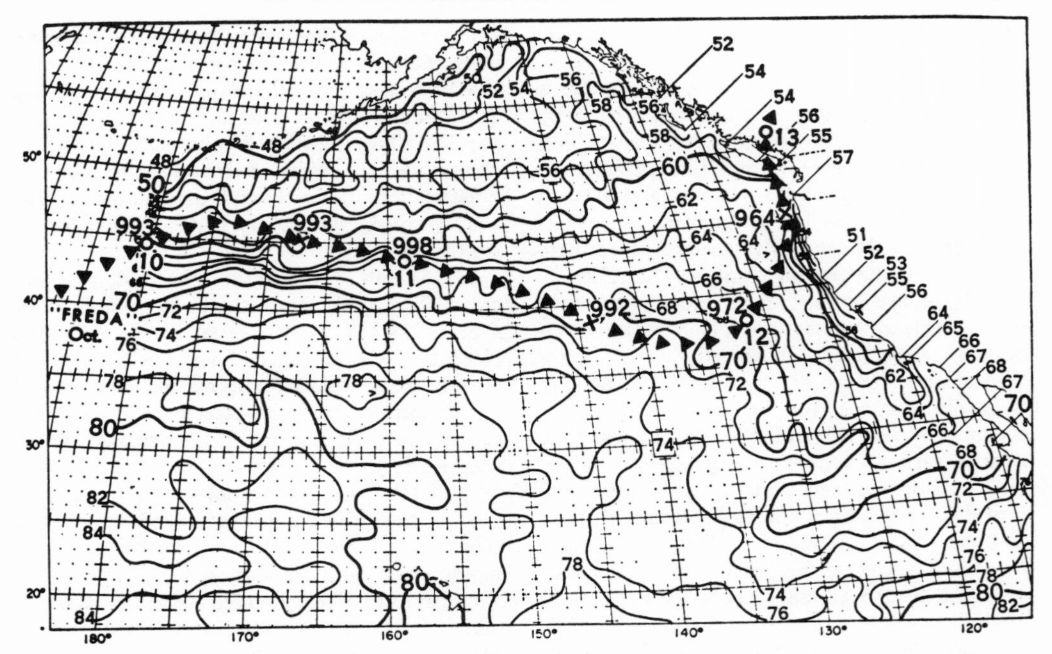

Fig. 5. Mean sea surface isotherms (°F) for September 1962 (solid, from *Renner* [1962]) and path of storm Freda (heavy arrows). 12-hr positions are indicated by circles and crosses. Central intensities, in millibars, are written above each 12-hr position. Dates, from Oct. 10 to Oct. 13, are written below the circles, which indicate positions at 1200 UT.

vailing current. In its northward course, the storm weakened considerably, partly owing to its trajectory over progressively colder water, from 30°C (85°F) in the area of formation to 13°C (56°F) at the northernmost point of its path; and by October 11 it seemed to have become an innocuous extratropical wave with a central pressure of 998 mb (see intensity and track in Figure 5). From 1200 UT on the 11th to 1200 UT on the 12th, the cyclone plunged southward and deepened 26 mb, of which 20 mb occurred during a 12-hour period.

It is believed that this rapid deepening, which runs counter to experience in southward-moving

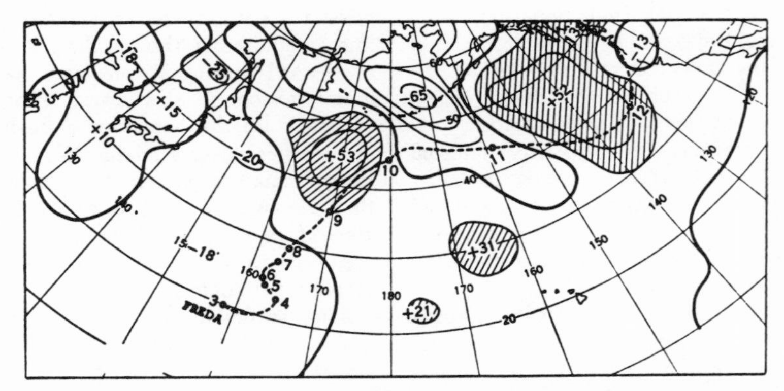

Fig. 6. Thirty-six-hour errors (in tens of feet) made by numerical (baroclinic) 500-mb prognosis made for 1200 UT, October 12, 1962.

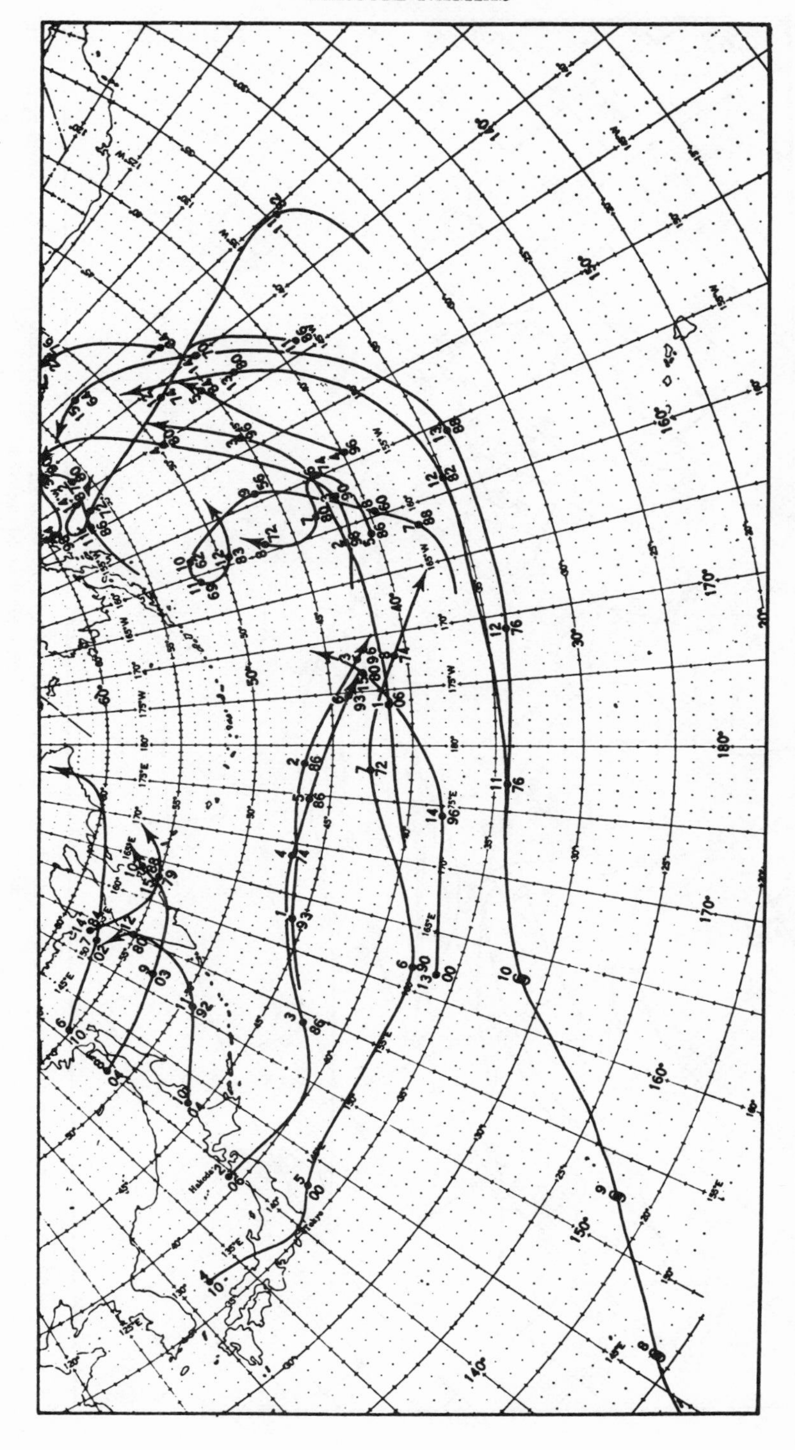

180°
170°
160°
150°
140°
30°

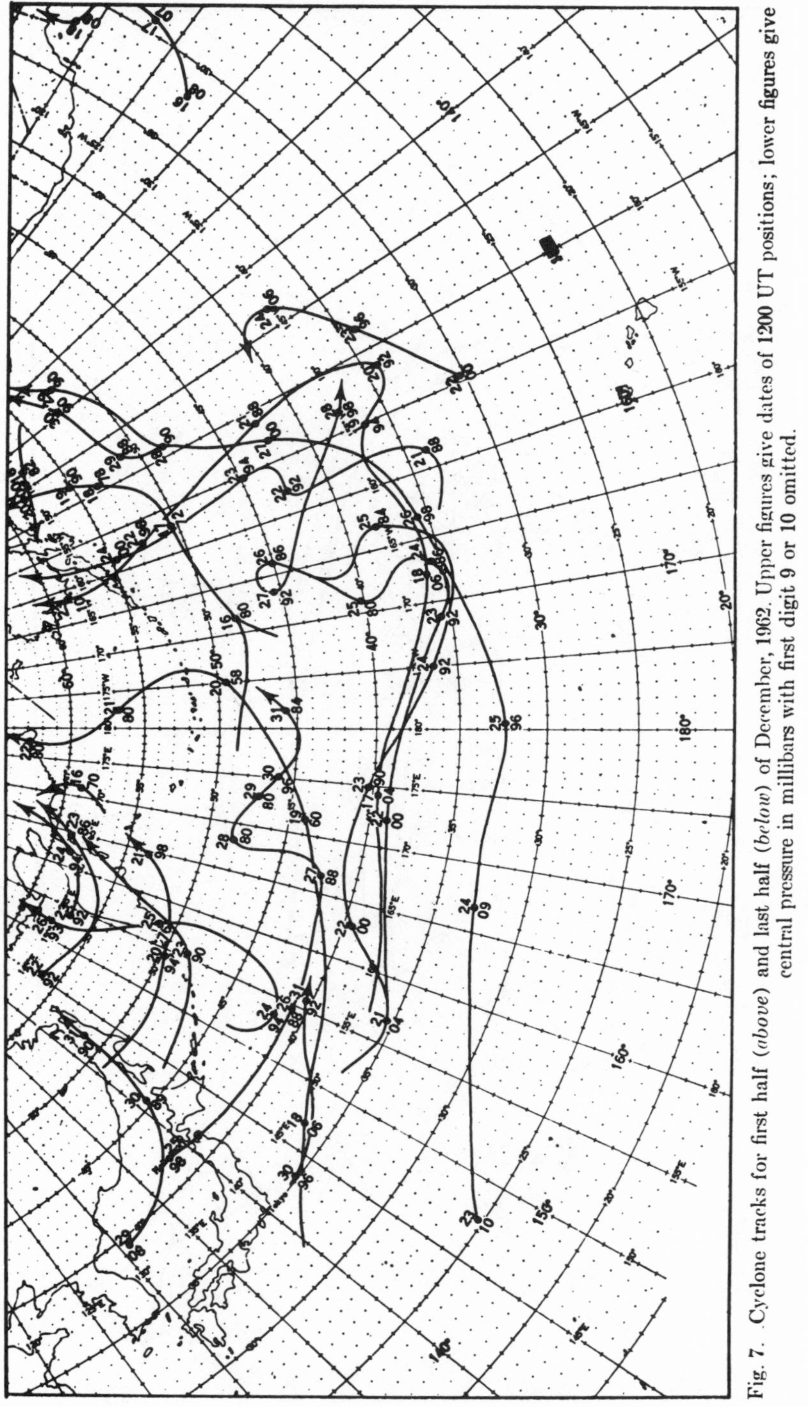

Fig. 7. Cyclone tracks for first half (*above*) and last half (*below*) of December, 1962. Upper figures give dates of 1200 UT positions; lower figures give central pressure in millibars with first digit 9 or 10 omitted.

storms, can be explained by the following arguments: Along its southward track the storm was forced to move into areas of increasing warmth where it picked up additional heat and moisture. This additional energy was apparently enough to resuscitate a storm which still retained some features of a typhoon aloft (perhaps in large values of moisture, vertical instability, and cyclonic vorticity at high levels) while appearing as an extratropical storm along the northern branch of its path. It is impossible to determine exactly how much of the storm's intensification was due to baroclinic deepening associated with the wave disturbance, but the errors of the 36-hour baroclinic numerical predictions at 500 mb made in the U. S. National Meteorological Center in Suitland (Figure 6) indicate failure to predict the amount of deepening by some 500 feet or 19 mb.

The southward track taken by the storm was also related to the mean circulation as indicated by the contours in Figure 4*B*, and to the anomalous gradient of water temperatures shown in Figure 2*B*. (The mean flow patterns for the first half of October were essentially the same as for September.) This relationship was complex, because the anomalies of water temperature were induced primarily by the same atmospheric circulation abnormalities (Figures 4*A* and 4*B*) which steered the storm vortex. In this case the cooler water probably resulted from the greater than normal cloud and vertical stirring of water associated with increased cyclonic activity in the Gulf of Alaska. But it must also be noted that the anomalous water temperatures, by imparting heat to or extracting it from the overlying air, favored similar temperature anomalies in the lower troposphere. Since pressure decreases more rapidly with height in air colder than normal, this helped to create a direct circulation in the same sense as the atmospheric flow pattern which generated the water anomalies in the first place.

To recapitulate, it appears that the formation growth, decay, and subsequent redevelopment of typhoon Freda along a most peculiar path were probably prescribed well in advance by interactive large-scale patterns of temperature and circulation in the ocean and atmosphere.

Subsequent short-period climatic fluctuations. It will be recalled that the positive anomaly of surface water temperature over much of the east-central Pacific continued well into winter. The maintenance of such warmth poses another complex problem wherein the coupling of atmosphere and ocean are inextricably entwined. Through October and November the westerlies continued to blow across the central North Pacific at latitudes farther north than normal, and cyclone paths for the most part continued to evade the warm pool of water. But a change began late in November and became consolidated in early December when the westerlies over the North Pacific dipped southward 10° of latitude and cyclones began to invade the domain of the warm pool. This is clear from Figure 7, where it is seen that after two cyclones plunged into the area on December 3 and 6, a series of intense storms moved over a route farther south than normal. The deepening of these cyclones as they moved eastward is indicated in Table 1. Indeed, this characteristic of the cyclonic activity—farther south than normal and appreciably deeper than normal—was to dominate the winter in the North Pacific. The statistical impact on the mean 700-mb circulation patterns for the winter is shown in Figure 8 where a departure from normal of 400 feet or 122 m (and 15 mb at sea level) was observed.

TABLE 1. Central Pressure of December Storms That Moved South of Latitude 45°N

	165°E	175°E	175°W
	988	975	973
	999	996	983
	985	977	976
	999	990	992
	1005	999	993
	1009	1001	997
Mean	998	990	986

We shall advance the thesis that the intense cyclogenesis was partly due to the abnormal warm pool of water and that the water was kept warm because of factors associated with the increased cyclonic activity.

First, the southward migration of the westerlies and prevailing cyclone tracks from fall to winter is a dependable feature of the climatology of the North Pacific, although the extent of the displacement of the cyclones as well as their degree of intensification varies from year to year. In November, zonal wind speeds over the North

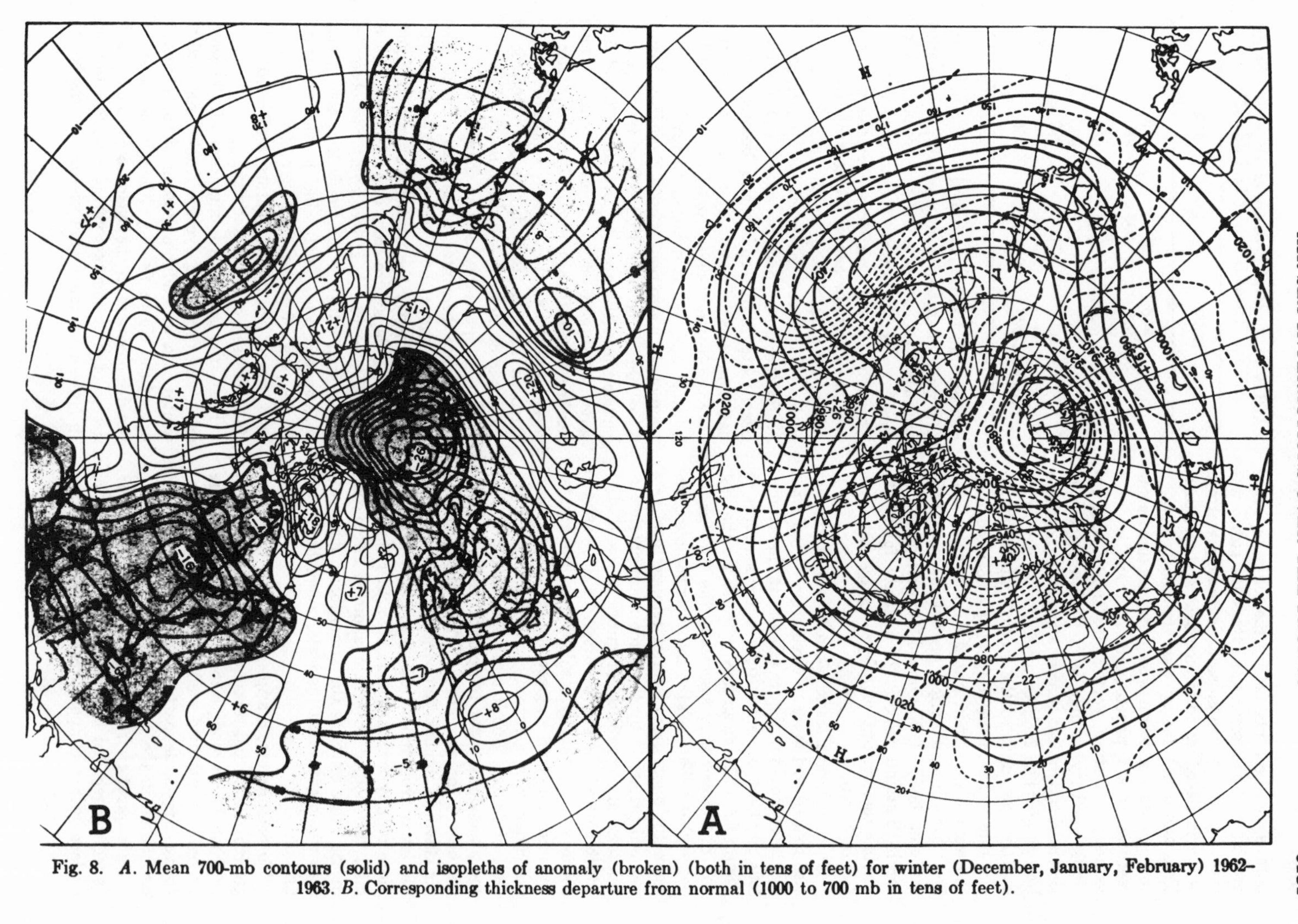

Fig. 8. *A*. Mean 700-mb contours (solid) and isopleths of anomaly (broken) (both in tens of feet) for winter (December, January, February) 1962–1963. *B*. Corresponding thickness departure from normal (1000 to 700 mb in tens of feet).

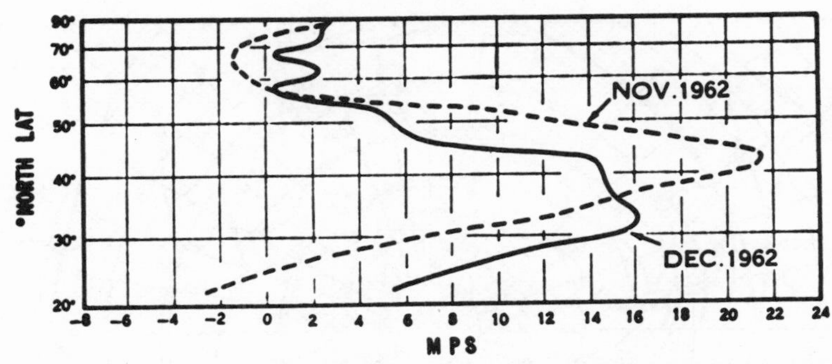

Fig. 9. Mean geostrophic zonal wind speed profiles over the North Pacific (135°E to 130°W) for November and December 1962.

Pacific increased strongly in midlatitudes, as indicated in Figure 9, with strong lateral shear north and south of the peak. In December, the zone of maximum speed shifted southward, and the low-latitude maximum continued in January and February but with still greater strength.

When some cyclones were forced to invade the area of relatively warm water (during the first week of December, for example), they probably received additional energy by virtue of the added (relative to normal) sensible and latent heat. Thus it is postulated that the cyclones, gaining energy from the excess heat and moisture supplied by the warm water, provided the vertical ascent necessary to release the latent heat of vaporization, which, in turn, further deepened the storms.

There is even more concrete evidence for this hypothesized chain of events. Again we refer to the numerical predictions of contour patterns produced by the 3-layer baroclinic model at the National Meteorological Center. It is important to note that such predictions, which are regularly extended to 36 hours, are based upon only the initial state of the atmosphere at the time the prediction is made; they do not include any effects of nonadiabatic heating. The mean of the 36-hour predictions for all days in December is shown in Figure 10*A*; the mean of the *observed* values for which the predictions were made is shown in Figure 10*B*. Obviously, the numerical predictions failed to capture the pronounced cyclogenesis in the east-central Pacific indicated by the strong trough observed there. Subtracting the observed from the forecast mean values brings into sharp focus (Figure 10*C*) the large algebraic errors (640 feet or 195 m) of the numerical prognoses—particularly in the vicinity of the warm pool of surface water.

Although this correspondence could be a coincidence, it must be pointed out that the usual place where errors of this sort (heights forecast too high) are found in other winters lies just off the Asiatic coast where the greatest heating normally occurs [*Jacobs*, 1951]. The other areas of positive error in Canada and Novaya Zemlya are perhaps related to the large one in the Pacific, but before getting into this phase of the problem we shall discuss further the feedback of the Pacific cyclones on the water temperatures.

The reservoir of anomalously warm water developed in summer and fall might not have been sufficient to persist through the ensuing winter. That is, the excess temperature of the surface water, relative to normal, might have been quickly depleted by latent and sensible heat transfer and radiation losses. It is clear, however, that the pool of warmer than normal water did, in fact, persist not only through the fall but also through the entire winter. Before exploring the heat budget associated with this retention of warmth we must obviously consider the depth of water involved.

It was indicated that the thermocline at 40°N, 170°W, is normally found at about 30 m in the month of September. However, in December it is normally found at about 70 m in the same area [*Berson*, 1962]. If we assume that the warming observed at the surface extended through the 70-m mixed layer, the additional heat being supplied in months other than September, we can estimate how long a positive

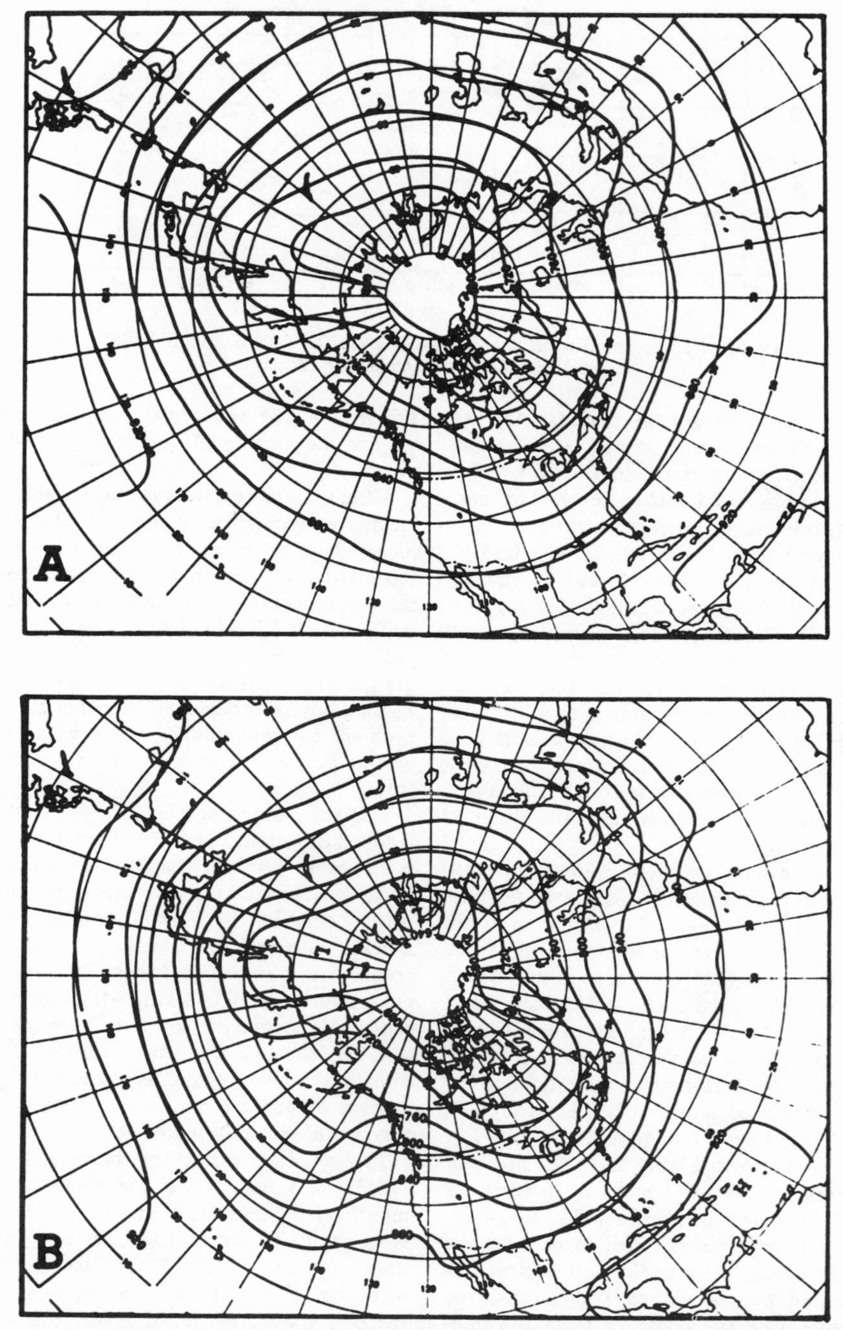

Fig. 10. *A*. Mean contours (in tens of feet minus 1000) of 36-hour numerical (baroclinic) 500-mb prognoses for all days in December 1962. *B*. Observed mean 500-mb contours for December 1962.

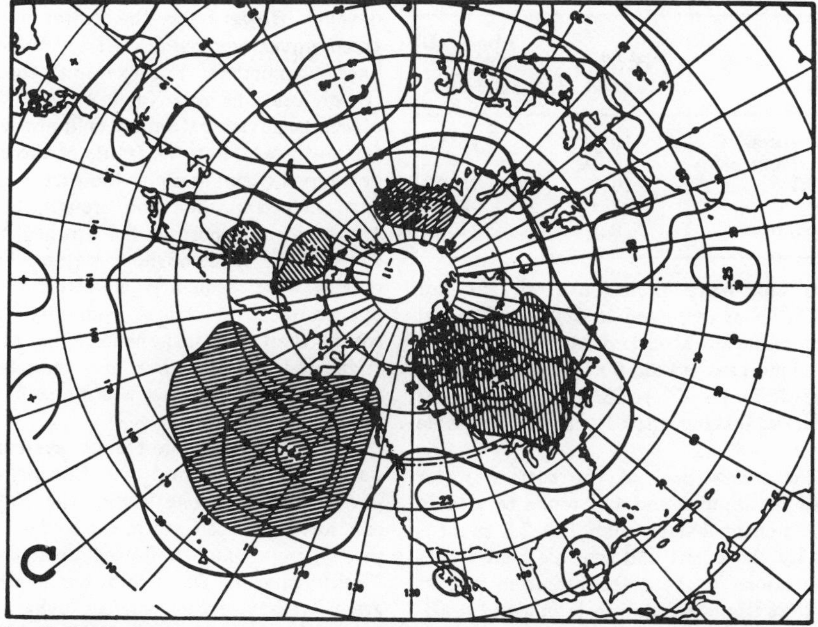

Fig. 10. *C*. Isopleths of mean error (drawn for 200-foot intervals) of numerical 36-hour prognoses for all days in December 1962, or *A* minus *B*.

anomaly of 2.3°C (4°F) would last if the heat were depleted by (*a*) sensible heat transfer, (*b*) latent heat exchange (evaporation), and (*c*) radiation imbalance, if we make the estimates shown in Table 2 for winter at 40°N, 170°W.

Using *Jacobs'* [1951] formula, we find that the normal transfer of *sensible heat* from the ocean surface is 57 ly/day or 1710 ly/month. For the abnormal case of the 1962–1963 winter, it is 4380 ly/month. Therefore, the anomalous removal of sensible heat is 2670 ly/month. To cool a 70-m layer originally 2.3 centigrade degrees above normal down to normal temperature would take about 5.8 months.

Using Jacobs' formula for the transfer of *latent heat* we find a normal extraction of heat of 6270 ly/month and an extraction of 9960 ly/month during the winter. Thus for the abnormal case, 3690 ly/month more than normal were extracted from the water. Consequently, the ocean surface will lose heat relative to normal at the greater rate of 6360 ly/month when account is taken of both latent and sensible heat loss. The time required to cool the 70-m layer is about 2.3 months.

In approximating the factors entering in the radiation balance it is immediately clear that the relative importance of the role played by each factor varies in accordance with the time of the year. Hence at 40°N, 170°W, cloudiness in summer, which screens insolation, results in less than normal heating of the water, whereas in winter it might result in less than normal cooling of the water because of its effect in trapping the long-wave radiation from the surface. If we make some computations for 40°N, 170°W, for December like those made for September, we find that with clear skies 7700 ly are received from sun and sky radiation, and, with the albedo of 12 per cent, 6700 ly are absorbed by the water. (Computations of short-wave absorption were made using the Savino-Ångström formula [*Budyko*, 1958].) With the normal 0.6 cloudiness in this area in December [*Landsberg*, 1945], about 4400 ly are absorbed. The long-wave back radiation from 0.6 cloudiness gives 2850 ly (using *Adem*'s [1962, p. 105] formula with an average cloud temperature of 261°K). If we interpose complete cloudiness instead of 0.6, a reasonable assumption in view of the persistent cyclonic activity, the absorbed incoming short-wave radiation is reduced to 2240 ly. The

TABLE 2

	Normal	Abnormal (Winter 1962–1963)
Sea temperature, °C	10.8	13.1
Sea temp. minus air temp., °C	−16.7	−15.6
Dew point, °C	7.1	8.9
Wind speed, mps	9.1	11.8

downward long-wave radiation from the low cloud at 280°K is increased to 6820 ly. The net downward radiation absorbed by the water for complete cloudiness is then 9060 (2240 + 6820) instead of 7250 (4400 + 2850) ly/month for 0.6 cloudiness, leaving 1810 ly/month available for heating.

Thus, the excess heat of the surface water established in summer and fall tends to remain about 1.2 months longer than the 2.3 months suggested by the latent and sensible heat transfer computations. In fact, the radiation budget indicates that the losses due to latent and sensible heat exchange (1650 ly/month) are more than offset by the radiant energy excess (1810 ly/month), but some of this (about 1000 ly) is lost to the water because of its higher than normal radiating temperature and the lower than normal radiating temperature of the air.

For these reasons, an anomalous reservoir of heat in the surface waters in fall, under special conditions, may not be depleted during the winter, a fact perhaps partly explaining the great persistence of abnormal surface temperature over the free ocean frequently cited in the literature [e.g., *Neumann et al.*, 1958].

Finally, we shall make a few remarks about possible influences well removed from the North Pacific. Returning to Figure 4*B* or Figure 8, we note the existence of four major troughs in the planetary atmospheric circulation: off Japan, in the central Pacific, over eastern North America, and in Europe. It is evident from the anomalies in Figure 8 that these wave systems are of much greater amplitude than normal. The central Pacific trough usually leads to a response of the ridge and trough downstream through the well-known barotropic redistribution of vorticity as first demonstrated by *Rossby* [1939]. In fact, with only one disturbance in the westerlies (in the central Pacific), it can be shown that a series of standing waves, about as observed, will develop. However, in the winter of 1962–1963, these waves were amplified by numerous feedback phenomena. For example, the ridge and trough locations over North America were ideally placed to advect extremely cold arctic air masses far southward, and the fronts of these air masses developed cyclones along the east coast which in turn helped deepen the trough, induce more arctic air, and increase the upward heat flux to the cold air from offshore waters. The cold air masses were frequently advected over an eastern snow cover whose boundary lay farther south than normal; hence, the air was refrigerated and the zone of strong temperature contrast along the coast was intensified [*Namias*, 1962].

Similarly, the ridge-trough system over the northeast Atlantic and over Europe, compatible in terms of wavelength with the North American wave, induced arctic air to flow southward into Europe, and deflected cyclones into the Mediterranean. The frigid arctic air produced great areas of frozen surface over Europe and led to frequent snows in areas where snow is uncommon. The cold and refrigerated air fed into the Mediterranean cyclones was rapidly heated by the Mediterranean, again producing greater than normal cyclogenesis there, and the cold east winds to the north of these storms assisted in keeping the ice and snow from melting, thus favoring anticyclogenesis.

In this manner, a self-aggravating or at least self-perpetuating system was established during the phenomenal winter of 1962–1963.

Acknowledgments. I express thanks to my colleagues Dr. Julian Adem and Mr. Philip Clapp for their sympathetic and stimulating discussions.

REFERENCES

Adem, J., On the theory of the general circulation of the atmosphere, *Tellus, 14,* 102–115, 1962.

Berson, F. A., On the influence of variable large-scale wind systems on the heat balance in the active layer of the ocean, *Tech. Mem. 25,* 75 pp., National Meteorological Center, U. S. Weather Bureau, 1962. (Available without charge from U. S. Weather Bureau, Washington 25, D. C.)

Budyko, M. I., *The Heat Balance of the Earth's Surface (Teplovi Balans Zemnoi Poverkhnosti),* 254 pp., Hydrometeorological Publishing House, Leningrad, 1956. (Translation from the Russian by N. A. Stepanova, U. S. Weather Bureau, 1958; in U. S. Weather Bureau Library, Suitland, Maryland.)

Decker, F. W., O. P. Cramer, and B. P. Harper, The Columbus Day 'big blow' in Oregon, *Weatherwise, 15,* 238–245, 1962.

Jacobs, W. C., The energy exchange between sea and atmosphere and some of its consequences, *Bull. Scripps Inst. Oceanog., 6,* 27–122, 1951.

Landsberg, H., Climatology, *Handbook of Meteorology,* pp. 927–997, McGraw-Hill Book Company, New York, 1945.

Namias, J., Recent seasonal interactions between North Pacific waters and the overlying atmospheric circulation, *J. Geophys. Res., 64,* 631-646, 1959.

Namias, J., Influences of abnormal surface heat sources and sinks on atmospheric behavior, *Proc. Intern. Symp. on Numerical Weather Prediction in Tokyo,* pp. 615–627, Meteorological Society of Japan, Tokyo, 1962.

Neumann, G., E. Fisher, J. Pandolfo, and W. J. Pierson, Jr., Studies in the interaction between ocean and atmosphere with application to long-range weather forecasting, *Final Rept. AFCRC-TR-58-236,* contract AF 19(604)-1284, (ASTIA-AD-152 555), Department of Meteorology and Oceanography, New York University, 1958.

Renner, J. A., Sea-surface temperature charts, eastern Pacific Ocean, *California Fishery Market News Monthly Summary, June through December 1962. Part II, Fishing Information,* U. S. Dept. of the Interior, Bureau of Commercial Fisheries, Biological Laboratory, San Diego, California, 1962.

Riehl, H., *Tropical Meteorology,* 392 pp., McGraw-Hill Book Company, New York, Toronto, and London, 1954.

Rossby, C. G., Relation between variations in the intensity of the zonal circulation of the atmosphere and the displacement of the semi-permanent centers of action, *J. Marine Res., 2,* 38–55, 1939.

(Manuscript received May 13, 1963;
revised August 26, 1963.)

Part III

TIDES

Editor's Comments on Papers 13 Through 19

Primitive people living by the seashore knew that the tides were linked to the passage of the moon (Bridges 1948, p. 166). In Greek science, however, mention of tides comes only after Aristotle's time since, in the Mediterranean, there was little opportunity for their observation. Following wider exploration, the later Grecian and early Roman writers described the twice-monthly springs and neaps; they related the daily tides to the lunar rather than to the solar day, and somewhat later (Pliny), gave an account of the lunitidal interval, the delay that occurs between the passage of the moon overhead and high water (Deacon 1971, ch. 1; Duhem 1913–1959; Harris 1898). A philosopher of the second century B.C. described the diurnal inequality of the tides in the Persian Gulf, the phenomenon in which one tide a day is higher than the other, and showed that it was related to and varied with the moon's position north or south of the equator (G. H. Darwin 1898, pp. 76–77). This tradition of observation and discussion was continued by Arab scientists and navigators (Aleem 1968; Duhem 1913–1959).

Even in medieval Europe, where science was not flourishing, there is some evidence of accurate knowledge of tides for which sailors and fishermen must have been responsible. In the early eighth century, Bede gave a well-informed description of the tides of the North Sea (Jones 1943); and four hundred years later, Gerald of Wales did the same for the Irish Sea (Deacon 1971, ch. 2). Bede's work shows he knew of the principle of the "establishment of a port," that is, that high water will always occur at a particular place when the moon is in the same position in the sky. This formed the basis of medieval tide tables and, at the time of the Renaissance, of printed tables like the one in William Cuningham's *Cosmographical Glasse* of 1559.

Most people who wrote about tides in medieval times were, however, satisfied to speculate on their causes without bothering too much about facts. This continued to be so into the seventeenth century, though a few were more careful; for example, the Yugoslav writer Nikola Sagroevic (Dadić 1968) and the English astronomer, Jeremiah Horrox, who made observations in the Mersey in 1640 (Wallis 1673). By this time the many causes of tides suggested in earlier times (Sarton 1927–1948) had largely been abandoned in favor of the theories of Galileo and Descartes. Galileo recognized that the earth was not only rotating round the sun but also on its own axis. Tides were, he thought the result of the effect on the oceans of the interference between these two motions which sometimes acted in concert, sometimes in opposition, at different points on the earth's surface (Aiton 1954; Drake 1961; Shea 1970. Descartes (1897–1909, vol. 9, pp. 227–230) believed that the spaces between the heavenly bodies were filled with ether. The passage of the moon overhead compressed the ether and the pressure was communicated to the surface of the sea, causing tides.

Galileo's ideas were adopted by an English mathematician John Wallis (1666), who added a third movement, that of the earth and moon round a common center of gravity (Deacon 1971, ch. 5). He made certain predictions about the behavior of tides which, he said, if found to be true, would prove his theory. The Royal Society accordingly called for volunteers to make observations at different places on the coast of Britain. Sir Robert Moray [1608–1673] drew up instructions for the observers (given in Paper 13). He did not think independent watchers would give sufficient information and suggested the foundation of a tidal observatory, equipped with a mechanical tide gauge. Those in charge of the observatory should not only make detailed observations of the rise and fall of the water but keep meteorological records as well.

The study of tides unfortunately did not inspire such a sense of urgency as the problem of finding longitude at sea which led to the founding of the Royal Greenwich Observatory in 1675 (Forbes 1975). However, some observations were made at different places. One of the people to whom Sir Robert Moray applied for information was John Winthrop, Governor of Connecticut (Winthrop 1878; Deacon 1971, pp. 100–108). These observations tended to contradict Wallis's predictions but revealed such things as the existence of a slight diurnal inequality off British coasts which Newton (1687) used as examples when applying his gravitational theory to tides.

Several attempts had been made, in the sixteenth and early seventeenth centuries, notably by Stevin (Sarton 1934), Kepler (Deacon 1971, p. 51), and William Gilbert (1600), to explain tides as the result of magnetic or other attraction; but such ideas appeared to the scientific rationalist spirit of the seventeenth century too much like the beliefs in magic or the occult of a previous, less enlightened age. It was only when Newton showed that tides could be explained as a consequence of his law of gravitation, that attraction seemed admissible. In fact, even those who were, like Wallis, opposed to such an idea had found it difficult to deal with the subject without actually admitting something of the kind. In spite of this, Newton's ideas on tides were not at once accepted and both popular and scientific works supporting Descartes' model, together with less well-known ideas, continued to appear into the eighteenth century (Deacon 1971, pp. 109–111).

Edmond Halley (1656–1702) had been instrumental in persuading Newton to publish the *Principia,* and, wishing to make his ideas better known, wrote a summary in English, which was presented to King James II; it appears here as Paper 14. Newton's work was at first taken up mainly on the continent. In France, systematic observations made during the early years of the eighteenth century were discussed by the astronomer Jacques Cassini. Further research was done by Bernoulli, Lalande, and the German mathematician Leonard Euler. The French astronomer Laplace instituted new observations at the port of Brest early in the nineteenth century. He first applied a hydrodynamical theory to the tides and stated that by analyzing tidal data from a port it would be possible to predict tides for the future, by relating the heights observed to the different positions of the sun and moon (Cartwright 1972; Harris 1898).

Laplace's initiative was followed in Britain by J. W. Lubbock, 1803–1865, and William Whewell. Lubbock was first of all concerned to produce new tide tables for London but the work was extended far beyond this. Self-registering tide gauges (Matthäus 1972), one of which is described in Paper 16, were gradually set up in naval dockyards and at the principal ports. With the data these provided and with existing records, Lubbock and Whewell began studying tides not only in harbours but also in the open sea (Deacon 1971, ch. 12).

Astronomers in the eighteenth century had, for the sake of convenience, treated the tide as a wave with two crests, 180° apart, traveling around the globe. In reality, the existence of the continental masses made the situation considerably more complex. In common with other scientists of their time, Lubbock and Whewell believed that the tide wave traveled freely around the Southern Ocean and sent subsidiary waves northward into the other major oceans (Lubbock 1832). They attempted to map the progress of the Atlantic tides by drawing cotidal lines, joining places where high tide arrived at the same time. Information was limited so they organized simultaneous observations on British coasts in June 1834. In 1835, a fortnight's simultaneous records were made at points on the western coasts of Europe and on the east coast of the United States and Canada.

Instead of showing a simple progression of the tide wave, the results for the North Sea showed what Whewell believed to be two rotating tidal systems. He predicted that at the center of each system there would be an area where the tidal rise and fall was very small, an amphidromic point. The existence of an amphidromic regime in the North Sea was demonstrated by Captain William Hewett of H.M. Survey Ship *Fairy* and his report to Beaufort, the Hydrographer of the Admiralty, is given in Paper 17. Whewell came to believe as the information available grew more plentiful, that amphidromic systems might also exist in the oceans and that this was why small tides were experienced on mid-ocean islands. His explanation was not, however, well received by other scientists. G. B. Airy, the Astronomer Royal, preferred to interpret the situation in the North Sea as a standing wave, produced by tidal waves entering the sea from north and south. He and other mathematicians considered tides on the basis of theoretical knowledge of the behavior of waves in canals. It was not until the twentieth century that the idea was reintroduced of amphidromic regimes in different parts of the ocean, caused by oscillations due to gravitational forces and influenced by the rotation of the earth (Cartwright 1969).

The other major development of the second half of the nineteenth century was the invention of harmonic analysis. Lord Kelvin, among others, showed that it was more convenient to represent the complex movements of the tides at any one place as a number of simpler movements related to particular motions of the sun and moon. This knowledge was then used to mechanize the process of making tide predictions for different places.

Professor Proudman's George Darwin Lecture "Tides of the Atlantic

Ocean," Paper 18, delivered in 1944, shows how the subject developed throughout this era, in which Darwin and Proudman, with his associate Doodson, were themselves prominent. It marks the end of a period in which great advances had been made in measurement, analysis of tidal records, and tide prediction. The tide-predicting machines of those days, still used in some places, but largely now relegated to museums, were masterpieces of engineering. They were in fact sophisticated mechanical computers, by means of which the amplitude, time, and phase of the response of water level to all the relative movements of earth, moon, and sun, that contribute to variations in the gravitational, tide-raising forces could be incorporated with great accuracy.

Following World War II, developments in digital computation and spectrum analysis of complex records allowed the analysis of tidal records to be extended to other responses, as well as those due to known variations in the gravitational attractions. W. H. Munk and D. E. Cartwright (1966) were pioneers in the new approach. It is necessarily a highly technical subject which, in a paper (Paper 19) read to a meeting of statisticians, Cartwright puts as simply as possible. Munk and Cartwright are now extending their tidal measurements into deep water so that they can study how the main tidal oscillations, small in the great ocean basins themselves, are modified as the tides cross the continental shelves and enter shallow water.

There has been little opportunity in dealing at this length with such a complicated subject, to say much about the tides as experienced on coasts, where their greatest effect is felt due to changes in the tide wave as it enters shallow water. When a tidal stream meets with some constriction, rough conditions, or overfalls as they were known in the seventeenth century, may result. When the stream enters the mouth of a river in certain cases it may create a bore which travels upstream at considerable speed (Tricker 1964). The Scottish scientist and clergyman John Fleming, 1785–1897, gave an early account of what happens when the tide comes upriver in conditions where the extreme circumstances of a bore do not apply (Paper 15). The sea water, because of its higher density, formed a wedge-shaped layer below the fresh water of the river which continued to run downstream at the surface. Fleming made these observations in the Firth of Tay, which he overlooked from his manse at Flisk, during the early years of the nineteenth century.

Now an integral part of oceanography, tidal studies developed very largely in isolation from the rest of marine science. This was partly due to their great complexity and partly to their close links with other disciplines. Thus, from the seventeenth to the nineteenth centuries, tides were studied as a branch of physical astronomy and then as an application of wave theory. Until the arrival of harmonic analysis, the theoretical picture of tidal movements often bore little relation to the experience of nautical surveyors—see, for example, the objections by Fitzroy (1836) to Lubbock and Whewell's version of the equilibrium theory. Modern work has provided explanations for the vagaries in the tides that puzzled previous generations. The subject is developing fast with the help of new equipment, such as the pressure recorders developed in the United States for use in the deep ocean, but

> There is still a large gap between our knowledge of the tidal forces as pioneered by Sir Isaac Newton and our knowledge and understanding of the way in which they generate the very complex tides found in the actual seas and oceans. (G. E. R. Deacon 1972)

13

Reprinted from the *Royal Soc. London Philos. Trans.* 1:298–301 (Sept. 1666)

CONSIDERATIONS AND ENQUIRIES CONCERNING TIDES

Sir Robert Moray

In regard that the High and Low waters are obſerved to increaſe, and decreaſe regularly at ſeveral ſeaſons, according to the Moons age, ſo as, about the *New* and *Full Moon*, or within two or three daies after, in the Weſtern parts of *Europe*, the *Tides* are at the *higheſt*, and about the *Quarter-Moons*, at the *loweſt*, (the former call'd *Spring-tides*, the other *Neap-tides*;) and that according to the height and exceſſes of the *Tides*, the *Ebbes* in oppoſition are anſwerable to them, the heigheſt Tide having the loweſt Ebbe, and the loweſt Ebbe, the higheſt Tide; the Tides from the *Quarter* to the *higheſt Spring-tide* increaſing in a certain proportion; and from the *Spring-tide* to the *Quarter-tide* decreaſing in like proportion, as is ſuppoſed: And alſo the *Ebbes* riſing and falling conſtantly after the ſame manner: It is wiſhed, that it may be inquired, in what proportion theſe Increaſes and Decreaſes, Riſings and Fallings happen to be in regard of one another?

And 'tis ſuppoſed, upon ſome Obſervations, made in fit places, by the above-mentioned Gentleman, though, (as himſelf acknowledges) not thoroughly and exactly performed, that the Increaſe of the Tides is made in the *Proportion* of *Sines*; the firſt Increaſe exceeding the loweſt in a ſmall proportion; the next in a greater; the third greater than that; and ſo on to the mid-moſt, whereof the exceſs is greateſt, diminiſhing again from that, to the higheſt Spring-Tide; ſo as the proportions, before and after the *Middle*, do greatly anſwer one another, or ſeem to do ſo. And likewiſe, from the *higheſt Spring-tide*, to the *loweſt Neap-tide*, the *Decreaſes* ſeem to keep the like proportions; the *Ebbes* riſing and falling in like manner and in like proportions. All which is ſuppoſed to fall out, when no Wind or other Accident cauſes an alteration.

And whereas 'tis obſerved, that upon the main Sea-ſhore the Current of the Ebbings and Flowings is ſometimes ſwifter, and ſometimes ſlacker, than at others, ſo as in the beginning of the Floud the Tide moves faſter but in a ſmall degree, increaſing its ſwiftneſs conſtantly till towards the *Middle* of the Floud; and then decreaſing in velocity again from the *Middle* till to the top of the High-water; it is ſuppoſed, that in Equal ſpaces of Time, the Increaſe and Decreaſe of velocity, and conſequently the degrees of the Riſings and Fallings of the ſame, in Equal ſpaces of time, are performed according to the *Proportion* of *Sines*.

But 'tis withall conceived, that the ſaid *Proportion* cannot hold *exactly* and *preciſely*, in regard of the *Inequalities*, that fall out in the *Periods* of the *Tides*, which are commonly obſerved and believed to follow certain *Poſitions* of the *Moon* in regard of the *Equinox*, which are known not to keep a *preciſe* and *conſtant* Courſe: ſo that, there not intervening equal portions of Time between one New Moon and another, the Moons return to the ſame *Meridian*, cannot be alwaies perform'd in the ſame Time; and conſequently there muſt be a like Variation of the Tides in the Velocity, and in the Riſings and Fallings of the Tides, as to equal ſpaces of time. And the Tides from New-moon to New-moon being not alwaies the ſame in number, as ſometimes but 57, ſometimes 58, and ſometimes 59, (without any certain order of ſucceſſion) is another evidence of the difficulty of reducing this to any great exactneſs. Yet, becauſe 'tis worth while, to learn as much of it, as may be, the *Propoſer* and many others do deſire, That Obſervations be conſtantly made of all theſe Particulars for ſome Months, and, if it may be, years together. And becauſe ſuch Obſervations will be the more eaſily and exactly made, where the Tides riſe higheſt, it is preſumed, that a fit *Apparatus* being made for the purpoſe, they may be made about *Briſtol* or *Cheap-ſtow*, beſt of any places in *England*, becauſe the Tides are ſaid thereabout to riſe to ten or twelve fathoms; as upon the coaſt of *Britanny* in *France*, they do to thirteen and fourteen.

In order to which, this following *Apparatus* is propoſed to be made uſe of. In ſome convenient place upon a Wall, Rock, or Bridge, &c. let there be an *Obſervatory* ſtanding, as neer as may be to the brink of the Sea, or upon ſome wall; and if it cannot be well placed juſt where the Low water is, there may be a Channel cut from the Low water to the bottom of the Wall, Rock, &c The Obſervatory is to be raiſed above the High-water 18. or 20. foot, and a Pump, of any reaſonable dimenſion, placed perpendicularly by the Wall, reaching above the High water as high as conveniently may be. Upon the top of the Pump a Pulley is to be faſtned, for letting down into the Pump a piece of floating wood, which, as the water comes in, may riſe and fall with it. And becauſe the riſing and falling of the water amounts to 60. or 70. foot, the Counterpoiſe of the weight, that goes into the Pump, is to hang upon as many Pulleys, as may ſerve to make it riſe & fall within the ſpace, by which the height of the Pump exceeds the height of the Water. And becauſe by this means the Counterpoise will rise and fall slower, and consequently by less proportions, than the weight it self, the first Pulley may have upon it a Wheele or two, to turn *Indexes* at any proportion required, so as to give the minute parts of the motion, and degrees of risings and fallings. All which is to be observed by *Pendulum-Watches*, that have *Minutes* and *Seconds*, with *Checks*, according to Mr. *Hugens's* way.

And because if the Hole, by which the water is let into the Pump, be as large as the Bore of the Pump it self, the weight that is raised by the water, will rise and fall with an Undulalation, according to the inequality of the Sea's Surface, 'twill therefore be fit, that the Hole, by which the water enters, be less than half as bigg as the Bore of the Pump; any inconvenience that may follow thereupon, as to the Periods and Stations of the Floud and Ebb, not being considerable.

And to the end, that it may appear the better, what are the *particular* Observations, desired to be made, near *Bristol* or *Cheap-stow* bridg, it was thought not amiss, to set them down distinctly by themselves.

1. The degrees of the Rising and Falling of the water every quarter of an hour (or as often as conveniently may be) from the Periods of the Tides and Ebbs; to be observed night and day, for 2 or 3 months.

2. The degrees of the velocity of the Motion of the Water every quarter of an hour for some whole Tides together; to be observed by a second *Pendul*-watch; and a logg fastened to a line of some 50 fathoms, wound about a wheel.

3. The exact measures of the Heights of every utmost High-water and Low-water, from one Spring-tide to another, for some Months or rather Years.

4. The exact Heights of Spring-tides and Spring-Ebbs for some Years together.

5. The Position of the Wind at every observation of the Tides; and the times of its Changes; and the degrees of its Strength.

6. The State of the Weather, as to Rain, Hail, Mist, Haziness, &c. and the times of its Changes.

7. At the times of observation of the Tides, the height of the *Thermometer*; the height of the *Baroscope*; the height of the *Hygroscope*; the Age of the Moon, and her *Azimuths*; and her place in all respects; And lastly the *Sun's* place; all these to *minutes*.

And it would be convenient, to keep *Journal Tables*, for all these Observations, each answering to its day of the Month.

For the *Apparatus* of all these observations, there will be particularly necessary.

A good *Pendulum*-watch.

A *Vane* shewing *Azimuths* to minute parts.

An *Instrument* to measure the strength of the Winde.

A large and good *needle* shewing *Azimuths* to degrees.

Thermometers, Barometers, *Hygroscopes.*

These Observations being thought very considerable as well as curious, 'tis hoped, that those who have conveniency, will give encouragement and assistance for the making of them; and withall oblige the publick by imparting, what they shall have observed of this kind: The ***Publisher*** intending, that when ever such observations shall be communicated to him, he will give notice of it to the ***publick***, and take care of the improvement thereof to the best use and advantage. A ***Pattern*** of the ***Table***, proposed to be made for observing the ***Tides***, is intended to be published the next opportunity, God permitting.

14

Reprinted from *Royal Soc. London Philos. Trans.* 19:445–457 (March 1697)

THE TRUE THEORY OF THE TIDES, EXTRACTED FROM THAT ADMIRED TREATISE OF MR. ISAAC NEWTON, INTITULED, *PHILOSOPHIAE NATURALIS PRINCIPIA MATHEMATICA*

Edmund Halley

IT may, perhaps, ſeem ſtrange, that this Paper, being no other than a partile Account of a Book long ſince publiſhed, and whereof a fuller Extract was given in Numb. 187. *of theſe Tranſactions, ſhould again appear here; but the Deſires of ſeveral honourable Perſons, which could not be withſtood, have obliged us to inſert it here, for the ſake of ſuch, who being leſs knowing in Mathematical Matters; and therefore, not daring to adventure on the Author himſelf, are notwithſtanding, very curious to be informed of the Cauſes of Things; particularly of ſo general and extraordinary* Phænomena,*as are thoſe of the Tides. Now this Paper having been drawn up for the late King* James's *Uſe, (in whoſe Reign the Book was publiſhed) and having given good Satisfaction to thoſe that got Copies of it; it is hoped the Savans of the higher Form will indulge us this liberty we take to gratifie their Inferiours in point of Science; and not be offended, that we here inſiſt more largely upon Mr.* Newton's Theory of the Tides, *which, how plain and eaſie ſoever we find, is very little underſtood by the common Reader.*

The ſole Principle upon which this Author proceeds to explain moſt of the great and ſurpriſing Appearances of Nature, is no other than that of *Gravity*, whereby in the Earth all Bodies have a tendency towards its Centre;

as is moſt evident: and from undoubted Arguments its prŏved, that there is ſuch a Gravitation towards the Centre of the Sun, Moon, and all the Planets.

From this Principle, as a neceſſary Conſequence, follows the Sphærical Figure of the Earth and Sea, and of all the other Cæleſtial Bodies: and tho' the tenacity and firmneſs of the Solid Parts, ſupport the Inequalities of the Land above the Level; yet the Fluids, preſſing equally and eaſily yielding to each other, ſoon reſtore the *Æquilibrium*, if diſturbed, and maintain the exact Figure of the Globe.

Now this force of Deſcent of Bodies towards the Center, is not in all places alike, but is ſtill leſs and leſs, as the diſtance of the Center encreaſes: and in this Book it is demonſtrated,that this Force decreaſes as the Square of the diſtance increaſes; that is, the weight of Bodies and the force of their Fall is leſs, in parts more removed from the Center, in the proportion of the Squares of the Diſtance. So as for Example, a Ton weight on the Surface of the Earth, if it were raiſed to the height of 4000 Miles, which I ſuppoſe the ſemidiamiter of the Earth, would weigh but $\frac{1}{4}$ of a Ton, or 5 Hundred weight: if to 12000 Miles, or 3 ſemidiameters from the Surface, that is 4 from the Center, it would weigh but $\frac{1}{16}$ part of the Weight on the Surface, or a Hundred and Quarter: So that it would be as eaſie for the Strength of a Man at that height to carry a Ton weight, as here on the Surface a 100$\frac{1}{4}$. And in the ſame Proportion does the Velocities of the fall of Bodies decreaſe: For whereas on the Surface of the Earth all things fall 16 Foot in a ſecond, at one ſemidiameter above this Fall is but 4 Foot; and at 3 ſemidiameters, or 4 from the Centre, it is but $\frac{1}{16}$ of the Fall at the Surface, or but one Foot in a ſecond: And at greater Diſtances both Weight and Fall become very ſmall, but yet at all given Diſtances is ſtill ſome thing, tho' the Effect become inſenſible. At the diſtance of the Moon (which I will ſuppoſe 60 Semidiameters of the Earth) 3600 Pounds weigh but one Pound, and the fall of Bodies is but $\frac{16}{3600}$ of a Foot in a ſecond, or 16 Foot in a minute; that is, a Body ſo far off deſcends in a Minute no more than the ſame at the Surface of the Earth would do in a Second of Time.

As was ſaid before, the ſame force decreaſing after the ſame manner is evidently found in the Sun, Moon, and all the Planets; but more eſpecially in the Sun, whoſe Force is prodigious; becoming ſenſible even in the immenſe diſtance of ***Saturn:*** This gives room to ſuſpect, that the force of Gravity is in the Celeſtial Globes proportional to the quantity of Matter in each of them: And the Sun being at leaſt ten Thouſand times as big as the Earth, its Gravitation or attracting Force, is found to be at leaſt ten Thouſand times as much as that of the Earth, acting on Bodies at the ſame diſtances.

This Law of the decreaſe of Gravity being demonſtratively proved, and put paſt contradiction; the Author with great Sagacity, inquires into the neceſſary Conſequences of this Suppoſition; whereby he finds the genuine Cauſe of the ſeveral Appearances in the Theory of the Moon and Planets, and diſcovers the hitherto unknown Laws of the Motion of Comets, and of the Ebbing and Flowing of the Sea. Each of which are Subjects that have hitherto taken up much larger Volumes; but Truth being uniform, and always the ſame, it is admirable to obſerve how eaſily we are enabled to make out very abſtruſe ***and difficult Matters,*** when once true and genuine Principles are obtained: And on the other hand it may be wondred, that, notwithſtanding the great facility of truth, and the perplexity and nonconſequences that always attend erroneous Suppoſitions, theſe great Diſcoveries ſhould have eſcaped the acute Diſquiſitions of the beſt Philoſophical Heads of all paſt Ages, and be reſerved to theſe our Times. But that wonder will ſoon ceaſe, if it be conſidered how great Improvements Geometry has received in our Memory, and particularly from the profound Diſcoveries of our incomparable Author.

The Theory of the Motion of the primary *Planets* is here ſhewn to be nothing elſe, but the contemplation of the Curve Lines which Bodies caſt with a given Velocity, in a given Direction, and at the ſame time drawn towards the Sun by its gravitating Power, would deſcribe. Or, which is all one, that the Orbs of the Planets are ſuch Curve Lines as a Shot from a Gun deſcribes in the Air, being caſt according to the direction of the Piece, but bent into a crooked Line by the ſupervening Tendency towards the Earths Centre: And the Planets being ſuppoſed to be projected with a given Force, and attracted towards the Sun, after the aforeſaid manner, are here proved to deſcribe ſuch Figures, as anſwer punctually to all that the Induſtry of this and the laſt Age has obſerved in the Planetary Motions. So that it appears, that there is no need of ſolid Orbs and Intelligences, as the Ancients imagined, nor yet of *Vortices* or Whirlpools of the Celeſtial Matter, as *Des Cartes* ſuppoſes; but the whole Affair is ſimply and mechanically performed, upon the ſole Suppoſition of a Gravitation towards the Sun; which cannot be denied.

The Motion of *Comets* is here ſhewn to be compounded of the ſame Elements, and not to differ from Planets, but in their greater ſwiftneſs, whereby overpowering the Gravity that ſhould hold them to the Sun, as it doth the Planets, they flie off again, and diſtance themſelves from the Sun and Earth, ſo that they ſoon are out of our ſight. And the imperfect Accounts and

Obſervations Antiquity has left us, are not ſufficient to determine whether the ſame Comet ever return again. But this Author has ſhewn how Geometrically to determine the Orb of a Comet from Obſervations, and to find his diſtance from the Earth and Sun, which was never before done.

The third thing here done is the Theory of the Moon, all the Inequalities of whoſe Motion are proved to ariſe from the ſame Principles, only here the effect of two Centers operating on, or attracting a projected Body comes to be conſidered ; for the Moon, tho' principally attracted by the Earth, and moving round it, does, together with the Earth, move round the Sun once a Year, and is according as ſhe is, nearer or farther from the Sun, drawn by him more or leſs than the Center of the Earth, about which ſhe moves ; whence ariſe ſeveral Irregularities in her Motion, of all which, the Author in this Book, with no leſs Subtility than Induſtry, has given a full Account. And tho' by reaſon of the great Complication of the Problem, he has not yet been able to make it purely Geometrical, 'tis to be hoped, that in ſome farther Eſſay he may ſurmount the difficulty : and having perfected the Theory of the Moon, the long deſired diſcovery of the Longitude (which at Sea is only practicable this way) may at length be brought to light, to the great Honour of your Majeſty and Advantage of your Subjects.

All the ſurprizing Phenomena of the Flux and Reflux of the Sea, are in like manner ſhewn to proceed from the ſame Principle ; which I deſign more largely to inſiſt on, ſince the Matter of Fact is in this caſe much better known to your Majeſty than in the foregoing.

If the Earth were alone, that is to ſay, not affected by the Actions of the Sun and Moon, it is not to be doubted, but the Ocean, being equally preſſed by the force of Gravity towards the Center, would continue in a perſect ſtagnation, always at the ſame height, without ever Ebbing or Flowing; but it being here demonſtrated, that the Sun and Moon have a like Principle of Gravitation towards their Centers, and that the Earth is within the Activity of their Attractions, it will plainly follow, that the Equality of the preſſure of Gravity towards the Center will thereby be diſturbed; and tho' the ſmallneſs of theſe Forces, in reſpect of the Gravitation towards the Earths Center, renders them altogether imperceptible by any Experiments we can deviſe, yet the Ocean being fluid and yielding to the leaſt force, by its riſing ſhews where it is leſs preſt, and where it is more preſt by its ſinking.

Now if we ſuppoſe the force of the Moons attraction to decreaſe as the Square of the Diſtance from its Center increaſes (as in the Earth and other Celeſtial Bodies) we ſhall find, that where the Moon is perpendicularly either above or below the Horizon, either in Zenith or Nadir, there the force of Gravity is moſt of all diminiſhed, and conſequently that there the Ocean muſt neceſſarily ſwell by the coming in of the Water from thoſe parts where the Preſſure is greateſt, *viz.* in thoſe places where the Moon is near the Horizon: but that this may be the better underſtood, I thought it needful to add the following Figure, where *M* is the Moon, *E* the Earth, *C* its Centre, and *Z* the place where the Moon is in the Zenith, *N* where in the Nadir.

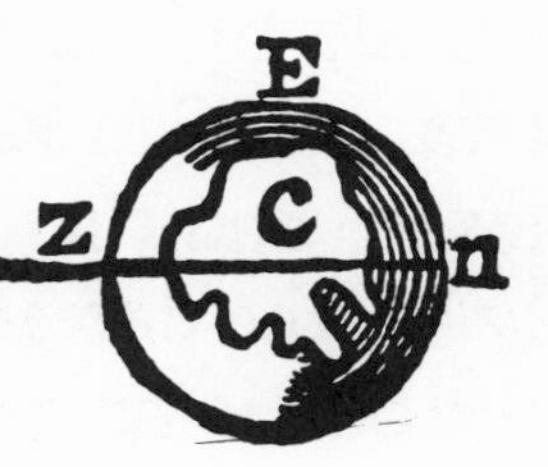

Now by the Hypothesis it is evident, that the Water in Z, being nearer, is more drawn by the Moon, than the Center of the Earth C, and that again more than the Water in N, wherefore the Water in Z has a tendency towards the Moon, contrary to that of Gravity, being equal to the Excess of the Gravitation in Z above that in C: And in the other case, the Water in N, tending less towards the Moon than the Center C, will be less pressed, by as much as is the difference of the Gravitations towards the Moon in C and N. This rightly understood, it follows plainly, that the Sea, which otherwise would be Spherical, upon the Pressure of the Moon, must form it self into a Spheroidal or Oval Figure, whose longest Diameter is where the Moon is Vertical, and shortest where she is in the Horizon; and that the Moon shifting her Position as she turns round the Earth once a day, this Oval of Water shifts with her, occasioning thereby the two Floods and Ebbs observable in each 25 Hours.

And this may suffice as to the general Cause of the Tides; it remains now to shew how naturally this Motion accounts for all the Particulars that has been observed about them; so that there can be no room left to doubt, but that this is the true cause thereof.

The Spring Tides upon the new and full Moons, and Neap Tides on the Quarters, are occasioned by the attractive Force of the Sun in the New and Full, conspiring with the Attraction of the Moon, and producing a Tide by their united Forces: Whereas in the Quarters, the Sun raises the Water where the Moon depresses it, and the contrary; so as the Tides are made only by the difference of their Attractions. That the force of the Sun is no greater in this case, proceeds from the very small Proportion the Semidiameter of the Earth bears to the vast distance of the Sun.

It is alſo obſerved, that *cæteris paribus*, the Æquinoctial Spring Tides in *March* and *September*, or near them, are the Higheſt, and the Neap Tides the Loweſt; which proceeds from the greater Agitation of the Waters, when the fluid *Sphæroid* reſolves about a great Circle of the Earth, than when it turns about in a leſſer Circle; it being plain, that if the Moon were conſtituted in the Pole and there ſtood, that the Sphæroid would have a fixt Poſition, and that it would be always high Water under the Poles, and low Water every where under the Æquinoctial: and therefore the nearer the Moon approaches the Poles, the leſs is the agitation of the Ocean, which is of all the greateſt, when the Moon is in the Æquinoctial, or fartheſt diſtant from the Poles. Whence the Sun and Moon, being either conjoyned or oppoſite in the Æquinoctial, produce the greateſt Spring Tides; and the ſubſequent Neap Tides, being produced by the Tropical Moon in the Quarters, are always the leaſt Tides; whereas in *June* and *December*, the Spring Tides are made by the Tropical Sun and Moon, and therefore leſs vigorous; and the Neap Tides by the Æquinoctial Moon, which therefore are the ſtronger: Hence it happens, that the difference between the Spring and Neap Tides in theſe Months, is much leſs conſiderable than in *March* and *September*. And the reaſon why the very higheſt Spring Tides are found to be rather before the Vernal and after the Autumnal Equinox, *viz.* in *February* and *October*, than preciſely upon them, is, becauſe the Sun is nearer the Earth in the Winter Months, and ſo comes to have a greater Effect in producing the Tides.

Hitherto we have conſidered ſuch Affections of the Tides as are Univerſal, without relation to particular Caſes; what follows from the differing Latitudes of places, will be eaſily underſtood by the following Figure.

Let *ApEP* be the Earth covered over with very deep Waters, *C* its Center, *P*, *p*, its Poles, *AE* the Æquinoctial, *Ff* the parallel of Latitude of a place, *Dd* another Parallel at equal distance on the other side of the Æquinoctial, *Hh* the two Points where the Moon is vertical, and let *Kk* be the great Circle, wherein the Moon appears Horizontal. It is evident, that a Spheroid described upon *Hh*, and *Kk* shall nearly repre-

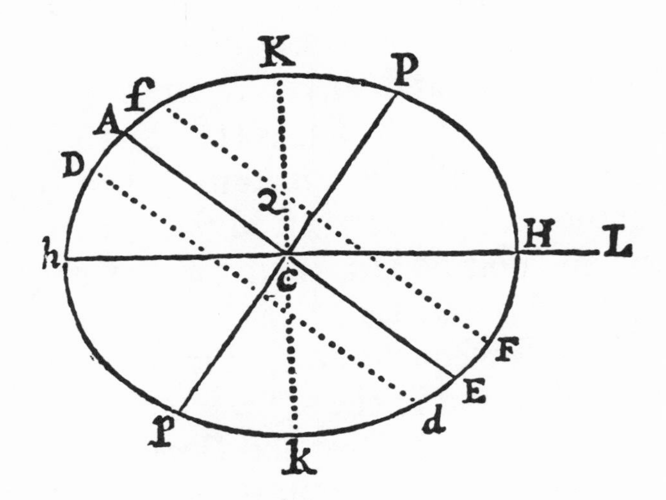

sent the Figure of the Sea, and *Cf*, *CD*, *CF*, *Cd* shall be the hights of the Sea in the places *f*, *D*, *F*, *d*, in all which it is High-water: and seeing that in twelve Hours time, by the diurnal Rotation of the Earth, the point *F* is transferred to *f*, and *d* to *D*: the hight of the Sea *CF* will be that of the High-water when the Moon is present, and *Cf* that of the other High water, when the Moon is under the Earth: which in the case of this Figure is less than the former *CF*. And in the opposite Parallel *Dd* the contrary happens. The Rising of the Water being always alternately greater and less in each place, when it is produced by the Moon declining sensibly from the Æquinoctial; that being the greatest of the two High-waters in each diurnal Revolution of the Moon, wherein ſhe approaches neareſt either to the Zenith or Nadir of the place: whence it is that the Moon in the Northern Signs, in this part of the World, makes the greateſt Tides when above the Earth, and in Southern Signs, when under the Earth; the Effect being always the greateſt where the Moon is fartheſt from the Horizon, either above or below it. And this alternate increaſe and decreaſe of the Tides has been obſerved to hold true on the Coaſt of *England*, at *Briſtol* by Capt. *Sturmy*, and at *Plymouth* by Mr. *Colepreſſe*.

But the Motions hitherto mentioned are ſomewhat altered by the Libration of the Water, whereby, tho' the Action of the *Luminaries* ſhould ceaſe, the Flux and Reflux of the Sea would for ſome time continue: This Conſervation of the impreſſed Motion diminiſhes the differences that otherwiſe would be between two conſequent Tides, and is the reaſon why the higheſt Spring Tides are not preciſely on the new and full Moons, nor the Neaps on the Quarters; but generally they are the third Tides after them, and ſometimes later.

All theſe things would regularly come to paſs, if the whole Earth were covered with Sea very deep; but by reaſon of the ſhoalneſs of ſome places, and the narrowneſs of the Streights, by which the Tides are in many caſes propagated, there ariſes a great diverſity in the Effect, and not to be accounted for, without an exact Knowledge of all the Circumſtances of the Places, as of the Poſition of the Land, and the Breadth and Depth of the Channels by which the Tide flows; for a very ſlow and imperceptible Motion of the whole Body of the Water, where it is (for example) 2 Miles deep, will ſuffice to raiſe its Surface 10 or 12 Feet in a Tides time; whereas, if the ſame quantity of Water were to be conveyed upon a Channel of 40 Fathoms deep, it would require a very great Stream to effect it, in ſo large Inlets as are the Channel of *England* and the *German* Ocean; whence the Tide is found to ſet ſtrongeſt in thoſe places where the Sea grows narroweſt; the ſame quantity of Water being to paſs through a ſmaller Paſſage: This is moſt evident in the *Streights*, between *Portland* and *Cape de Hague* in *Normandy*, where the Tide runs like a Sluce; and would be yet more between *Dover* and *Calis*, if the Tide coming about the Iſland from the North did not check it. And this force being once impreſſed upon the Water, continues to carry it about the level of the ordinary height in the Ocean, particularly where the Water meets a direct Obſtacle, as it is at St. *Malo's*; and where it enters into a long Channel, which running far into the Land, grows very ſtreight at its Extremity; as it is in the *Severn-Sea* at *Chepſtow* and *Briſtol*.

This ſhoalneſs of the Sea and the intercurrent Continents are the reaſon, that in the open Ocean the time of High-water is not at the Moons appulſe to the Meridian, but always ſome Hours after it; as it is obſerved upon all the Weſt-Coaſt of *Europe* and *Africa*, from *Ireland* to the *Cape of Good-Hope*: In all which a S. W. Moon makes High-water, and the ſame is reported to be on the Weſt ſide of *America*. But it would be endleſs to account all the particular Solutions, which are eaſie Corollaries of this *Hypotheſis*; as why the *Lakes*, ſuch as the *Caſpian Sea*, and *Mediterranian Seas*, ſuch as the *Black Sea*, the *Streights* and *Baltick*, have no ſenſible Tides: For *Lakes* having no Communication with the Ocean, can neither increaſe nor diminiſh their Water, whereby to riſe and fall; and Seas that communicate by ſuch narrow Inlets, and are of ſo immenſe an Extent, cannot in a few Hours time receive or empty Water enough to raiſe or ſink their Surface any thing ſenſibly.

Laſtly, to demonſtrate the excellency of this Doctrine, the Example of the Tides in the Port of *Tunking* in *China*, which are ſo extraordinary, and differing from all others we have yet heard of, may ſuffice. In this Port there is but one Flood and Ebb in 24 Hours; and twice in each Month, *viz.* when the Moon is near the Æquinoctial there is no Tide at all, but the Water is ſtagnant; but with the Moons declination there begins a Tide, which is greateſt when ſhe is in the Tropical Signs: only with this difference, that when the Moon is to the Northward of the Æquinoctial, it Flows when ſhe is above the Earth, and Ebbs when ſhe is under, ſo as to make High-water at Moons-ſetting, and Low-water at Moons-riſing: But on the contrary, the Moon being to the Southward, makes High-water at riſing and Low-water at ſetting; it Ebbing all the time ſhe is above the Horizon. As may be ſeen more at large in the *Philoſophical Tranſaction*, Num. 162.

The Cauſe of this odd Appearance is propoſed by Mr. *Newton*, to be from the concurrence of two Tides; the one propagated in ſix Hours out of the great *South-Sea* along the Coaſt of *China*; the other out of the *Indian-Sea*, from between the Iſlands in twelve Hours, along the Coaſt of *Malacca* and *Cambodia*. The one of theſe Tides, being produced in North-Latitude, is, as has been ſaid, greater, when the Moon being to the North of the Equator is above the Earth, and leſs when ſhe is under the Earth. The other of them, which is propagated from the *Indian-Sea*, being raiſed in South Latitude, is greater when the Moon declining to the South is above the Earth, and leſs when ſhe is under the Earth: So that of theſe Tides alternately greater and leſſer, there comes always ſucceſſively two of the greater and two of the leſſer together every day; and the High-water falls always between the times of the arrival of the two greater Floods ; and the Low-water be tween the arrival of the two leſſer Floods. And the Moon coming to the Æquinoctial, and the alternate Floods becoming equal, the Tide ceaſes and the Water ſtagnates : but when ſhe has paſſed to the other ſide of the Equator, thoſe Floods which in the former Order were the leaſt, now becoming the greateſt, that that before was the time of High-water now becomes the Low-water, and the Converſe. So that the whole appearance of theſe ſtrange Tides, is without any forcing naturally deduced from theſe Principles, and is a great Argument of the certainty of the whole *Theory.*

15

Reprinted from *Royal Soc. Edinburgh Trans.* 8:507-513 (1818)

OBSERVATIONS ON THE JUNCTION OF THE FRESH WATER OF RIVERS WITH THE SALT WATER OF THE SEA

John Fleming

It is possible, that the following observations may contain little that is new to those who are familiarly acquainted with the details of the science of Hydrostatics. But as I have not met with any remarks on the subject, in the course of my limited reading, the experiments which were performed, and the conclusions to which they lead, are here submitted to the consideration of the Royal Society.

When the flux of the tide obstructs the motion of a river, the wave has been supposed to produce its effects in the same manner as a dam built across a stream. This popular opinion, however, appears to have been adopted without sufficient consideration, as it can only hold true, in those cases, where the opposing fluids are of equal density, but never at the junction of opposite currents of fresh and salt water, which are of different densities. In this last case, where currents of fresh and salt water come in opposition, the lighter fluid, or the fresh water, will be raised upon the surface of the denser fluid, or the salt water, and when the stronger current of the tide has reversed the direction of the stream, the salt water will be found occupying the bottom of the channel, while the fresh water will be suspended or diffused on the surface. This view of the matter occurred to me in 1811; but it was not until the 29th of September 1813, that I had an opportunity of verifying the conjecture, by an examination of the waters of the Frith of Tay.

Flisk Beach, opposite to which the experiments were made, is situated a considerable way up the Frith, being upwards of sixteen miles from Abertay and Buttonness, where the Frith of Tay actually joins the German Ocean. The channel of the Frith at this place is about two miles in breadth; but upwards of a mile and a half of this extent consists of sand-banks, left dry at every ebb of the tide, and during flood, covered with from three to ten feet of water. These banks are separated from one another by deep pools, or *lakes* as they are termed, which occasion great irregularities in the motion of the currents. The channel of the *river* is near the south side. It is about half-a-mile in breadth, having in the deepest part about eighteen feet of water, when the tide has ebbed, and upwards of thirty feet during flood.

The apparatus which I employed was very simple: It consisted of a common bottle, with a narrow neck, having a weight attached to it. Besides the cord by which the bottle was lowered, there was another connected with the cork, in such a manner, that I could pull it out when the bottle had sunk to the place of its destination. The weather was favourable, and, on the day of the experiment, there was no wind to disturb the surface of the stream.

With this apparatus, I proceeded to the middle of the channel of the river, at *low water*, when the current downwards had ceased to be perceptible in the boat at anchor; and I obtained water from the bottom, the middle, and the surface of the stream. The water taken from the surface of the stream, was fresh, and tasted like ordinary river water. The water taken from the middle, was not perceptibly different; but that which was brought from the bottom was sensibly brackish. The water from the surface did not contain any salt, as a thousand grains of it, when evaporated with care on a sand-bath, left only a grain and a half of residue, apparently mud, which, when applied to the tongue, communicated no impression of saltness. The water from the middle of the stream yielded two grains of residue, when the same quantity was evaporated, of a whiter colour than the former, and having a perceptibly salt taste. The water from the bottom, which was saltish even to the taste, yielded four grains of saline matter. According to these experiments, the layers of water were arranged according to their densities, the heaviest water occupying the bottom of the stream, and the lightest floating on the surface.

At *half-flood*, I repeated the experiments on the waters obtained from the same situations as before. The water at the surface had now become very sensibly salt to the taste, and thus gave decided proofs of the progress of the tide. The three bottles of water now obtained, yielded results, not in unison with those already taken notice of. The arrangement of the different strata of water, according to their densities, as observed at ebb-tide, was in some degree reversed; for here the water at the surface was salter than that which was obtained from the bottom, and the water from the middle was salter than either. A thousand grains of water from the bottom, yielded by evaporation only ten grains of saline matter, while the water from the surface yielded eleven grains, and from the middle twelve grains, by the same process.

This anomaly is easily accounted for. Were the current of the tide confined entirely to the channel of the *river*, an arrangement of the waters, similar to that which existed in the first experiments, would have prevailed. But during the flowing of the tide, the sea-water soon occupies more than the channel of the river, and spreads itself in various streams among the hollows of the sand-banks. These streams reunite at different places with the principal current, and, in this manner, prevent the salt and fresh waters from gaining their natural relative position. But as soon as these sand-banks are covered with water, the tide proceeds with regularity in its course, so that the different layers of water can then arrange themselves according to their specific gravities.

A thousand grains of water obtained from the bottom, at the *height of flood*, yielded by evaporation twenty-three grains of salt, while the same quantity of water from the middle yielded only eighteen grains; and from the surface only seventeen grains. This was a difference of no less than six grains, and seemed to afford a decisive result.

In order, however, to complete the series of observations, I examined the conditions of the currents at *half-ebb*. The same irregularities prevailed, as before observed at half-flood. A thousand grains of the water, from the bottom, yielded after evaporation eleven grains of salt; from the middle, nine grains, and from the surface, twelve grains. At this time the densest water was at the surface, and the lightest occupied the middle. The cause of this was obvious. Extensive portions of the sand-banks had already been left dry by the receding tide, and various currents of water, disjoined from the main stream by the inequalities of these banks, were now re-uniting with it, through various channels, and disturbing the natural arrangement which had prevailed during the time of flood.

Although the Frith of Tay is very ill calculated for experiments of this kind, from the circumstances already taken notice of, still the premises which we have stated seem to warrant the conclusion, that when the wave of the tide obstructs the motion of a river, and causes it either to become stationary, or to move backwards, the effect is produced by the salt water presenting to the current of the river an inclined plane, the apex of which separates the layer of fresh water from the bed of the channel, and suspends it buoyant on the surface *.

It may here be observed, that this inferior current of salt-water, will never reach that point of the bed of the river, which is intersected by a line drawn perpendicular to the altitude of the wave of the tide, in the ocean, at the mouth of the river. This point is undoubtedly the place at which the salt-water would arrive, at every flood, were there no fresh-water current, as has been demonstrated with regard to the waters of the Tay, by the accurate observations of Mr JAMES JARDINE. But as the motion of the current of salt-water is retarded by the opposite current of the fresh-water, and the apex of the wedge which it forms, also washed away by the same agent, the point which the salt-water reaches will be considerably lower than the summit of the tide-wave with which it is connected.

The surface of the higher part of the river, whose elevations and depressions are influenced by the movements of the tide, will necessarily attain a higher level than the summit of the tide-wave, in consequence of the lower specific gravity of the river-water, when compared with the denser column of sea-water

* I understand that my friend Mr ROBERT STEVENSON has made similar observations at the mouth of the Dee, near Aberdeen, and also on the Thames, and that his conclusions and my own nearly coincide.

which it counterbalances; and this, independent of the progressive motion of the tide in the river.

If the view which we have taken of this subject, in reference to the progress of the salt water, be considered as just, it will enable us to explain some of the phenomena of nature, at present rather perplexing, and may even be useful in its practical application.

In examining the vegetable productions of the banks of rivers, at their junction with the sea, we are sometimes surprised to witness the growth of plants, considered as the natural inhabitants of the sea-shore. But our surprise will cease when we reflect, that the sea-water proceeds farther up the river at every flood-tide than the sensible qualities of the water at the surface indicate; so that the plants, which we hastily conclude to be out of the reach of the salt-water, are still within the sphere of its influence. Thus, at the Beach of Flisk, and even farther up the river, the *Fucus vesiculosus* (the species commonly cut for making kelp) not only vegetates, but in its season appears in fructification.

But that which proves in a still more decisive manner, the action of the inferior stratum of salt-water at the place, is the growth of the coralline termed *Tubularia ramosa* (ELLIS's Corallines, Tab. xv. fig. A.), and another of a different genus, closely resembling the *Sertularia gelatinosa* of PALLAS. There are likewise some traces of *Flustræ*.

A knowledge of the facts which we have already stated, may be of use to those who are engaged in the erection of salt-works at the mouths of rivers. In such situations, the openings of the pipes for obtaining the salt water, should be placed as near the bottom, or as deep in the water as possible; and water ought only to be drawn during the height of flood-tide, when the fresh-water is diffused over the surface.

Even to navigators, an acquaintance with this subject may sometimes be of use. Thus, for example, when entering a creek in an unknown coast, they may easily ascertain whether any streams of fresh water flow into it, by examining the comparative density of the water taken from the surface and from below.

These experiments appear to give countenance to the opinion which supposes that the water at the surface of the sea contains less salt than the water at the bottom. This may be expected to take place in the neighbourhood of continents, at least, whatever may be the case in the open ocean. During winter, the difference is probably very considerable, as at that season the rivers incessantly pour vast quantities of fresh water into this great reservoir, while but a small portion is abstracted by evaporation. In the Frith of Forth, the difference between the dense water of summer and the diluted water of the winter season, is as eighteen to sixteen, and that even as far down as Prestonpans.

MANSE OF FLISK,
11*th March* 1816.

16

Reprinted from *Naut. Mag.* 1(9):401–404 (Oct. 1832)

The Tide Gauge at Sheerness

The drawing at the commencement of this number presents a plan of the Tide Gauge in H. M. Dock-yard at Sheerness, which, being the most perfect machine of the kind that we know of, a description of it may not prove unacceptable to our readers :—

A—Is the elevation in profile of the tide-gauge house, which stands on a small wharf at the S.W. angle of the Boat Basin, open to the sea.

B—A front view of the same, shewing as much of the machinery as can be seen from the door-way.

C—A wooden trunk, two feet square, and twenty-six feet long, reaching to the bottom of the basin; the lower end bored with holes, to admit freely the ebb and flow of the tide, without being affected by the waves.

D—A long wooden tube, six inches square, and twenty-six feet high, from the top of the wharf, to protect the gauge-rod.

E—A copper float, or buoy, fixed at the lower end of

F—A moveable gauge-rod, of light wood, one foot and a quarter square, and twenty-six feet long, guided by friction rollers at the top, and sliding inside the trunk against a fixed scale, divided to tenths of an inch; the divisions on the scale being numbered upwards, on the right-hand side of it, to shew the rise of the tide, and on the left-hand side, downwards, to shew its fall. Between the scale and the moveable rod, on each side is a groove, containing a sliding vernier bearing a catch, G and H, both of which project over the face of the rod: the catch on the right hand,

G—Is caught by a projecting nail, I, on the lower part of the rod, and is carried up during the rise of the tide: while that on the left,

H—Is brought down (by another nail, K, fixed on the upper part of the rod) as the water falls; the distance between the two nails, or I and K, being exactly fourteen feet.

When the sea is at its mean level at Sheerness, which has been found from the mean of several years' observations, (and which corresponds exactly with the mark "eighteen feet" at the entrance of the basin,) the O mark, or zero, on the rod is made to agree with seven feet on the two scales, as is represented in the plan. Now, should the tide flow to cause the nail at I to raise the catch G with its vernier, to 8 ft. on the scale; and again fall, till the nail K on the upper part of the rod bring the catch H, with its vernier, on

[*Editor's Note:* The plate accompanying this article has not been reproduced because of limitations of space.]

the left-hand side down to 8 ft. it is evident that the tide has ranged sixteen feet; which, with the time, the observer must note down, and set the verniers afresh.

Again, supposing the rising tide to carry the catch G up to 11 ft. on the right-hand scale, and the falling tide to bring H down to 5 ft. on the left-hand scale, the tide has still ranged sixteen feet; but it is plain that the rise has far exceeded the fall, and that either a strong wind or some other cause has prevented the ebb of the water. This arrangement of the indices is perfectly simple, and no mistake in reading off can be made by the commonest observer.

Yet, notwithstanding the correctness of this gauge in giving the rise and fall of tide, it requires an observer at high and low water to watch for the time. By night, no attendant is on the spot; by day, too, he is sometimes absent; and even when present, the most watchful observer often cannot tell the precise time from five to twenty minutes; as the water is at times stationary, or nearly so, for more than half an hour; moreover, sometimes it falls a few inches, and rises again. To meet these difficulties, it occurred to the civil engineer at that dock-yard, Mr. Mitchell, to cause the tide-gauge to register itself, which is done in the most ingenious, and at the same time in the simplest manner, by the application of a little machinery, as follows:

A sheet of paper is divided by lines into twenty-four equal parts, for hours, on the scale of 0·5 of an inch to an hour, and these lines are crossed at right angles by twenty-four other lines, to shew the range of the tide in feet, on the scale of 0·4 of an inch to a foot.* The paper (S) so prepared is wrapped round a cylinder or roller of wood, the circumference of which is exactly equal to the twenty-four divisions of time; the roller is then supported horizontally, and so connected with a clock, (R,) that it makes one revolution in twenty-four hours.

By means of three wheels, (LNO,) the vertical motion of the gauge-rod is communicated horizontally to a sliding bar of brass, Q, (in which is fixed a pencil, P,) causing it to traverse across the paper as the tide rises and falls parallel to the axis of the cylinder;† the pencil at the same time tracing a line indicated by the

* The principle of this tide-guage is the same as that mentioned in Mr. Lubbock's valuable paper on "tides" in the Companion to the British Almanack for 1830; but the plan and execution are solely those of Mr. Mitchell, of the Dock-yard.

† The vertical motion of the guage-rod is thus communicated:
L—Is a wheel twelve inches in diameter, with a grooved edge to receive M, a small cord which passes round L, in the direction of the arrows, and is attached to a spring at each end of the gauge-rod; which springs, as well as the tightening pulley, T, always keep the cord at an equal tension; as the rod rises, the cord will make the wheel revolve in the direction of the arrows; as, also, another wheel, N, of four inches diameter, on the same axis which communicates motion by an endless cord to O, a wheel of twelve inches diameter, on the axis of which is a brass pinion which causes the rack, or toothed bar, Q, with the pencil in it, to move to and fro, as the tide rises or falls, in proportion as the divisions on the paper are to each foot of the gauge-rod.

combined motions of the clock and the gauge-rod: this line is the tidal curve, and, as far as the nature of the materials employed will admit of, is strictly correct;* entirely obviates the necessity of an observer, excepting once a fortnight, at full and change of moon, to replace the sheet of paper, and shews, on inspection, not only the rise and fall, and time of high and low water, but also every, even the most trifling irregularity which may have occurred at any time during the lunation.

Such is the plan of the self-registering tide-gauge at Sheerness; a wind-gauge will shortly be added to it, that will shew both the force and direction of the wind, and thus render it the most complete thing of the kind in Europe. May we not hope, now that such attention has been shewn to the subject of "*Tides;*" in proof of which we may allude to the able "Paper on Tides" read by Mr. Lubbock at the late "*Solemn Session of Science*" at Oxford; may we not hope, that ere long we may see such establishments formed at all our principal sea-ports; the diagrams and observations from which would enable us, by comparing them, to throw some light on the irregularities of the tides, irregularities which are proved by the following instance not to depend solely upon the direction of the wind: *e.g.* the greatest spring-tide yet measured at Sheerness, viz., 20 feet, and the least range of neap, 7 feet 9½ inches, both occurred with a strong S.W. wind.

The expense of an establishment similar to that at Sheerness would not be great; from 50 to £100 would complete the tide-gauge at Portsmouth, and erect one at Plymouth: and surely such a sum is not worth consideration, more especially in a country rendered by its sea-girt shores so essentially maritime as Great Britain. Indebted to the ocean, as we are, for our very existence as a nation, surely it ill becomes us to be indifferent, or inattentive, to the striking phenomena which that ocean daily presents to us, in the unceasing ebb and flow of its tides.

There are other means by which persons who, having the opportunity, might with great facility be enabled to register the phenomena of the tides, and the following is perhaps as simple and certain as any.

Let a leaden pipe be obtained, at one end of which a rose similar to that of a watering pot is to be fixed. Let this, as the outer end, be placed in the water, sufficiently deep to be always well below the level of the lowest spring tides. In placing it, care should be taken that no weeds, sand, or mud should prevent the water from passing through the rose into the pipe; and a situation where the water is deep should be preferred, so that the rose, with its face downwards, while it is below the level of the low-water of spring-tides, should be at the same time well above the bottom. The pipe

* A check is always kept by an observer, and in no case has this machine ever been at fault since it first came into use in September, 1831.

may be continued to any distance from the sea-side to the place where it is intended to make the observations in the apartment of a building, but great care must be taken that in this extent it be lower than the level of lowest springs, so that the water may always have a free passage through it. Having led the inner end of the pipe to the place desired, it should then be turned up in a vertical direction, and the end of it left open at any convenient height above the level of the highest spring-tides. A long thin rod of light wood, one end of which should be fixed in a piece of cork to answer as a float, being introduced into this end of the pipe as the cork rests on the surface of the water, the rod will always rise and fall with the motion of the tide. A scale, with two sliding verniers, one for high and the other for low water, similar to those adopted at Sheerness, to be acted upon by two catches in the rod, may then be placed close to it, or the rod may be graduated instead, into feet and inches. The exact time must be observed when the verniers reach their highest and lowest places on the scale, or that of the greatest rise and fall of the rod, and the verniers should always be displaced from the position on the scale in which they may be left by the last high and low water, in order to be ready for the next. The times of high and low water being noted, with the direction and force of the wind, the gauge will give the actual rise and fall in feet, and a series of valuable observations will be thus obtained by very simple means. Should there be any vibratory motion in the water within the pipe, so as to produce an unsteady jumping movement of the rod, it may be readily overcome in the same manner as that of the mercury in a marine barometer, by merely flattening it in any part between the lowest range of the rod and the rose at the outer end of the pipe. Any sudden motion in the water, is thus counteracted, while it has a free access to the inner end, although through a smaller channel. Several other little details will readily present themselves to an ingenious mind when the gauge is in operation.

There are many convenient situations on our coasts, where naval officers and other persons are residing, at which a tide-gauge of this nature might be placed, the expense of which would not be more than the value of the leaden pipe and the rose. And we will venture to say, that to many it would afford a most interesting amusement, while the information thus obtained by them would bring to light many secrets arising from general and local causes, and shortly enable us to explain all the laws which influence and regulate this most interesting phenomenon.

17

Reprinted from *Rep. Br. Ass. Advmt Sci.* 11(2):32-35 (1841)

Letter from the late Capt. Hewett to Capt. Beaufort, R.N. (referred to in a communication by Professor Whewell).

H.M. Ship Fairy, Harwich, August 31st, 1840.

Sir,—On the 24th inst., being in lat. 52° 27′ 30″ N., long. 3° 14′ 30″ E., with light breezes and smooth water, I deemed it a fitting opportunity for making a further trial on the rise and fall of tide in the middle of the North Sea; and although I was then many miles both to the northward and eastward of the spot near which Mr. Whewell had previously expressed his wishes that the experiment should be made, yet I thought that if good observations by any means could be obtained at the above position, they would at the least serve to show, in some measure, the truth or error of that gentleman's theory; either in the one case by a sensible diminution of the vertical movement of the tide, when compared with the known rise and fall on the shores of England and Holland, or in the other by ascertaining the rise and fall, beyond a doubt, to be so great as to throw some doubt on the correctness of the theory in question. But as I apprehend that Mr. Whewell's theory is founded mainly upon the fact, that the tide waves, to make high water on the opposite coasts of England and Holland, come from different directions, namely, on the former, round the northern extreme of Great Britain, and so working its way along the eastern coast; and on the latter, through the straits of Dover, and running thence along the coasts of France, Belgium and Holland; and that it might reasonably be inferred, that these waves gradually diminish in importance as they recede from their respective shores, or approach each other, there would be left a broadish space about the middle of this part of the North Sea, where no rise and fall of tide exists, and that therefore the waters between the two opposite shores would assume a convex form at low water by the shores, and a concave one at high water.

Allowing this view of the foundation of Mr. Whewell's theory to be correct, (and I have not his book at present near me to refer to,) this line, or more properly speaking, "broad belt" of no rise and fall, would doubtless run for a considerable distance in the north-easterly direction into the North Sea, from the point where it may commence on the North-Sea-side of the straits of Dover. It would therefore follow, that the fact of my being to the northward of Mr. Whewell's position would of itself be of no material importance, and by reference to the Chart, it will be seen that the longitude places me not many miles to the eastward of the "broad belt" above alluded to. Having thus reflected, I came to the conclusion, that if Mr. Whewell's views were correct, true observations made in this position would exhibit some indications thereof, and I accordingly made the necessary dispositions.

A rise and fall by the shore is a case which falls immediately on the conviction, by the sense of sight; but to ascertain the fact of a vertical motion of five or six feet in the middle of a great sea, and out of sight of land, is a problem of no small difficulty, and requires the exercise of many precautions to arrive at anything like true results. In making an observation of this description we find two important obstacles in the way of obtaining these, namely, the stream of tide and the undulating character of the surface of the ground. Under the influence of a strong stream of tide, it is utterly impossible, except in very shallow water, to take a strictly correct depth from the vessel, or a boat, at anchor, (and therefore a fixed point,) for the line *will* assume a curved form in the act of descent; and after all, from the want of perpendicularity in the line, a large allowance, in a depth of nearly twenty fathoms, is necessarily left to the exercise of the judgment; and both of these may amount to considerably more than the "rise and fall" sought for. On the other hand, the undulations of the surface render it essential that the depths should be always taken *over* some discovered elevated spot. The stream of tide and the undulations of the ground are therefore alternately opposed to the making of observations from which correct results can be derived. I experienced on this, as on the former occasion, considerable difficulty in overcoming these obstacles; but I soon found myself compelled to resort to the former plan, (with the addition of such precautions as experience then gave me,) namely, that of mooring one boat and taking the depths in another.

The accompanying diagram will assist my account of the plan pursued.

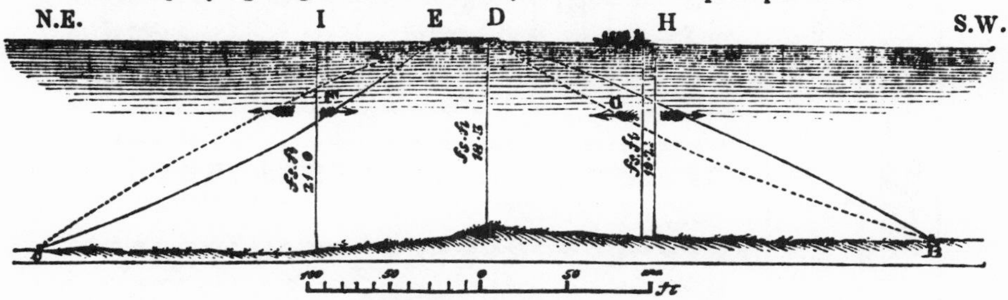

The ship was anchored in 21½ fathoms, and on searching, a convenient rise in the ground A was soon found near her, over which there was exactly 18 fathoms 3 feet, by a well-measured line. The second gig (of 26 feet) was then moored, "head and stern," in the direction of the strength of the stream (N.E. and S.W.), so that she should be as nearly as possible over the overfall A. This was accomplished thus:— I prepared a coil of 1½ inch rope, and fastened a grapnel at either end. The first grapnel was let go at B, the whole of the line was then veered away, and the second grapnel was let go at C; the gig was hauled along the bight of the rope, until it was found, by repeated trial, that the summit of the overfall was exactly abreast the foremast row-lock of figure D, at about six feet from the boat, while the N.E. stream was running. She was there secured. At the turn of the tide to the S.W., it was found that the weight of the stream F had operated so powerfully on the bight of the N.E. line, as to draw the boat from D to E, so that the summit of the overfall, which was before under the foremost row-lock, was found to be eight feet on her bow. On the return of the N.E. tide, its operation (G) upon the bight of the S.W. line again drew the gig ahead to her former position D, and the summit of the overfall was found as before under the foremost row-lock.

It will then be evident, that at each change of tide I knew exactly where the overfall was to be found, while taking the depths; and thus prepared, it only remained to get the *least* and *exact vertical* depth over the summit of the overfall at the intervals determined upon, and which were every half-hour. With the N.E. stream running, I dropped the lead from the other gig about the point H, and exactly in the stream, which I knew would drift her at the proper distance of six feet from the moored boat; the lead was constantly lifted off the ground, so that the line was perfectly straight and perpendicular, and the undulations of the ground carefully observed until the lead passed over the summit of the overfall, where the depths were strictly noticed, and recorded in the accompanying Table. The boat on this stream was allowed to drift to the point I. With the S.W. stream I began about

August 24th, 1840. Moon's Age, 26·6 days. } Latitude 52° 27′ 30″ N. Longitude 3° 14′ 30″ E.

Times. P.M.	Depths.	Direct. (Comp.)	Rate.	Winds. (Comp.)	Force.	Remarks.
h. m.	fms. ft.		Kts. $\frac{1}{10}$s.			
1 0	18 3¼	N.E. ¾ E.	1 5	S.W.	2	
1 30	18 3	N.E.	1 4			
2 0	18 2¾	N.E. ½ N.	1 2			
2 30	18 2¾	N.E. ¾ N.	1 0			
3 0	18 2¾	N.E. by N.	0 7			
3 30	18 2	N.E.	0 5			
4 0	18 1	N.E. by E.	0 3			
4 30	18 1	East.	0 3			
5 0	18 1½	Slack.	0 0	S.S.W.		
5 30	18 2	W. by S.	0 3			
6 0	18 2½	S.W. by W.	0 7			
6 30	18 3	S.W. ½ W.	1 4	S. by W.		
7 0	18 3¼	....	1 5	South.		
7 30	18 3½	....	1 5	S.S.E.		
8 0	18 3½	....	1 6	S.E. by S.		
8 30	Too dark for observations.					
9 0						
9 30						
10 0						
10 30						
11 0						
11 30						
12 0						
12 30						
1 0						

August 25th, 1840. Moon's Age, 27·6 days. } Latitude 52° 27′ 30″ N. Longitude 3° 14′ 30″ E.

Times. A.M.	Depths.	Direct. (Comp.)	Rate.	Winds. (Comp.)	Force.	Remarks.
h. m.	fms. ft.		Kts. $\frac{1}{10}$s.			
5 30	18 3	Slack.	0 0	Calm.	0	
6 0	18 3	S.W. by S.	0 5			
6 30	18 3	S.W.	0 7			
7 0	18 3	....	1 0	W. by N.	1	
7 30	18 3	....	1 3	....	2	
8 0	18 3	....	1 5			
8 30	18 3	....	1 3			
9 0	18 3	....	1 2			
9 30	18 3	....	0 9			
10 0	18 3	....	0 5			
10 30	18 3	S.W. by W.	0 2	....	......	Tide slack from 10·45 to 11·0.
11 0	18 3	Slack.	0 0	W. by S.		
11 30	18 3	N.E.	0 3			
12 Noon	18 4	....	0 9			
P.M. 30	18 4	....	1 3			
1 0	18 4	....	1 6			
1 30	18 4	....	1 6			
2 0	18 4	....	1 6			
2 30	18 4	....	1 6			
3 0	18 4	....	1 3			
3 30	18 4	....	1 0			
4 0	18 3½	....	0 5			
4 30	18 4	....	0 3	W.S.W.		
5 0	18 4	Slack.	0 0	S.W.		
5 30	18 4	S.W.	0 2			

(Signed) WILLIAM HEWETT, *Captain of H.M.S. Fairy*, *August 31st*, 1840.

the point I and terminated at H, using the same observances and precautions until 5h. 30m. P.M. of the 25th, when the appearance of the weather required my removing.

It will be seen that the observations recorded on the afternoon of the 24th are not so regular as those of the following day. I attribute this to some degree of *uncertainty* on account of a long swell, perhaps of one and a half or two feet rise, interrupting the observation at the moment of passing over the overfall; but this little swell had nearly subsided on the 25th, and the depths were then recorded with much satisfaction. It will also be noticed, that at the turn of the stream about noon of the latter day, the depth had increased to eighteen fathoms four feet, and went on uniformly so; but I investigated the cause of this on the spot, and found that the wind having increased to 2 from W. by S., and therefore operating upon the starboard bow of the boat, had sidled her a few feet to the S.E., so as to bring the eighteen fathoms three feet immediately under her; and that by observing the same distance from the boat while drifting past her (and which was always on her larboard side), I obtained eighteen fathoms four feet instead of eighteen fathoms three feet.

From the care and pains taken in these observations, and *that* under favourable circumstances, I do not entertain a doubt of the correctness of any one of the depths over the summit of the overfall as recorded on the 25th; but as this interesting result of observations on an unexpected theory may no doubt give rise to a strong desire for further observations as corroboratives, I shall not fail to make such when I find myself in a position and circumstances to do so with any prospect of success. It is a difficult observation, and can be made but seldom. In the mean time I would offer my congratulations to Mr. Whewell on these results, should they prove in any degree gratifying to him. I have the honour to be, &c.

(Signed) WILLIAM HEWETT, *Captain.*

18

Reprinted from *Royal Astron. Soc. Monthly Notices* **104**(5):244-256 (1944)

THE TIDES OF THE ATLANTIC OCEAN

(*George Darwin Lecture, delivered by Professor J. Proudman, F.R.S., on* 1944 *October* 13)

It is appropriate, I think, that I should lecture on a subject connected with the tides of the ocean, for Sir George Darwin himself gave a great deal of his attention to this subject. From about 1882 until the end of his life in 1912, he was generally regarded as the greatest living authority on ocean tides.

I shall be mainly concerned with the discovery of the distribution of the tides over the open Atlantic Ocean, by the application of the principles of dynamics. This will involve a large measure of dynamical explanation of the tides as they have been observed on the coasts of the continents and islands. Owing to the great difficulty of measuring tides in mid-ocean, it is at present only by theoretical methods that we can expect to discover their distribution over the open oceans. A method depending on the tidal analysis of barometric observations has been put forward by E. Barkow (1911) and J. Bartels (1926), but has not been successfully employed, and to me the difficulties seem very great. Up to the present time the leading features of the tides have not been determined in any really convincing way. I propose to give a progress-report on an attempt which I have recently been making.

It was Newton who laid down the fundamental principles on which the tides are to be explained, and thus reduced investigations to problems in mathematics. A great deal of attention has been paid to these problems, some of it by the ablest of mathematicians, including Laplace and Poincaré. Much progress has been made in the determination of the tides in ideal geometrically simple basins, notably in recent years by A. T. Doodson and G. R. Goldsbrough. I shall show that some progress has been made with the actual basin of the Atlantic, but practically none has been made with the basins of the Pacific and Indian Oceans.

A question often asked is: Why are the tides of the Mediterranean so small? The answer is: The tides of the Mediterranean are just about of the magnitude that the simplest dynamical theory indicates. The much more interesting question is: Why are the tides of the Atlantic so large?

In order to introduce a fair amount of precision into the argument, I am going to restrict myself very largely to the principal harmonic constituent of the tides. This is the oscillatory motion with a period of 12 hours 25 minutes which is commonly denoted by M_2. For most places in the Atlantic it may be regarded as the average tide. In this constituent only the generating forces of the Moon are taken into account, and the Moon may be taken to remain in the plane of the Earth's equator and at a fixed distance from the Earth.

From the results of the analysis of observations the amplitude of this constituent has been evaluated at a large number of stations both on the continental coasts and on islands. A very convenient standard for the order of magnitude is provided by the equilibrium-form of the tides. This equilibrium-form gives what the tides would be if the water could immediately take up the position corresponding to the tide-generating forces of the Moon. Now the amplitude of the equilibrium-form of the tides is a function only of latitude and is given by $25 \cos^2 \phi$ cm., ϕ denoting the latitude, the mean value of which along a meridian is 12·5 cm. It is often not realised that the tide-generating forces are so small. Taking the mean amplitude of the equilibrium-form as a standard, there is at Brest a 16-fold magnification, at Boston, Mass., an 11-fold magnification, and at the oceanic islands a magnification of 4-, 3-, and 2-fold. This degree of magnification indicates that there is some form of resonance in the basin of the Atlantic.

The longer curves of Fig. 1 show the distribution of the tidal elevation along the continental coasts, including a line just outside the West Indies. The full lines show the Greenwich lunar time of high water and the broken lines the amplitude. The data for this figure have been obtained by taking mean values along stretches of the open coasts, each covering 5° of latitude. We notice that there is a progression northwards

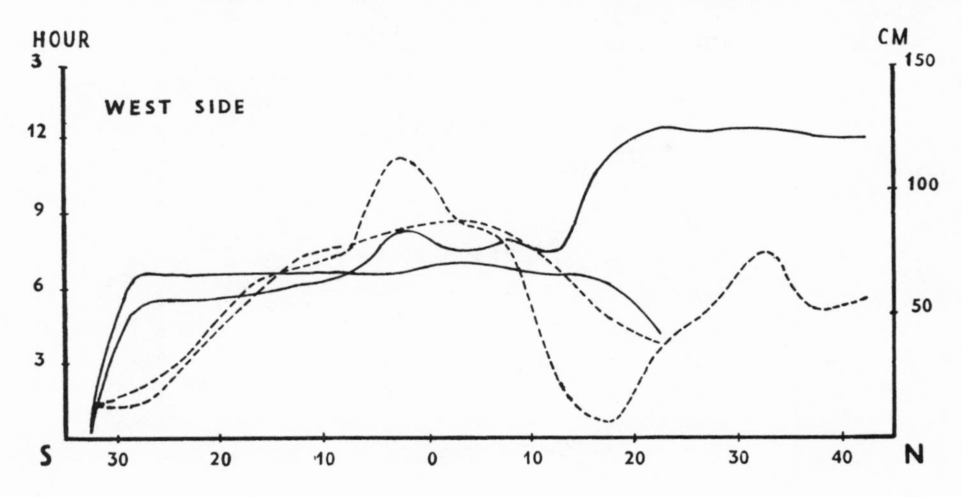

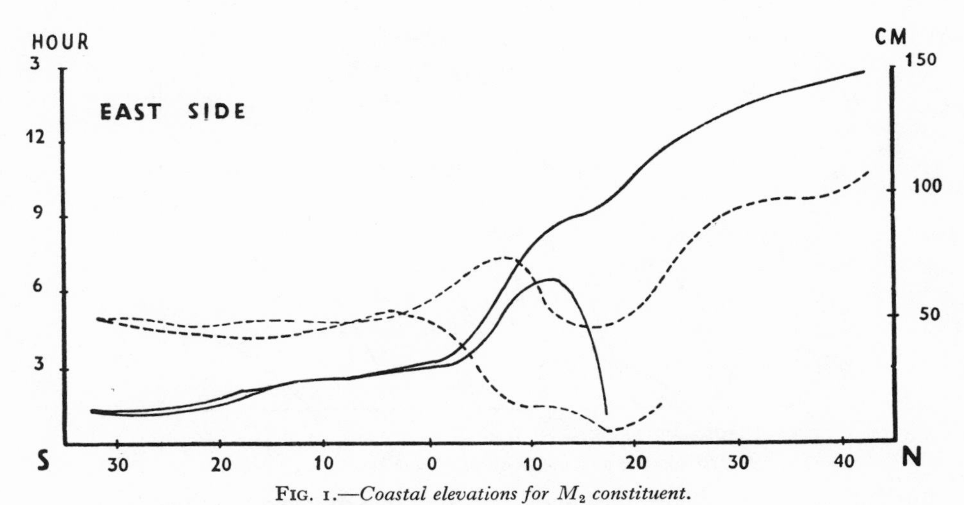

FIG. 1.—*Coastal elevations for M_2 constituent.*
Observation (longer lines) and synthesis (shorter lines).
Full lines give Greenwich lunar hour of high water.
Broken lines give amplitude in cm.

up both coasts of the South Atlantic and along the eastern coast of the North Atlantic. On the west side, at about 32° S. and 17° N. the range is very small and the time changes very rapidly; from about 2° S. to 14° N. and again along the U.S.A. from 22° N. to 42°·5 N. the times change very little, the hours being 7 and 12 respectively. Our problem is to find the theoretical distribution of the tides over the ocean which agrees with these coastal observations, and then we can check the theory by comparing the theoretical distributions with observations on the islands.

I shall now give a short historical review of the attempts that have been made to account for the chief features of the semi-diurnal tides of the Atlantic. In 1833, William Whewell published the first *co-tidal charts*, and Fig. 2 shows his chart for the Atlantic.* The idea of a co-tidal chart is that along each line high water occurs at the same time; in Whewell's case the numbers give the solar times of high water on days of New Moon and Full Moon. The reasoning behind this chart is largely non-dynamical. Only the

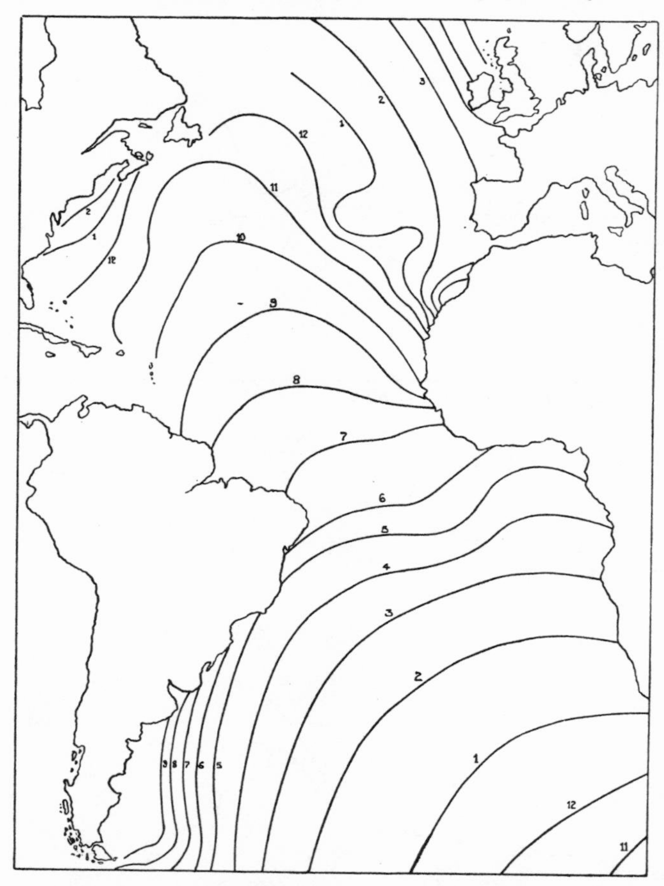

FIG. 2.—*Whewell's chart for the tides on days of New Moon and Full Moon. The numbers give the Greenwich solar hour of high water.*

forward curvature of the co-tidal lines is derived from dynamics, and this derivation is wrong. The lines just connect up the various points on the coasts at which high water has been observed to occur at the times indicated. In order to produce his charts Whewell organised an extensive series of observations. On the west, he got the simultaneity along the coast of the U.S.A., but he did not get the crowdings of co-tidal lines near 32° S. and 17° N. This particular chart led to the hypothesis that the tides of the Atlantic are waves derived from the Southern and Pacific Oceans. In 1842, Sir George Airy stated that he could not accept this hypothesis, and by 1848 Whewell himself had very largely abandoned it; but it lived for a long time in textbooks. Whewell's hypothesis raised the question as to whether the tides of the Atlantic are generated mainly within or mainly without the basin of the Atlantic itself. According to the hypothesis

* W. Whewell, *Phil. Trans. Roy. Soc.*, 1833, 147.

they would be generated mainly without the basin of the Atlantic. The question requires more precise statement, and when this is done it is seen that the problem of obtaining a complete answer is more difficult than that which I have already formulated. Of course, such a complete answer has not yet been given.

In 1874, William Ferrel calculated, on dynamical principles, what the tides would be in canals along parallels of latitude and stretching from one side of the Atlantic to the other.* For a canal between Ireland and Newfoundland he thought that there might be an approach to the conditions of resonance. He considered that if there were a dyke extending from the Cape of Good Hope to the coast of South America, the tides of the North Atlantic would most probably be very nearly the same as they actually are. In Ferrel's case there were no meridian components of tidal current, but I have considered zonal bands in which there is no lateral constraint. I have taken into account both longitudinal and transverse components of generating force, the variations over each band in the direction of the meridian being determined by the variation of the generating forces. Some of the amplitudes are quite large; larger than those actually observed. This indicates that large tides are generated in the basin of the Atlantic itself. On the coast of the United States high water in these bands occurs at about 6 hours in lunar time. But there is so much discontinuity between the tides in two contiguous bands that it is a very difficult problem to find the connecting links.

In 1900, R. A. Harris introduced a series of hypotheses on the basis of which, in 1904, he constructed charts of co-tidal lines for all the oceans and seas.† These hypotheses are:

1. That the tides are dominated by the forms of simple free oscillations which have the period of 12 hours 25 minutes.
2. That these forms can be determined without paying attention to the absence of lateral constraint.
3. That these free oscillations can be determined without reference to the rotation of the Earth.

For the tides of the Atlantic, Harris used four simple free oscillations. The times of these oscillations are supposed to follow from the times of the various phases of the generating forces of the Moon, but in this I believe Harris was three hours in error. In combining the oscillations of overlapping areas, we see that at the intersection of two nodal lines there is still no rise and fall of the water, and in this way we arrive at an amphidromic point. Also, the openings cause progressive waves. Fig. 3 shows Harris's co-tidal chart, and the times on this do agree with observations. The nodal lines have been replaced by crowdings of co-tidal lines ; in particular there are crowdings of co-tidal lines on the west at 32° S. and 17° N. We notice that there is a progression from east to west across the north of the ocean. Harris's work may be subjected to some serious criticisms:

1. For the South Atlantic he neglects the resonance of transverse oscillations.
2. He gives no adequate consideration of the effect of the water in neighbouring regions.
3. He gives no adequate consideration of the dynamical effects of the Earth's rotation.
4. His work is very largely only qualitative; he gives no absolute amplitudes of tide.
5. His times do not follow from the generating forces.

His chart was not accepted by Sir George Darwin.

In 1910, Poincaré discussed the energy implications of progressive waves. An influx of energy into any region must be balanced by:

1. The dissipation of energy by the friction of tidal currents on the bottom of the ocean and connected seas.

* W. Ferrel, *Tidal Researches*, 237, 1874. † R. A. Harris, *Manual of Tides*, Part IVB, 1904.

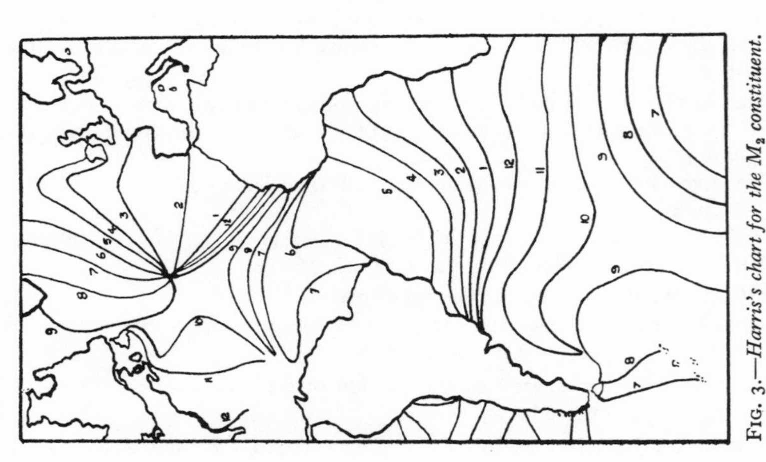

Fig. 3.—*Harris's chart for the* M_2 *constituent. The numbers give the Greenwich lunar hour of high water.*

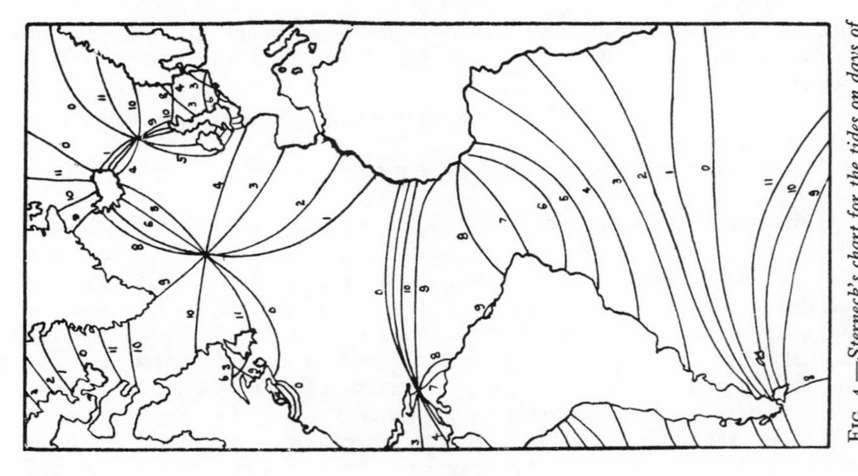

Fig. 4.—*Sterneck's chart for the tides on days of New Moon and Full Moon.*

The numbers give the Greenwich time of high water in hours which are one-twelfth of the period.

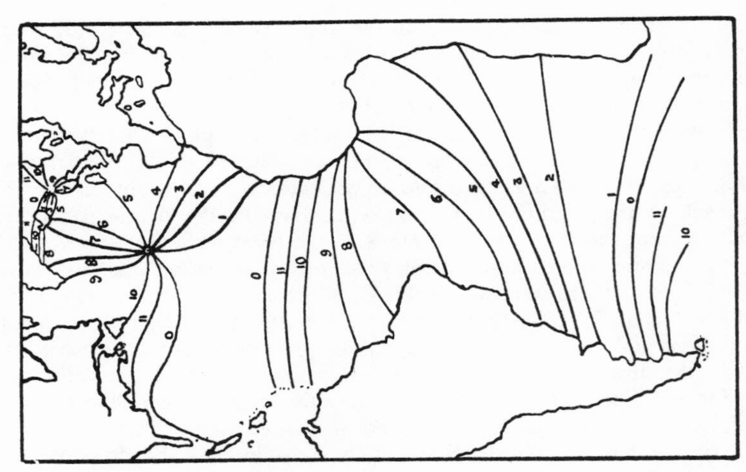

Fig. 5.—*Defant's chart for the semidiurnal tides. The numbers give the Greenwich hour of high water.*

2. The work done against the generating forces.
3. The flow of energy out of the region.

In applying this principle to the Arctic seas, Poincaré assumed No. 2 to be negligible on account of the smallness of the generating forces in the Arctic, and No. 3 to be negligible on account of the smallness of Bering Strait. He assumed that No. 1 is always negligible. He thus deduced that the Arctic tides do not consist of a wave progressing away from the Atlantic. Incidentally, it would also follow that the energy passing into the Atlantic from the Southern Ocean must be given up to the tide-generating forces acting on the water of the Atlantic. But friction is not negligible, as was shown in 1918 by G. I. Taylor; and the progression of co-tidal lines does not necessarily involve a progressing of energy. For the M_2 constituent in the northern hemisphere and for a northerly progression of co-tidal lines, if the range increases towards the east sufficiently rapidly then the current at the time of high water may be directed towards the south, and in this case there will be a mean progression of energy towards the south.

In 1920, H. Jeffreys continued Taylor's work on the dissipation of energy in shallow seas, and in 1923 he showed that this friction must have very important effects on the distribution of the tides. There is another strong argument for the important effects of friction on the distribution of the tides, but for this we have to go beyond M_2 and consider also the corresponding solar harmonic constituent, S_2. One of the most striking facts about the actual tides is that, almost universally on the open coasts of the ocean, spring tides occur sometime *after* New and Full Moon. This can be accounted for by the dissipation of energy in marginal seas. It was in fact one of the points in favour of Whewell's hypothesis, for the S_2 constituent having a shorter free wave-length than the M_2 constituent, there would be an increasing lag in the time of S_2 high water behind the time of M_2 high water as we proceed northward up the ocean. But apart from this hypothesis it is clear that marginal dissipation must form a part of any adequate dynamical explanation of the observed tides.

In 1920, Sterneck published co-tidal charts for all the oceans.* These charts are not based on dynamical principles, but only on observations. He took all the observations available, resolved them into two phases differing by a quarter period, and then estimated the positions of lines of zero elevation in each phase. The intersection of a line of zero elevation in one phase with a line of zero elevation in the other phase gives an amphidromic point in the combination. By joining these amphidromic points to various points on the coasts he produced his co-tidal charts. For the semidiurnal tides of the Atlantic, his co-tidal chart, Fig. 4, shows the same general features as that of Harris, but the co-tidal lines for the North Atlantic have a more natural shape. He has the crowding of co-tidal lines on the west near 17° N. but not near 32° S.

In 1924, A. Defant gave a dynamical treatment of the spring tides of the Atlantic and Arctic regarded as one basin.† He applied to the combined basin of the Atlantic and Arctic the same methods which had been successful for such narrow seas as the Red Sea and the Adriatic. He considered two phases of a longitudinal oscillation. In each phase the transport of water is assumed to vanish at the northern end and the elevation to agree with observations at the southern end. All the generating forces are assumed to be allowed for in one of these phases, so that the other phase is a co-oscillation with the water of the Southern Ocean. Defant makes no allowance for the dissipation of energy. With one exception, the nodes of one phase are very near to the nodes of the other phase, so that their combination is practically a standing oscillation. There can be very little progression either northwards or southwards. When the rotation of the Earth is allowed for, transverse surface gradients are introduced, except quite near to the equator. In the North Atlantic the surface rises to the right of the current, while in the

* R. Sterneck, *Sitz. Akad. Wiss. Wien*, II*a*, **129**, 131, 1920.
† A. Defant, *Ann. Hydrog. Marit. Meteorol.*, **52**, 153, 1924.

South Atlantic the surface rises to the left of the current, and of course in both cases it oscillates with the current. We thus have a series of amphidromic regions, the co-tidal lines going round in the anti-clockwise direction in the North Atlantic and Arctic, and in the clockwise direction in the South Atlantic. This means that a wave progresses northwards up the coast of South America and of Western Europe, and that a wave progresses southwards down the coast of North America and West Africa. There is, of course, no south-going wave down the coast of West Africa. But then Defant has not yet allowed for the transverse generating forces, and this difficult calculation he does not do adequately. His forced transverse oscillations are chosen so as to make the combination of all his oscillations agree with observations on the coasts. Fig. 5 shows Defant's co-tidal chart, but it may be subjected to some very serious criticisms:

1. He assumes that longitudinal generating forces produce only longitudinal currents, and that transverse generating forces produce only transverse currents. Owing to the Earth's rotation, these assumptions are not valid.
2. His total neglect of dissipation is very serious.
3. His forced transverse oscillations are without value.
4. The large range which he obtained on the arctic coast of Canada is not in agreement with observations.
5. The small range which he obtained at the Azores is not in agreement with observations.

In 1932, Defant abandoned his work of 1924 and confined his attention to the neighbourhood of the central line of the ocean and also to the part of the Atlantic south of Iceland.* He assumed:

1. That the rotation of the Earth does not affect tidal conditions on the central line.
2. That the transverse generating forces do not affect tidal conditions on the central line.
3. That there is no dissipation of energy to the south of Iceland.

His method does allow for the dissipation of energy outside the region which he considered, and in fact he stressed the importance of friction in the Arctic Sea. He considered two phases, differing by a quarter period, such that they agree with observations at the Azores and at Tristan da Cunha. This time the node of one phase coincides with the loop of the other, and their arrangement indicates that there is a northerly progression throughout the region considered. His results show good agreement with the observations at the islands. Defant's assumptions cannot be more than rough approximations, but it does seem as if conditions on the central line can be largely accounted for by means of longitudinal oscillations. Defant's longitudinal current-phases also increase from south to north. They appear to be in good agreement with the results of the "Meteor" expedition of 1925–27. But his longitudinal current-amplitudes are only about one-third of those observed by the "Meteor". I am not disposed to attach very much importance to this disagreement, because it is very difficult to obtain an adequate depth-mean of current when the observations are much affected by the presence of internal tides. But if we discredit the disagreement shown by the amplitudes we must not attach much importance to the agreement shown by the phases, because the accurate determination of current-phases from observations is subject to the same difficulty.

I now propose to describe an attack on the subject which I started some years ago, but which is still only in its early stages. As a preliminary, let us examine a series of precisely formulated mathematical problems. Consider a region of the Atlantic bounded by two parallels of latitude, namely those for 45° N. and 35° S. Suppose that we are given either:

* A. Defant, *Wiss. Ergeb. "Meteor"*, 7 (1), 273, 1932.

1. The general conditions to be satisfied along the coasts and the actual meridian-components of currents across the bounding parallels; or
2. The general conditions to be satisfied along the coasts and the actual tidal elevations on the bounding parallels; or
3. The general conditions to be satisfied at the coasts and the actual meridian-components of currents on one bounding parallel and the actual elevations on the other bounding parallel;

then the tidal elevations and currents are mathematically determinate all over the region considered, including their values on the coasts. For example, we might imagine the meridian-components of currents at 45° N. and 35° S. to be produced by suitably moving flexible pistons, and then in the first case it is clear that we have a definite physical problem. But we do not know either the actual currents or the actual elevations on the bounding parallels, whereas we do know the coastal elevations. Hence we consider the problem the other way round. Given the general conditions along the coasts and the actual values of the coastal elevations, the tides are mathematically determinate over the whole region considered, including the elevations and currents on the bounding parallels. This is my problem, but I propose to solve it by considering a series of other subsidiary problems of still different type.

Given the general conditions along the coasts and both the meridian-components of currents and the elevations along *one* bounding parallel, the tides are still mathematically determinate over the whole region considered.

This double condition along one bounding parallel has great advantages from a computational point of view. We can work steadily away from that parallel until we cover the whole region. Of course, in each subsidiary problem we have to take arbitrary distributions of current and elevation along the bounding parallel from which we start, but the object is to take such a series of distributions that a suitable linear combination of them will represent the actual distributions, whatever these may turn out to be.

I consider then a number of subsidiary oscillations. Each of these oscillations is a possible motion of the waters of the Atlantic under the conditions prescribed. These prescribed conditions include the general coastal conditions in all cases, and in each case they also include sufficient conditions along the parallel of latitude 35° S. to make the motion completely determinate.

I have not, as yet, introduced any dissipation of energy within the region considered, but this is quite consistent with any amount of dissipation of energy outside the region considered. My reasons for this neglect within the region are as follows:—

1. Jeffreys' total dissipation for the whole ocean did not include any contribution from parts of this region.
2. Defant's latest investigation did not involve any dissipation in this region.
3. It is desirable to keep the first attempt as simple as possible, and the introduction of coastal dissipation does complicate the details, though the general method allows for any amount of this.

I do not propose to describe the mathematical details by which these individual oscillations have been calculated. We proceed from one parallel of latitude to another 5° to the north of it, and much use has been made of an expansion-theorem, specially evolved for the purpose, which is being published by the London Mathematical Society.

It is possible that my treatment of the great irregularities of the basin formed by the Gold Coast and the north-east coast of Brazil may not yet be adequate. It is also possible that I have not yet taken sufficient terms in the series which represent the individual oscillations. I have still to examine these questions.

Fig. 6 shows the co-tidal and co-range lines of a forced oscillation. In this oscillation both the longitudinal and transverse generating forces are taken into account, and both the elevations and the meridian-components of currents are taken to be zero on latitude

35° S. The oscillation is thus independent of the tides to the south of latitude 35° S. The conditions satisfied along the southern bounding parallel secure that there is no passage of energy into or out of the Atlantic. But they put on a much bigger restriction than that, and keep the whole oscillation very small for some distance north of 35° S. This does not mean that only a small part of the tides of the Atlantic are generated within the Atlantic itself. If we put a barrier along any parallel of latitude, the cutting out

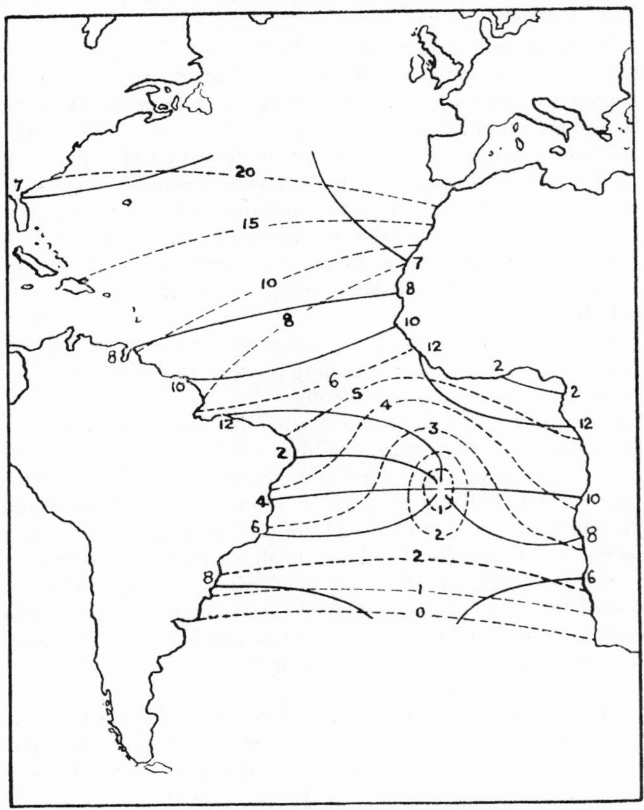

Fig. 6.—*Forced oscillation for* M_2 *constituent.*
Full lines give Greenwich lunar hour of high water.
Broken lines give amplitude in cm.

of the currents of this forced oscillation would require large multiples of the other oscillations.

The other subsidiary oscillations could be maintained by tides to the south of latitude 35° S. and are independent of the generating forces acting on the water of the Atlantic itself.

Fig. 7 shows an oscillation which I have called a north-going Kelvin-wave. It corresponds to a certain mean rate of northward progression of energy. As there are no generating forces and as there is no dissipation, this mean rate of progression of energy must be the same for all parallels of latitude. The particular type of motion assumed along the parallel 35° S. is that first investigated in 1879 by Sir W. Thomson for a canal on a flat Earth. The currents are directed along the meridians and both the elevations and the currents increase exponentially towards the west. No complication is produced by the curvature of the Earth along the parallels of latitude, but much complication is produced by the curvature of the Earth along the meridians, by variations in the width of the ocean and by variations of depth. It is these complications which make the distribution shown in Fig. 7 vary from the simple conditions along the parallel of latitude 35° S. Continuous reflection along the coasts is allowed for, but if a complete barrier were put across the Atlantic, then of course the wave would arrive at an impossibility. We notice that there are large amplitudes in the south-west and in the north-east.

Fig. 8 shows the co-tidal and co-range lines of a south-going Kelvin-wave. This time there is a certain mean rate of southward progression of energy, which is the same for all parallels of latitude. Again, the currents in latitude 35° S. are along the meridian, but this time both the elevations and currents increase exponentially towards the east. This time it is not a question of following up the wave, but of finding what it must have been on any parallel of latitude in order to reach given values 5° to the south. We notice that there are large amplitudes in the north-west and the south-east.

Fig. 9 shows the co-tidal and co-range lines in what I have called a north-going Poincaré-wave. The type of motion along the parallel of 35° S. is next in simplicity to that of a Kelvin-wave; it was first studied by H. Poincaré in 1910 for the same canal as was considered by Sir W. Thomson. Instead of the transverse currents being zero, they now follow a sine-distribution from coast to coast. For a uniform flat canal there is a line of zero range parallel to the sides of the canal; the co-tidal lines run straight across the canal, but the time of high water changes by six lunar hours on crossing the line of zero range. These simple conditions hold in our case only along the parallel of latitude 35° S. On account of the curvature along the meridians and of the variations of width and depth, a line of zero range is too artificial and becomes replaced by a special form of amphidromic region. We notice that in the centre of the ocean the amplitudes remain small.

Fig. 10 shows the co-tidal and co-range lines of a south-going Poincaré-wave. This corresponds to a certain mean rate of southward progression of energy. The conditions in latitude 30° S. are similar to those for the north-going Poincaré-wave.

These Poincaré-waves are quite different from anything in Defant's work. They are the two simplest of an infinite series of oscillations, and it will probably be necessary to consider more. When in 1919 G. I. Taylor gave his determination of tidal motions in a rectangular gulf, he used both Kelvin-waves and the infinite series of Poincaré-waves.

It is now a question of assigning an amplitude and a time-origin to each Kelvin- and Poincaré-wave, so that when the four are added to the forced oscillation, the coastal elevations in the synthesis will agree with observations as far as is possible. As all the oscillations have been computed from prescribed conditions at the southern boundary of the region considered, it is desirable to make the agreement between synthesis and observation as good as possible in the southern part of the region, and then to discover where it begins to break down. As there are four free waves, I have made the sum of all the waves agree with observations at four points on the coasts, two on the east coast and two on the west coast. Two of these points I have chosen in latitude 32°·5 S. and the other two in latitude 7°·5 S., *i.e.* south of the Gulf of Guinea and Cape San Roque where the major irregularities of the basin begin.

I then find, as expected, that the north-going Kelvin-wave is the most important. I calculate that this wave must have an amplitude of 52 cm. and a high-water time of 1·8 hour at the south-west corner of the region considered, where it is greatest. The next most important wave is the south-going Poincaré-wave. This must have an amplitude of 35 cm. and a high-water time of 9·1 hours at the south-west corner, where it also is greatest. The other two free waves have each an amplitude of 22 cm. at the south-east corner, where they both are greatest in the south. The high-water time of the south-going Kelvin-wave at this point is 10·8 hours, and that of the north-going Poincaré-wave 0·5 hour.

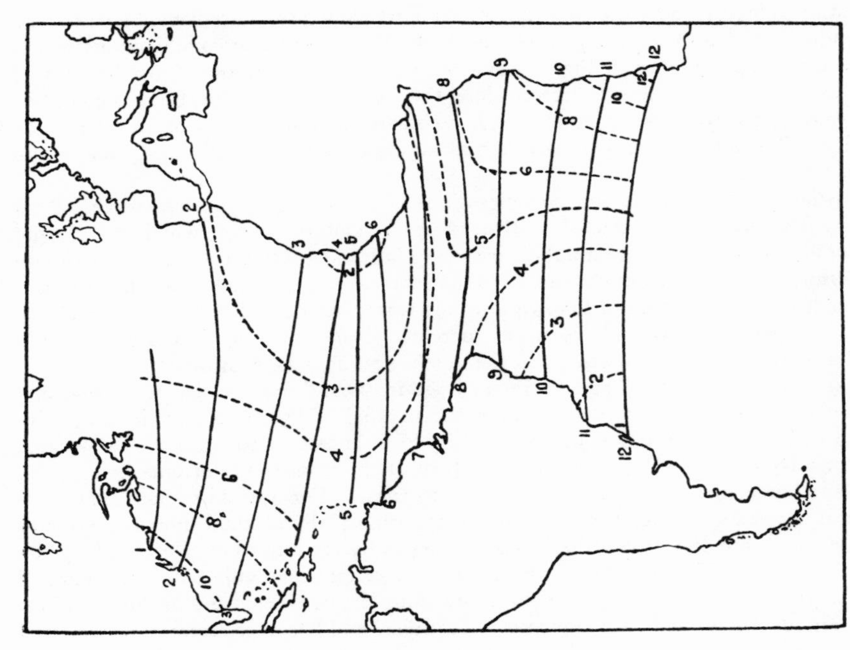

FIG. 8.—*South-going Kelvin-wave for M_2 constituent.*
Full lines give lunar hour of high water, origin arbitrary.
Broken lines give amplitude, scale arbitrary.

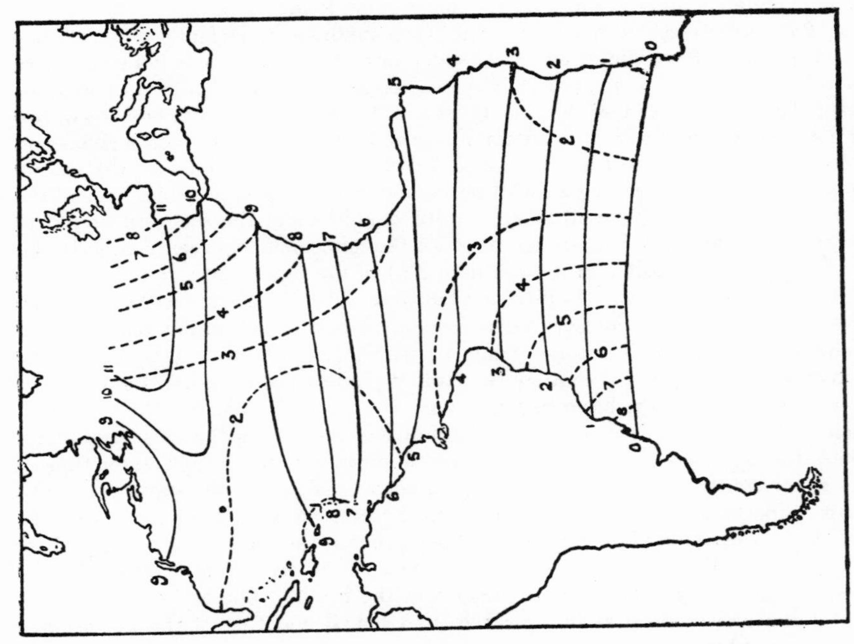

FIG. 7.—*North-going Kelvin-wave for M_2 constituent.*
Full lines give lunar hour of high water, origin arbitrary.
Broken lines give amplitude, scale arbitrary.

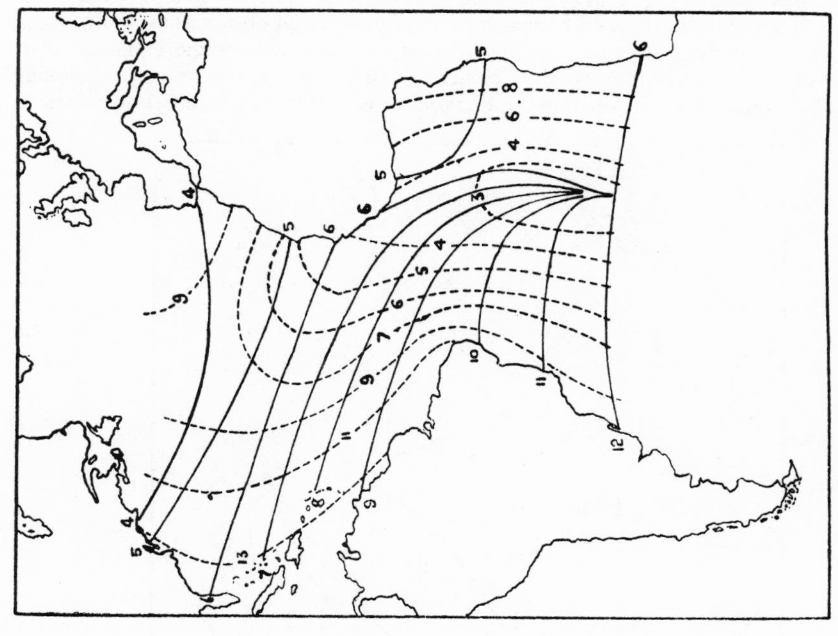

FIG. 10.—*South-going Poincaré-wave for M_2 constituent. Full lines give lunar hour of high water, origin arbitrary. Broken lines give amplitude, scale arbitrary.*

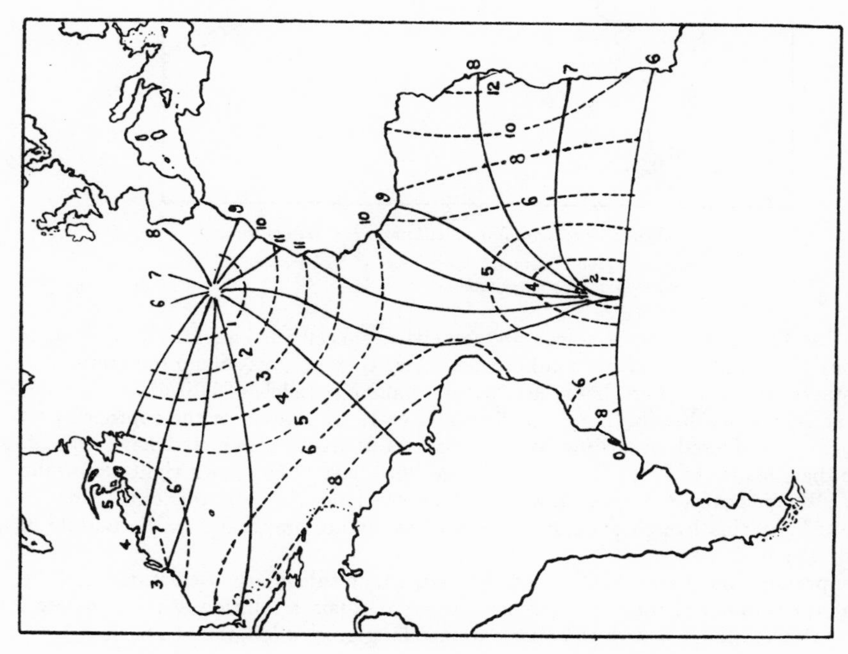

FIG. 9.—*North-going Poincaré-wave for M_2 constituent. Full lines give lunar hour of high water, origin arbitrary. Broken lines give amplitude, scale arbitrary.*

The shorter lines of Fig. 1 show the lunar times of high water and amplitudes in the synthesis. They show fairly good agreement with the results of observations to the south of the equator, but they diverge very considerably to the north of the equator. Part of this divergence may be due to the deficiencies of the individual waves, as already indicated, but it may also be due to my not having taken a sufficient number of Poincaré-

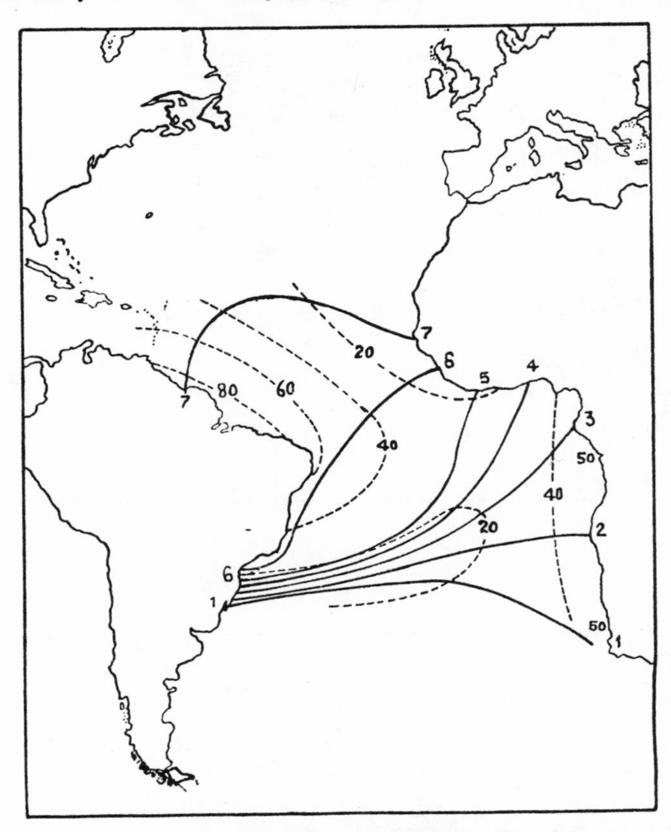

FIG. 11.—*Synthesis of subsidiary oscillations for M_2 constituent.*
Full lines give Greenwich lunar hour of high water.
Broken lines give amplitude in cm.

waves. The first two Poincaré-waves are somewhat special in that, for the particular dimensions of the Atlantic, at the southern boundary they progress along the meridians. All the others, at the southern boundary, progress along parallels of latitude.

Fig. 11 shows the distribution of co-tidal and co-range lines over the southern part of the region considered, according to the synthesis of waves at which I have arrived. I believe that this is the first time that co-range lines have been drawn right across the Atlantic. They show a trough of low amplitude reaching the coast of South America at about 32° S.; this trough contains a point of minimum amplitude in the middle of the South Atlantic.

By improving the determination of the individual subsidiary oscillations and by increasing the number of these oscillations, I hope to obtain a solution of the dynamical equations which will agree with the observations along greater lengths of coasts.

19

Reprinted from *J. Appl. Prob.* 4:103–112 (1967)

TIME-SERIES ANALYSIS OF TIDES AND SIMILAR MOTIONS OF THE SEA SURFACE

D. E. CARTWRIGHT, *National Institute of Oceanography, Wormley, Godalming, Surrey, U.K.*

1. Introduction

This survey article, most of whose results are described in greater detail in Munk and Cartwright (1966), which will hereafter be abbreviated to MC, describes methods which aim to separate the response of the sea level at a given place due to various exciting forces such as gravity, solar radiation, non-linear effects, and weather. In so doing, it provides predictors for sea level which are formally simpler and somewhat more accurate than those given by the classical methods.

We are concerned with the variations in time of the height of the sea surface $\zeta(t)$ above a fixed point, for which data are available in the form of time series sampled at hourly intervals from tide-gauge records of many years' extent. A typical spectrum of density of variance (i.e. a "power spectrum", or, in oceanographic contexts, "energy spectrum") of $\zeta(t)$, such as that illustrated in Figure 1, occupies a range in frequency of about 0 to 10 cycles per day (c/d). It may be described as a complicated line spectrum due to tides superposed on a smooth continuum which rises steadily with decreasing frequency. The continuum can be regarded as "random noise" for many purposes, but, as will be shown later, certain parts of it may be related coherently to other variables describing the weather, and to interactions between the continuum itself and the tides. The weather-dependent part is the frequency transform of what are commonly called "storm surges", whose prediction is important in flood warning.

The frequencies at which the tidal lines are situated are known very precisely from the motions of the Moon and the Sun. They are expressible in the general form:

$$f = K_1F_1 + K_2F_2 + \cdots + K_6F_6, \tag{1}$$

where

Paper read at the European Meeting of Statisticians, London, 6 September 1966.

$$F_1 = 1 \text{ cycle/lunar day} = 0.9661 \text{ c/d},$$

$$F_2 = 1 \text{ cycle/month(c/m)} = 0.0366 \text{ c/d}$$

$$F_3 = 1 \text{ cycle/year (c/y)} = 0.0027 \text{ c/d},$$

and F_4, F_5, F_6 are known lower frequencies. The K_n are small integers, mostly within the range $+ 4$ to $- 4$. The intensities of the lines vary greatly, and lines to an order of 10^2 are usually sufficient to define the purely tidal variance to within a few cm^2. In practice, the number is reduced to about 40 by the use of modulating factors.

2. The harmonic method of tide prediction

In the classical (so-called "harmonic") method of tide analysis, due essentially to G.H. Darwin, with later improvements by A.T. Doodson and others, about a year's hourly series of data, say ζ_r, $r = - N(1)N$, is subjected to a number of complex spectral filters $\Phi_r^{(k)}$ whose characteristics are narrow peaks centred on selected tidal frequencies, f_k. After some minor adjustments to allow for modulations and imperfect filtering, one thereby arrives at a number of amplitudes $|H_k|$ and phase lags $\text{Arg}(H_k)$, where

$$H_k = \sum_{r=-N}^{N} \Phi_r^{(k)} \zeta_r. \tag{2}$$

The adjusted amplitudes and phases are thereafter used to represent the selected tidal lines for prediction purposes.

A typical figure for the continuum level is $100 \text{ cm}^2/(\text{c/d})$, so, since the bandwidth of $\Phi_r^{(k)}$ cannot be less than 1 c/y in the case considered, H_k is a normal estimator with variance $100/365 = 0.3 \text{ cm}^2$ for the "true" amplitude, H_k' say. The probability distribution of $|H_k|$ is however a non-normal function, well known to statisticians, such that the expectation

$$E(|H_k|) = |H_k'| \; [1 - O(0.3/|H_k'|^2)]$$

is positively biassed, giving a consistent error to the smaller amplitudes. Apart from this, the somewhat arbitrary selection of lines and the not very satisfactory modulations employed make the harmonic method less than ideal, though still very serviceable for many purposes.

3. The response method

The method of tidal analysis and prediction evolved in MC avoids these and other difficulties by correlating $\zeta(t)$ with noise-free input functions. The gravity potential which generates the tides may be expressed as:

$$V(\theta, \lambda; t) = g \sum_m \sum_n \; [a_n^m(t) U_n^m(\theta, \lambda) + b_n^m(t) V_n^m(\theta, \lambda)], \tag{3}$$

where $U_n^m + iV_n^m$ is a standard complex spherical harmonic of geographical latitude θ and longitude λ, and $a_n^m(t)$ and $b_n^m(t)$ are orthogonal functions of time which can be computed by a straightforward procedure. The summations are not at all extensive, since only degrees $n=2$ and 3 have appreciable magnitude, and m is limited to $0 \leqq m \leqq n$. The order m determines the leading frquency K_1F_1 in the time harmonic expansion (1), so that, apart from a slight complication in separating the effects of degrees 2 and 3, the specification of m and n also specifies the zone of the frequency spectrum we are dealing with. We therefore drop the spherical harmonic designation m, n for convenience in most of what follows.

We may compute the time series $a_r(t)$ for the same times as the sea level $\zeta_r(t)$, and apply the same filter functions $\Phi_r^{(k)}$ to obtain another spectral sequence

(4) $$G_k = \sum_{r=-N}^{N} \Phi_r^{(k)} a_r.$$

The complex quantity[1] $G_k H_k^*$ is then a sample of the "cross-spectrum" between $a(t)$ and $\zeta(t)$ at around the frequency f_k. Further, since a_r is *a priori* a noise-free input series, then

(5) $$Z(f_k) = G_k H_k^* / G_k G_k^*$$

is an unbiassed estimate of the "admittance" of the sea level to the spherical harmonic of gravity potential considered at frequency f_k.

The important fact is that the admittance or "transfer function" $Z(f)$ is observed to vary only slowly with varying frequency f. Thus, the detailed structure of the tidal lines (1) may be largely ignored, and we obtain essentially the same estimate of $Z(f)$ from (5) by employing a filter with broader frequency characteristics than is necessary for the harmonic method. A broader filter uses a shorter time span, so we may for example obtain about 12 independent samples of $Z(f_k)$ for every k from a year's data by applying filters $\Phi_r^{(k)}$ of about 1 month's extent. Suppose that in general we estimate p samples of the cross-spectra at each relevant frequency, and form ensemble-averages, (denoted by $\langle \;\; \rangle$). Then it may be shown that $\langle G_k H_k^* \rangle / \langle G_k G_k^* \rangle$ is an unbiassed normal estimator for $Z(f_k)$ with sampling variance

(6) $$\sigma_p^2 = (|Z|^2 / 2p)(\gamma^{-2} - 1),$$

where γ^2 is the "coherence" at the frequency f_k for the given bandwidth.[2]

The obvious estimate for γ^2 is the ratio

(7) $$\gamma_p^2 = |\langle GH^* \rangle|^2 / \langle |G|^2 \rangle \langle |H|^2 \rangle,$$

---

(1) * is used throughout to denote the conjugate $x - iy$ of a complex quantity $x + iy$.

(2) Note that for line spectrum signals, the coherence increases with decreasing bandwidth, unlike a purely stochastic dynamic process.

but this is positively biassed, the more so the smaller the sample number p. For example, in the trivial case $p = 1$, the right hand side of (7) is identically unity, whatever the true coherence. The sampling distribution of γ_p^2 is somewhat different from the usual case tabulated by Amos and Koopmans (1963), because the input intensity $|G|^2$ is approximately constant, not stochastic. A simplified analysis shows that the expectation of γ_p^2 is very nearly

$$\gamma^2 - (1 - \gamma^2)/p,$$

so that an unbiassed estimator for γ^2 is

$$(p\gamma_p^2 - 1)/(p - 1) \quad (p > 1),$$

which is the value most appropriate for use in (6). With this adjustment, we may usefully divide the spectral variance $S(f_k) = \langle H_k H_k^* \rangle$ into the "coherent variance" $\gamma^2 S(f_k)$, which is the part of the level spectrum directly related to the input, and the "incoherent variance" $(1 - \gamma^2)S(f_k)$, which is effectively "noise".

4. Parametric representation of $Z(f)$

We find, in cases not too distorted by nonlinear effects (to be considered later), that consecutive values of $Z(f)$ referring to a particular spherical harmonic appear to lie on a smooth curve, as might be expected from dynamical tide theory. This leads us to represent the tides at the given place in terms of the few parameters necessary to define $Z(f)$ with reasonable precision, instead of the heterogeneous collection of amplitudes and phases required by the harmonic method. The complete set of lines is of course embodied in the functions $a_n^m(t)$, so we are in fact employing them directly without recourse to the clumsy harmonic expansion.

To parametrise $Z(f)$ we use the Fourier series representation

$$Z(f) \cong w_0 + \sum_{s=1}^{S} w_s \exp(2\pi i f \tau_s), \tag{8}$$

which is exactly equivalent to the regression formula

$$\tilde{\zeta}(t) = \mathrm{Re}\left\{ w_0 c^*(t) + \sum_{s=1}^{S} w_s c^*(t - \tau_s) \right\}, \tag{9}$$

where $c(t) = a(t) + ib(t)$, and the weights w_s are complex. The symbol $\tilde{\zeta}(t)$ is used to denote that (9) is a "prediction" formula, as distinct from $\zeta(t)$ itself which includes noise.

The choice of the time lags τ_s and their number S require special consideration. Firstly, we choose an arithmetic series of form $\tau_s = s\Delta\tau$, since computational experiments to choose arbitrary sequence for τ_s giving a best "fit" failed to suggest any meaningful system. With this restriction, (8) has periodicity $(1/\Delta\tau)$ in frequency f. The periodicity is not real in nature, and in common physical systems it would therefore be necessary to make $\Delta\tau$ as small as possible (in fact to the sampling rate of $\zeta(t)$). However, we have here a special case of inputs $c_n^m(t)$ whose frequency bands are limited virtually (see Equation (1)) to about

$$f = mF_1 \pm 4F_2. \tag{10}$$

It is therefore convenient to take this range of 8 c/m as about half the frequency-period of (8), i.e. to make $\Delta\tau = 1/(16\text{c/m}) = 2$ days. The fact that $Z(f)$ is thereby almost certainly wrongly represented outside the range (10) is irrelevant, because there is no input spectrum there.

Choice of S has analogy with the "aliassing" problem in time sampling. From (1), the input spectrum is concentrated into line-groups separated by voids a little less than 1 c/m wide. Variations of period smaller than 2c/m are therefore "aliassed" (or redundant). In fact, other considerations lead us to limit the minimum frequency-period $1/S\Delta\tau$ to about 4c/m, which limits S to 3.

Finally, it can be shown that in this special case of regression with band-limited input spectrum, we need not be restricted to positive time lags τ_s, but may extend the summation range to $-S$ (1) $+S$. This increases the degrees of freedom available for "curve-fitting" from 8 to 14 (i.e., 7 complex pairs).

The technical procedure for estimating the regression weights w_s is to minimise either

$$\sum_k \left[\sum_s w_s \exp(-2\pi i f_k s\Delta\tau) - Z(f_k) \right]^2 / \sigma_{p,k}^2, \tag{11}$$

or, more conveniently in practice,

$$\sum_t \left[\mathrm{Re}\left\{ \sum_s w_s c^*(t-\tau_s) \right\} - \zeta(t) \right]^2. \tag{12}$$

Minimising of form (12) involves a familiar matrix of unbiassed covariances, whose relative variability tends to zero with increasing span of the summation in t.

5. Multiple inputs and nonlinear effects

The values of K_1 (Equation (1)) or of m, which determine the centres of the separate frequency bands, are known in tidal literature as "species". Thus we have tides of species 0, 1, 2, 3, etc. Because the covariance of two time series whose spectra are separated by an order of 1 c/d is negligible when evaluated over many days, the Equations (12) for the regression weights w_s may be conveniently separated into distinct tidal species.

However, there are distinct input functions with distinct dynamic responses which do fall within the same tidal species, and therefore cannot be evaluated by separate regression equations. These include

(i) the spherical harmonic coefficients $c_2^m(t)$ and $c_3^m(t)$;
(ii) similar coefficients related to solar radiation (as distinct from gravity);
(iii) nonlinear effects.

The difficulty here is that the line spectra of (i)–(iii), which are all of the same form as (1), not only differ by the small frequencies $F_3, \cdots, F_6$, but are in some cases identical in frequency. For example, the strong solar line at exactly 2 c/d, given by $K_1 = 2$, $K_2 = 2$, $K_3 = -2$, in general contains contributions from all three inputs, (although $c_3{}^2$ may probably be neglected here). Separate cross-spectra, say $G_k' H_k^*$, $G_k'' H_k^*$ $G_k'' H_k^*$, are meaningless, since the three inputs G', G'', G'', are themselves highly correlated. It is easily shown that the distinct admittances $Z'(f_k), Z''(f_k), Z'''(f_k)$ cannot be derived from these cross-spectra at a single frequency f_k. However, they may be derived from cross-spectra at a few adjacent frequencies, using our basic assumption that the admittances vary only slowly with frequency. The symbolism in the frequency domain becomes bulky, but we may conveniently revert to the regression form (12) by adding the relevant lagged input functions analogous to $c(t - \tau_s)$ with additional weights $w_s^{(r)}$.

The nonlinear input functions (iii) deserve further description. Except in very shallow water, the dynamics of tidal waves are only weakly nonlinear, and so may be described in terms of a dominant linear effect with smaller perturbations involving products of linear elements. In the frequency domain, as is well known, such a system gives rise to sums and differences of the frequencies of the linear part of the spectrum. Thus, linear tidal elements ζ_1 at frequency f_1 in species 1, and ζ_2 at frequency f_2 in species 2, produce by nonlinear interaction contributions $Z(f_1, f_2)\zeta_1\zeta_2$ in species 3, and $Z(-f_1, f_2)\zeta_1^*\zeta_2$ in species 1, where $Z(f, g)$ is a sort of bivariate transfer function, or "interaction coefficient". Similarly, a relatively important contribution to species 2 has the form $Z(f_2, f_2, -f_2)\zeta_2\zeta_2(\zeta_2)^*$, derived from a triple interaction caused by frictional effects. Fourth and higher order interactions appear to be negligible at most places, however. Note that the use of *complex* ζ is essential to this representation.

It will be appreciated that the nonlinear effect produces an unwieldy proliferation of tidal lines if tackled by the harmonic method, with the dubious compensation that some lines of identical frequency may be counted as single lines without distinguishing their origins. Indeed, numerical tests have shown that the advantage of our response method over the harmonic method increases with the importance of the nonlinear (shallow water) effects.

In practice, we deal with terms of type (i)–(iii) by first obtaining approximate linear forms $\tilde{\zeta}(t)$ as in (9) for the species concerned. The orthogonal or "imaginary" part of $\tilde{\zeta}(t)$ is easily derived at the same time from the complex form $c(t) = a(t) + ib(t)$. We then use products such as $\tilde{\zeta}_1\tilde{\zeta}_2$ as quasi-input functions in the complete regression formalism, and re-evaluate all weights, including those corresponding to $Z(f, g)$.

6. Storm surges and surge-tide interaction

A large part of the *non-tidal* spectrum can be ascribed to weather. Such variations often take the form of large waves of tidal proportions called "storm-surges", which when they coincide with high tide can cause serious flooding of low-lying territory, but they are nearly always present in $\zeta(t)$ as smaller, less obvious oscillations. There have been many elaborate studies of the response of enclosed seas to various fields of normal and tangential surface stress. Here we follow for example Rossiter (1959) in assuming that the surge can be dynamically related to local atmospheric pressure $P(t)$, its spatial gradients $P_x(t)$, $P_y(t)$ (which largely determine the wind field), and second derivatives $P_{xx}(t)$, $P_{xy}(t)$, $P_{yy}(t)$ (which allow for linear variation of the wind field).

Cross-spectral analysis of $\zeta(t)$ with the above variables, at all frequencies not dominated by the tides, enables one to evaluate new transfer functions $Z_n(f)$, from which the appropriate lags in an equivalent regression formula may be determined as before. Certain of the input functions, such as $P(t)$ and $P_{xx}(t)$, are fundamentally correlated ($\rho \simeq -0.5$), but their continuous spectra allow the admittances to be separated more easily than in the case of the tidal line spectra. Since the input spectra occupy a wide frequency range of about 0 to 2 c/d, the basic regression lag $\Delta\tau$ has to be much smaller than in the special tidal case; $\Delta\tau = 3$ or 6 hours is satisfactory.

Finally, a small but significant part of the continuum-spectrum of $\zeta(t)$ may be accounted for in terms of nonlinear interaction between surge and tide. In seas round northern Europe the tide is largely dominated by species 2, and in particular the strong lunar semi-diurnal line known as M_2 ($K_1 = 2$). Therefore this interaction can conveniently be studied spectrally in terms of the cross-spectrum of $H(f_2 \pm f)$ and $H(f_2)H(f)$ or $H(f_2)H(f)^*$, where f is a nontidal frequency and $f_2 = 1.9323$ c/d. Such analysis (essentially a "bispectrum" analysis) has been shown to give good coherences in certain parts of the spectrum for shallow water stations such as Southend, where nonlinear effects are relatively strong. It yields estimates of the bilinear interaction function $Z(f, f_2)$ as a continuous function of f, into which the coefficients obtained for the purely tidal regions fit reasonably well.

The resulting bivariate regression is of the form

$$\tilde{\zeta}(t) = \tilde{\zeta}_0(t) + \sum_{s=0}^{S} [w_s \tilde{\zeta}''(t) + W_s^+ \tilde{\zeta}''(t)\tilde{\zeta}''(t) + W_s^- \tilde{\zeta}''(t)\tilde{\zeta}''(t)^*]\tilde{\zeta}_0(t - s\Delta\tau),$$

where $\tilde{\zeta}_0(t)$ is the non-tidal sea level (surge) predicted from a linear formula, $\tilde{\zeta}''(t)$ is the predicted side of species 2, and $\Delta\tau = 3$ hours. The w_s are weights corresponding to the bilinear interaction mentioned above, and W_s^+ and W_s^- are weights derived from a trilinear analysis of "surge × tide × tide".

7. A typical variance budget

Table I shows a spectral breakdown of the total variance (cm^2) of a 3-year record of $\zeta(t)$ from Southend Pier into the tidal and non-tidal zones of interest, and the "predictable variances" due to the tidal and weather regressions discussed above. By "predictable variance" I mean the reduction in variance when a particular prediction formula is subtracted from $\zeta(t)$.

One notes that the tidal regressions (dominated by species 2) account for a very large proportion of the variance, 24888 cm^2 *in toto*. The harmonic method, using 36 pairs of standard prediction constants—compared with about 25 complex weights in our response method—gives a predictable variance of 24821 cm^2, about 67 cm^2 less. The response method is therefore superior in all important respects.

TABLE I

Spectral breakdown of sea level variance in cm^2 at Southend Pier

Frequency (c/d)	Total	Predictable		Residual
		Tide	Surge	
0.0—0.1	93	23	50	20
0.1—0.8	219	—	150	69
0.8—1.1	159	127	20	12
1.1—1.7	25	—	15	10
1.7—2.1	24668	24613	5	50
2.1—2.7	11	—	5	6
2.7—3.1	29	23	1	5
3.1—3.7	5	—	1	4
3.7—4.1	62	59	1	2
4.1-10.0	50	43	1	6
OVERALL	25321	24888	249	184

The predictable surge variances are tentative at the time of writing, but are of the right order numerically. As expected, they are mainly concentrated in the range (0, 2 c/d).

The "residual" variances (Total − Tide − Surge) represent the true noise, which is unpredictable in terms of the input functions considered. The overall total of 184 cm^2 is reduced by about a further 10 to 174 cm^2 by the surge-tide interaction formula. This noise is largely concentrated in the very low frequencies $\ll 1$ c/d, and near the species 2 tide zone. The distribution is perhaps more clearly seen in the sketch of spectral *density* included in Figure 1 (note that the scale is logarithmic).

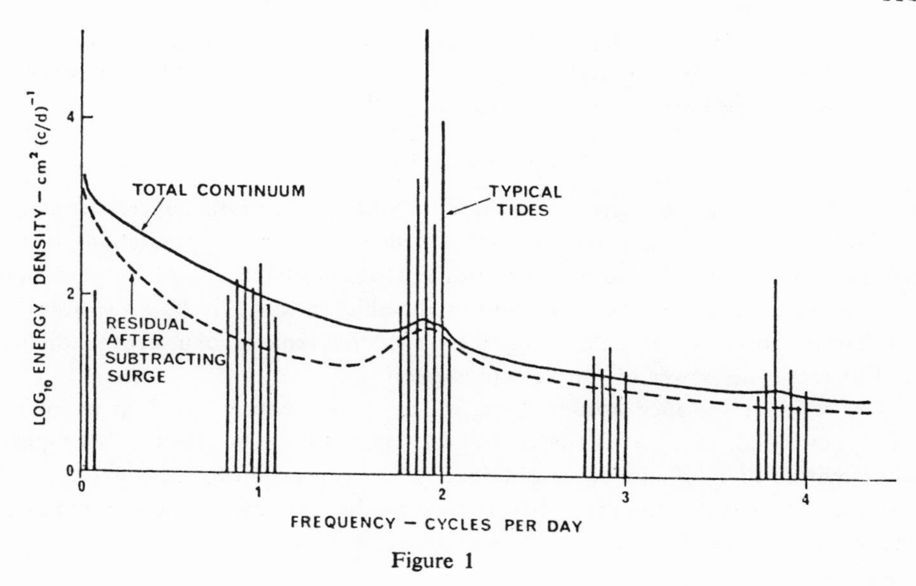

Figure 1

The low frequency residual is probably chiefly due to thermal and other slow fluctuations in conditions over the Atlantic Ocean as a whole. The residual in the species 2 tide zone is thought to represent nonlinear interaction between the tide and some low frequency oceanic variable. It was at first thought that this variable might be local "mean sea level", but tests show very low coherence close to the tidal lines. Also present in the residue are some actual tidal lines, which would presumably be removed by allowing more degrees of freedom to the regression scheme.

8. Auto-prediction

Prediction of tides alone, either by our method, or by the classical harmonic method, can be made almost indefinitely into the future, because of the constancy of the lunar and solar motions. Inclusion of the surge at once limits predictions to the few hours for which detailed weather forecasts are available, a situation quite acceptable for a flood-warning scheme. Within this time range we may also conceive of auto-regressive formulae for the residual noise. Indeed, much of the low frequency tidal variation such as the annual term is probably better treated as an auto-regression than in terms of a response to tidal (radiational) forces.

The low frequency concentration of the residual noise spectrum helps us. We apply a lowpass filter with cut-off at about 0.5 c/d to $\zeta(t)$, form a series of auto-covariances with the resulting series at time intervals of about 1 day. These are then used to compute optimum Wiener-type linear auto-regressions with various combinations of time lag, and their "predictable variances". We find that inclusion of more than one lagged term hardly improves the predictable variance; that is, low frequency sea level behaves statistically much like a Markov process.

The predictable variance of the single lag model

$$\tilde{\zeta}_0(t) = \rho_\tau \zeta_0(t - \tau)$$

naturally decreases with increasing τ. At Honolulu, where weather effects are small, we find (see MC) that half the total low frequency variance is predictable with τ as large as 15 days. At Newlyn, Cornwall, where we have not yet removed the considerable surge variance, the auto-predictable variance is half the total at $\tau = 2$ days. This is still acceptable, but it is hoped that removal of a surge prediction will increase the power of the auto-prediction.

The residual variance around 2 c/d, which is also concentrated in a narrow frequency band, is also amenable to auto-prediction. One uses a band-passs filter centred on 2 (c/lunar day), and applies auto-regression to the low frequency envelope. This technique may also be used in the last resort to compensate for inadequate tide predictions.

References

AMOS, D. E. AND KOOPMANS, L. H. (1963) Tables of the distribution of the coefficient of coherence for stationary bivariate Gaussian processes. *Sandia Corporation Monograph* SCR–483, U.S.A.

MUNK, W. H. AND CARTWRIGHT, D. E. (1966) Tidal spectroscopy and prediction *Phil. Trans.* A **259**, 533–581.

ROSSITER, J. R. (1959) Research on methods of forecasting storm surges on the east and south coasts of Great Britain. *Quart. J.R. Met. Soc.* **85**, 262–277.

Part IV

WAVES

Editor's Comments on Papers 20 Through 24

The waves of the sea are perhaps its most immediately obvious feature, yet it is scarcely thirty years since the breakthrough occurred that enabled them to be studied systematically. Naval officers of the nineteenth century, among them Captain Owen Stanley (1848) sometimes attempted to measure waves, but the confused patterns of the sea surface made visual observations difficult. When sailors reported waves 20 m in height it was regarded as just another sailors' yarn (Deacon 1971, pp. 285–286).

Scientists such as Froude, Stokes, and John Scott Russell realized that the confusion of the sea surface was due to the presence of two or more wave trains interacting with each other but they were unable to suggest methods as to how to reduce the data. In the early twentieth century, studies of waves and of their effect on coastal formations were carried out by Vaughan Cornish (1934); but the problem of measuring waves remained.

The breakthrough came in World War II, under pressure of the need for waves forecasts to facilitate the landing of troops on occupied beaches. Professors N. F. Barber and F. Ursell, while working at the Admiralty Research Laboratory, Teddington, England, found a way of breaking down the data obtained from a wave recorder into the component wave trains, by applying spectral analysis (Papers 20, 21, and

22; and Barber and Ursell 1948). Their work showed that, as expected from classical wave theory, the waves generated during a storm traveled away from the area of disturbance at a speed proportional to the square root of their wave lengths. A continuous spectrum of wave trains was set up and since the longer waves traveled faster, they would be the first to arrive on a distant coast. From the lapse of time between the arrival of the longer and the shorter waves it was possible to calculate the position of the storm. By 1945 it became clear that some waves arriving on the coasts of Britain had been generated by storms in the vicinity of Cape Horn.

Work on wave prediction was carried out in Britain and by H. Sverdrup and Walter Munk at the Scripps Institution of Oceanography (Schlee 1973). This work has been continued since, with useful results for routing ships to avoid the worst wave conditions and for the construction of coastal works and offshore structures such as oil-drilling rigs. Wave analysis has also been employed to improve ship design.

Since World War II, apparatus of different kinds have been invented for measuring waves at sea, for example the ship-borne wave recorder and the clover-leaf buoy developed at the National Institute of Oceanography. Measurements over long periods have shown that waves above average height do occur when different wave trains momentarily coincide and that the sailors of former days may not have been so guilty of exaggeration as commonly supposed. Further work has been done on the dispersion of waves. In the early 1960s, a team of American scientists led by Munk measured waves at stations spanning the Pacific Ocean, one of which was FLIP (the Floating Instrument Platform). They showed that waves generated in a storm in the Southern Ocean traveled as far as Alaska. This work and trends of research on how the wind forms waves is summarized by Cartwright in Paper 24.

When waves enter shallow water, their behavior changes. They slow down in depths of less than half a wave length and the accumulating energy makes them get higher. The direction and force of waves on different types of coastline has an important effect on the cycle of erosion or beach-building observed in different places. As well as influencing the long-term development of the coastline, waves can be dangerous to the casual visitor to the beach. Unexpectedly high waves can come in without warning where components of different wave-trains have reinforced one another. Surf breaking on a beach can raise the water level inside the breakers and the result is a localized rip current that can carry a swimmer away, running out to sea through the waves. The origin of these currents is explained by Dr. F. P. Shepard in Paper 23.

In 1941 a French scientist, P. Bernard, showed that the very small oscillations, or microseisms, recorded on seismographs were the result of waves at sea. This had been suspected before (Darbyshire 1962) but no account could be given of the way in which energy was transfered to the sea bed. Bernard showed that the period of the microseisms was about half that of the waves. Longuet-Higgins (1950) later demonstrated that they were due to standing waves, produced in the ocean by conflicting wave trains and on coasts where part of the swell is reflected back from the shore.

While it is generally accepted that the waves commonly observed at sea are generated by the wind, the process by which this is done is not yet fully understood. There are other forms of long waves, due to various causes, which are not visible at sea but which can have disastrous effects on the ocean margins. Tidal waves, now better known as seismic surges or tsunamis, are caused by submarine earthquakes or landslides. Because they are long waves, they travel at a speed determined by the depth of the water. In the deep ocean the seismic surge therefore travels extremely fast but causes only an imperceptible oscillation in sea level. In shallow water the movement is retarded, the accumulating energy goes into piling up the water to tremendous heights and great devastation is caused in low-lying coastal areas. These surges are particularly experienced in seismically active areas such as the Pacific Ocean and warning systems have been set up there to minimize casualties. In other areas, such as the North Sea, the Bay of Bengal, and the Caribbean, greater danger comes from storm surges, long waves created by fierce storms over the sea. These cause a rise in sea level at the coast and are made worse by accompanying high winds and wind-generated waves.

In lakes or partially enclosed seas, variations in atmospheric pressure can give rise to oscillations which result in periodic small-scale rise and fall of the water level at the shore. Aimé (1845) was one of the first to observe these phenomena, known as seiches. Later in the nineteenth century, detailed observations were made in Lake Geneva by the Swiss geographer F. A. Forel and a mathematical explanation was provided by George Chrystal, professor of mathematics at the University of Edinburgh (Fairbridge 1966).

Wave motion is not confined to the surface of the sea and some waves exist entirely in the deeper layers, between layers of water of differing densities. In 1762, Benjamin Franklin observed internal waves in a lamp on board ship. A layer of oil lay on top of a layer of water and he noticed that though the ship's motion did not disturb the surface of the oil, the interface between the oil and the water was forming waves. This led him to investigate the old tale of stilling waves at sea by using oil. His first trial was made on a pond. He relates:

> At length being at CLAPHAM where there is, on the common, a large pond, which I observed to be one day very rough with the wind, I fetched out a cruet of oil, and dropt a little of it on the water. I saw it spread it-

> self with surprizing swiftness upon the surface; but the effect of smoothing the waves was not produced; for I had applied it first on the leeward side of the pond, where the waves were largest, and the wind drove my oil back upon the shore. I then went to the windward side, where they began to form; and there the oil, though not more than a tea spoonful, produced an instant calm over a space several yards square, which spread amazingly, and extended itself gradually till it reached the lee side, making all that quarter of the pond, perhaps half an acre, as smooth as a looking-glas. (Franklin, Brownrigg, and Farish 1774, p. 449)

He made a further trial at sea and collected accounts of similar instances, including the following:

> A gentleman from Rhode-island told me, it had been remarked that the harbour of Newport was ever smooth while any whaling vessels were in it; which probably arose from hence, that the blubber which they sometimes bring loose in the hold, or the leakage of their barrels, might afford some oil, to mix with that water, which from time to time they pump out to keep the vessel free, and that same oil might spread over the surface of the water in the harbour, and prevent the forming of any waves. (p. 452)

The modern study of internal waves was begun by V. W. Ekman (1906) who was asked by Nansen to account for the phenomenon of dead water, frequently encountered by ships in the Norwegian fjords and which the *Fram* had experienced in the Arctic. Ekman showed that where there were layers of relatively fresh water from melting ice overlying sea water, a wave could be generated in the lower layer by the movement of the boat and have the effect of retarding its motion through the water. Modern work on the natural causes and properties of internal waves show that they occur widely in the ocean (LaFond and Cox 1962) and that one may have been responsible for the unexplained loss of the U.S. submarine *Thresher* in 1963.

20

Reprinted from *Occ. Pap. Challenger Soc.* 1:1–13 (April 1946)

OCEAN WAVES AND SWELL

by G. E. R. Deacon, D.Sc., F.R.S.

In view of the interest and possible applications of the study of waves it is doubtful whether the subject has received sufficient attention. Although much has been done, there are many questions of primary importance which cannot be satisfactorily answered. Thorade, probably the most authoritative German oceanographer, maintained in 1931 that theory and observation had not gone sufficiently hand in hand. Mathematicians have been too much occupied with theoretical problems, while many observers, unable to follow the language of the mathematicians, have worked without such precise formulation of their problems as would direct their attention to the observations most likely to further the investigation.

Sverdrup, Johnson and Fleming (1942) make similar remarks. While paying tribute to the work of Gerstner, Cauchy and Poisson, Stokes, Kelvin, Rayleigh, Lamb and other mathematicians, they observe that theoretical investigations have preceded the accumulation of exact information as to the nature of the phenomena, and describe some of the early work as more mathematically beautiful than practically applicable. They make special mention of the work of Jeffreys, who has done much to bridge the gap between the theoretical worker and the more general student in his notes at the end of the book on Ocean Waves by Cornish (1934).

The difficulties encountered in the study of ocean waves are familiar in any oceanographical problem. The controlling factors cannot be varied at will, and the effect of each has to be discovered gradually by observing it under a wide range of conditions. Because of their magnitude and complexity they are difficult to imitate in the laboratory. Expense is another obstacle ; sooner or later one needs a ship and costly equipment.

Like many other fundamental studies, the investigation of waves was notably advanced during the war, because reliable information was needed for military purposes, and adequate facilities were made available for research. In this paper it is intended to summarize the information published before the war, and to indicate the advances that have been made, or those that will be made in the next few years.

Waves in a Storm Area.

There is some difficulty in explaining the exact mechanism by which the wind begins to make waves on undisturbed water ; in particular, how it gives rise to oscillatory movements instead of simply driving the water before it. It has been demonstrated theoretically that the boundary surface between two fluids moving with different velocities, one above the other, is unstable while it is level, so that the wind sweeping over the surface of the sea is likely to deform it, but it is not clear how a particular type of oscillation grows.

Two types of movement are involved in the motion of a wave, the most obvious being the forward movement of the wave form. The water itself does not move continuously ; observation of a floating mark shows that the small advance made with a wave-crest is followed by a return in the trough to almost the original position. Theory, and studies of artificial waves, have shown that the paths of the water particles in waves in deep water are approximately circles, but not exactly, because the forward movement at the crest is slightly greater than the backward movement in the trough. At the surface the diameter of the circular orbit is the same as the height of the wave, but the amplitude of the movement falls off rapidly with depth.

The growth of waves once they are formed is attributed mainly to the greater pressure of the wind on their windward slopes. Jeffreys, who demonstrated this mathematically, concluded that the effect of the frictional drag of the wind, a second factor which might be expected to contribute to the development and growth of waves, was negligible. Thorade (p. 48) argues, however, that the evidence is not sufficient, and Sverdrup, in work not yet published, maintains that the effect of frictional drag must be taken into account. If the pressure of the wind is the only factor, waves cannot travel faster than the wind which makes them, but if the effect of frictional drag is considerable, the maximum velocity of the waves may be slightly greater than that of the wind. The evidence on this question is conflicting, and more observations are necessary to decide it ; Thorade emphasizes that in making such observations due regard must be paid to the sharp decrease in wind velocity which is known to exist very close to the surface.

The rate at which waves increase in height in a steady wind is not uniform, the growth being rapid at first and then slower. For a particular wave it has been shown that the limit is probably reached when the height grows to about 1/7 of the wave-length. When the wave is as steep as this the crest becomes a sharp ridge and curls over. Thus short waves will soon break and cover the sea with whitecaps, but longer waves, which can build up to a greater height without becoming unstable, continue to grow. Theoretically there is an upper limit when the waves travel as fast as the wind, or slightly faster if the effect of frictional drag is considerable ; this maximum velocity fixes a maximum wave-length. It is not known exactly how the long waves develop ; Jeffreys concludes that the tendency for them to predominate when a strong wind has been blowing for a long time is due to their being able to store more energy than shorter waves.

It is generally agreed that the most striking feature of waves in a storm area is their irregularity, and the predominance of short-crested waves (short ridges of water in comparison with the long ridge of a long-crested wave). Jeffreys (p. 138) has remarked that they may be the resultant of long-crested waves crossing at an angle " so that the whole of the waves produced by a wind can be regarded as a combination of swells of different lengths, whose directions are grouped about that of the wind, but not coinciding with it." He also believes that the long waves which predominate after a strong wind has been blowing for a long time tend to be long-crested, and sometimes, as recorded by Cornish (p. 3) one sees " a magnificent procession of storm waves sweeping across the sea from horizon to horizon."

When the wind falls, or the waves travel out of the storm area, they assume a rounder form and greater regularity, changing from waves to swell. The short and short-crested waves decay most rapidly and long-crested swells tend to predominate. One reason suggested by Jeffreys is that the short-crested waves will tend to separate into two long-crested parts, which will not go to the same distant observation point.

The Form of Deep Sea Waves.

The profile of a wave or swell has been found to approximate to the shape of a trochoid, the curve traced by a point on the spoke of a wheel imagined to roll along the underside of a level surface. Such a curve can vary from the limiting form of the straight line traced by the centre of the wheel to the sharp-crested cycloid traced by a point on the circumference, but it has been shown that waves depart from the shape of a trochoid before they reach such a sharp-crested form. Michell has calculated that a wave becomes unstable when the angle between the forward and rear slopes of the crest is 120°, and the ratio of height to length 1/7. In practice there is some evidence that the limit is reached sooner.

When the amplitude of the wave is small the trochoidal shape approximates to that of a sine curve, and using this simplification the speed, length and period of the wave are related by the formulae

$$\text{Speed in knots} = 3.1 \times \text{Period}$$
$$\text{Length in feet} = 5.1 \times (\text{Period})^2$$

where the period is measured in seconds. By means of these formulae, measurements of one of the variables can be used to calculate the other two, and if all three are measured the adequacy of the simple theoretical treatment can be tested. Such an investigation made by Krümmel (1911), using wave observations from many deep-water areas, showed that if reasonable allowance was made for variations in the reliability of the data, the agreement between theory and observation is very satisfactory. The comparison also suggested that the length of a wave is the most difficult of the three characteristics to estimate.

Forecasting Waves and Swell.

The only systematic swell forecasts made before the war were those of the French swell-prediction service on the coast of French Morocco, where heavy swell often makes communication between anchored vessels and the shore difficult, and sometimes impossible. With the help of observers stationed in the Azores, and on the coast of Portugal, and with some reports from ships, they were able to trace an outstanding swell on the coast of Morocco to its origin, generally in a depression moving across the North Atlantic Ocean, sometimes as far north as Iceland. Various types of depression were classified, and with the help of a number of rules, including an allowance for the effect of favourable or contrary winds in the path of the swell, they were able to make useful predictions of the state of the anchorages. The foundation of the predictions was mainly the reports from the Azores.

The wartime requirement was more comprehensive, being in general terms the prediction of waves and swell from meteorological data and forecasts. It was met by the Naval Meteorological Service, who made the best possible use of all previous data and conclusions, filling the gaps as quickly as possible with new observations. The empirical formulae relating wave-height, wind-strength and other characteristics, put forward during the past 100 years proved inadequate, but useful tables were compiled, and the first instructions for predicting waves and swell were issued in 1942. Similar work was undertaken in the United States with the active help of the Scripps and Woods Hole Oceanographical Institutions.

Detailed agreement was not reached, but the Admiralty Swell Forecasting Section, staffed with both United States and British officers, was able to make reasonably accurate forecasts when they had accurate meteorological data and forecasts. The starting point of a prediction is a meteorological chart on which one has to distinguish **generating areas** in which a strong wind blows towards the scene of the prediction, or in a direction inclined at not more than 30° to the direct line. The average wind speed in the generating area is estimated from the isobar spacing, some allowance being made for the lesser speeds at the beginning and end of the storm, when the wind was rising and falling. The length of the generating area, known as the **fetch,** which may be bounded by a coastline or change in wind speed and direction, is also important. The effect of the wind also depends on its **duration,** which includes an allowance for the waves already present when the wind rises. If the area for which the prediction is required is exposed to ocean swell, detailed information is needed from distant storm areas from which swell may have been travelling towards the area for several days, and an allowance has to be made for the effect of favourable or contrary winds during this long journey. If the prediction is to include an estimate of breakers and surf some general information is required of the depth contours and beach slope.

Using tables and graphs showing the effect of the different factors the Swell Forecasting Section began by making experimental predictions for the English coast, and then routine predictions for the Normandy beaches and approaches. A review of their work, when the landings were complete, showed that when the wind forecasts were correct to within 1 unit of the Beaufort Scale and 2 points of the compass 85 per cent. of the predicted wave-heights were correct to within 1 foot for waves less than 5 feet high, and 2 feet for waves higher than 5 feet. Such agreement is remarkable, and it can probably be attributed to some extent to the detailed experience gained by the forecasters in their limited area, which allowed them to modify the rules every now and then to suit their own ideas. Their task was made somewhat easier by the absence of swell from the Atlantic Ocean, partly due to the forecasts being limited to the summer months, and partly because the beaches were screened from the west.

On a coast which is exposed to ocean swell the effect of the swell from storms, even several thousands of miles distant, may be more important than the effect of local winds. An approximate rule formulated by Commander Suthons of the Naval Meteorological Service states that swell loses 1/3 of its height in travelling a distance which, measured in nautical miles, has the same numerical value as the wave-length measured in feet. Swells longer than 1,000 feet are common, and these will lose only 1/3 of their height in travelling 1,000 miles. The famous rollers of Ascension and St. Helena are very spectacular, and have caused great damage, in spite of the fact that they have travelled at least 3,000-4,000 miles.

As the swell enters coastal waters it undergoes more striking modifications. Its velocity becomes noticeably reduced at soon as the depth of water is less than a quarter of a wave-length, and if it approaches the coast at an angle this change in velocity causes it to turn more or less parallel to the shore. The phenomenon is very similar to the refraction of light or sound waves, and the swell is made to bend round headlands, and into bays according to similar laws. The height of the swell also changes as it moves into shallow water, a small decrease being followed by steeper and steeper growth, till the crest rises sharply above the general water level just before the wave breaks.

The height of the breakers on the beach depends on the height and length in deep water, and to some extent on refraction. A low long swell will rise to about twice its deep-water height, while a short steep wave will not perceptibly increase. Refraction may decrease the height of a wave by increasing the length of its crest, spreading the energy over a broader front. If part of a coastline and the adjacent sea-bottom contours are straight, the waves will be more or less the same height at all points along it ; but if the bottom contours are irregular the waves will be more refracted in some places than others, and although the coast is straight the waves may be twice as high at one part of the coast as at another. They tend to be focussed where a shoal extends out below the water, and dispersed, to give quieter conditions, above a submarine gully. The effect can be calculated if the bottom topography is fairly well known. Special problems arise where there are steep slopes or vertical walls.

The depth of water in which a wave breaks is generally between 3/2 and 2/3 of the wave-height. The movement of the water in the wave is then mainly forward, and the undertow and rip-currents associated with them are not fully understood.

Measurements of Waves.

Until recently measurements of waves have been made almost entirely by visual observations. In deep water the wave period is measured by timing the successive appearances of a patch of foam on the wave crests. Wave-length is estimated by comparing the distance between wave crests with the length of the ship, or by using a small float paid out astern on a measuring line ; and velocity by measuring the time taken by a wave to travel a measured distance along the ship's side. Estimation of height is the most difficult, the best method being to find some height above water level at which the wave crest can be seen in line with the horizon. All the measurements are made more easily when the ship is hove-to. The chief difficulty in devising any kind of recording instrument is that there is no stationary surface from which to start the measurements. Use has to be made of the rapid decrease in wave motion with depth ; Froude was the first to measure waves against a graduated pole, held upright and steady by a long wire which anchored it to a large drogue sunk well below the surface. Stereophotography has also been used, but it is a most laborious task to measure one picture.

From the shore the task is easier, wave heights can be measured against poles driven into the bottom or with a telescope and graticule. In the United States a float recorder installed at the end of the Scripps Institution Pier was used, and the waves were also photographed against the background of the pier.

More ambitious recording instruments, allowing routine measurements in exposed sites up to a mile or more from the coast, were ready in this country for use before the invasion of Normandy. A sensitive pressure-measuring instrument was placed on the sea bottom, and the pressure fluctuations below the waves were recorded electrically in a hut on shore through a cable. The pressure fluctuations depend on the wave-length and the depth of water as well as the wave-height, but the corresponding wave heights can be calculated from the record with reasonable accuracy. They were laid by scientists attached to the Admiralty's Mine Design Department, several of whom, notably C. H. Mortimer, continued to develop wave-recording techniques as part of their work when they joined a newly formed oceanographical group in the Admiralty Research Laboratory in the summer of 1944. The next advance was to use echo-sounding wave-recorders which stand on the sea bottom, pointing upwards, recording echo-profiles of the surface. The idea was not a new one, having been used with apparatus supplied by Messrs. Hughes to measure tides and waves at Dover in 1939, but considerable development was necessary before a more or less permanent installation could be laid at the end of a long cable. Two types are now available, a permanent-station type made by the Anti-Submarine Establishment, and a semi-portable type made in the Admiralty Research Laboratory.

Interference between Waves.

As often happens in the investigation of a natural phenomenon, better observations emphasize the complexity of the phenomenon, and then point the way to further, more satisfactory investigation. In this instance the new wave records emphasized the widespread interference between waves of different lengths. A clear example was seen by a small party making visual observations and wave-recordings off the north coast of Cornwall in July 1944. Three wave recorders laid at distances of approximately 1,000, 2,000 and 3,000 yards were used. It would have been very convenient if a particular wave could have been identified and measured as it passed each of the recorders in turn, but this was never possible either with single waves or with groups of waves. New crests formed and old ones disappeared between each pair of recorders.

The explanation is that the observations were not being made on a simple train of waves but on the combination of a number of waves of different lengths. The observed pressure fluctuations were the resultant of the pressure fluctuations in these different waves which sometimes add and sometimes subtract, and the actual appearance of the surface was the resultant of their combined vertical displacements. In such a combination the relation of the individual waves to each other, and therefore the surface profile, varies with distance, because the individual waves travel with different velocities. If one could follow a surface wave (crest to crest) it would be seen to be continually changing its length, period and velocity. When one has such an opportunity it is often seen that a wave crest cannot be followed very far before it loses its identity among other crests which rise in its place.

Similar considerations explain why waves often travel in groups, a sequence of high waves being followed by a sequence of low waves. This is often noticeable in a ship, two or three violent movements being followed by a quieter period. The intervals between the prominent groups depend on the range of wave-lengths present, so that it is not necessarily the fourth, fifth, seventh or tenth wave which is always the highest. Generally there are so many wave periods present that although some repetition is obvious the intervals are very irregular. Certain geographical regions may be found to have fairly characteristic rhythms, because the range of periods present depends on the intensity, size and distance of the storm, which may often be more or less the same for those regions.

The observations and wave-recordings were examined on the spot by Barber, Ursell and Darbyshire, whose names will go down in the history of the subject. Mathematicians and observers had come together, and although statistical analysis, using crest to crest measurements, was given a thorough trial they soon reached the conclusion that no fundamental advance could be made in the study of waves till a method was developed by which a complex wave-record could be resolved into its individual waves, in sufficient detail to allow the amplitude of each wave-length present to be measured.

One immediate benefit of such analysis would be the possibility of recognizing the low long swell which travels fastest and is the first to arrive from a distant storm, but on its arrival cannot usually be detected because it is obscured from visual observation by the shorter and higher waves caused by local winds. An interesting example was seen in Harlyn Bay, where a 14-second swell could be seen in a quiet cove sheltered from the west wind, but not among the short waves in every other part of the bay.

Frequency Analysis of Waves.

The first step in the separation of different wave lengths was to obtain the wave record in black and white as in Fig. 1. This was done by arranging the recording system to move a line of light backwards and forwards across the photographic recording paper, instead of the usual spot of light. At the same time a time-scale was printed along the edge of the record by interrupting a fixed lamp every 20 seconds.

Twenty minutes was chosen as the most suitable length of wave record for analysis, and this length of record, about 8 feet long, is fastened round the circumference of a flywheel which carries it past an optical system in which a steady light shines across the record through a narrow slit. The light which is scattered

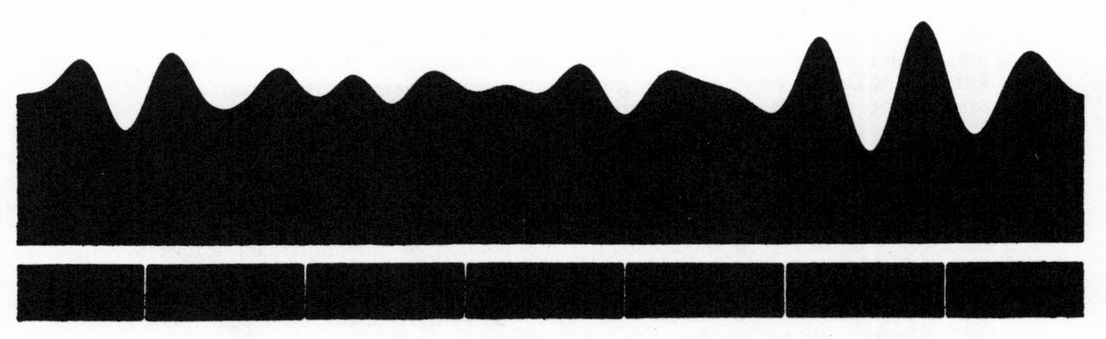

Fig. 1.—A pressure " wave record." Time scale: 20 seconds between white marks. Pressure scale: the largest fluctuation near right edge of record corresponds to three feet of water. The measurements were made at a depth of 100 feet.

back from the record is measured by photocells, and as the wheel rotates the electrical output of these photocells varies in accordance with the width of the white parts of the record passing the slit. In this way the wave profile is measured as a varying electrical current, the frequency of the variations being multiplied by the rotation of the wheel, one of the advantages of this multiplication being to bring the frequencies into a range for which a simple resonant selective system could be developed.

One of the simplest methods used to examine any particular wave-length in the record is to use the resonance of a tuning fork built into a system which picks out a certain frequency (in this case 128 cycles a second) which corresponds to a long wave length on the record when the wheel is rotating fast, and to a short one when the wheel is rotating slowly. In operation the wheel is turned by hand to about 4 revolutions a second, and then allowed to slow down under its own friction. As the speed of the wheel decays the tuning-fork system examines each wave-length in turn from the longest to the shortest and measures their amplitudes. The output is amplified and recorded on a moving-paper pen recorder, the result being a graph of the amplitudes of successive wave periods. There is no difficulty in identifying the wave periods in the analysis, because a period scale is written on the analysis by a second pen worked by a second analysing system working off the known, 20-second, time marks along the edge of the wave record.

Several ideas incorporated in the analyser are not new ; the great advance is the use of the continuously decaying speed of the wheel. With a receiving circuit of the right selectivity, an analysis can be made with more than sufficient detail in a reasonable time. The present models analyse a 20-minute record in about 10 minutes ; the record has to be put on and taken off, but the rest of the operation is automatic.

A good indication of the advantages that can be obtained by the frequency analysis of ocean waves is afforded by the analysis of 20-minute wave records taken automatically every 2 hours between 14 March and 19 March 1945, by a pressure recorder lying in 17 fathoms, 4,000 feet N.W. of Pendeen lighthouse near Land's End.

Early on 14 March calm weather extended over a wide area west of the British Isles, and the wave analysis showed only small waves and swell with a maximum period of about 14 seconds (Fig. 2). During the afternoon a much

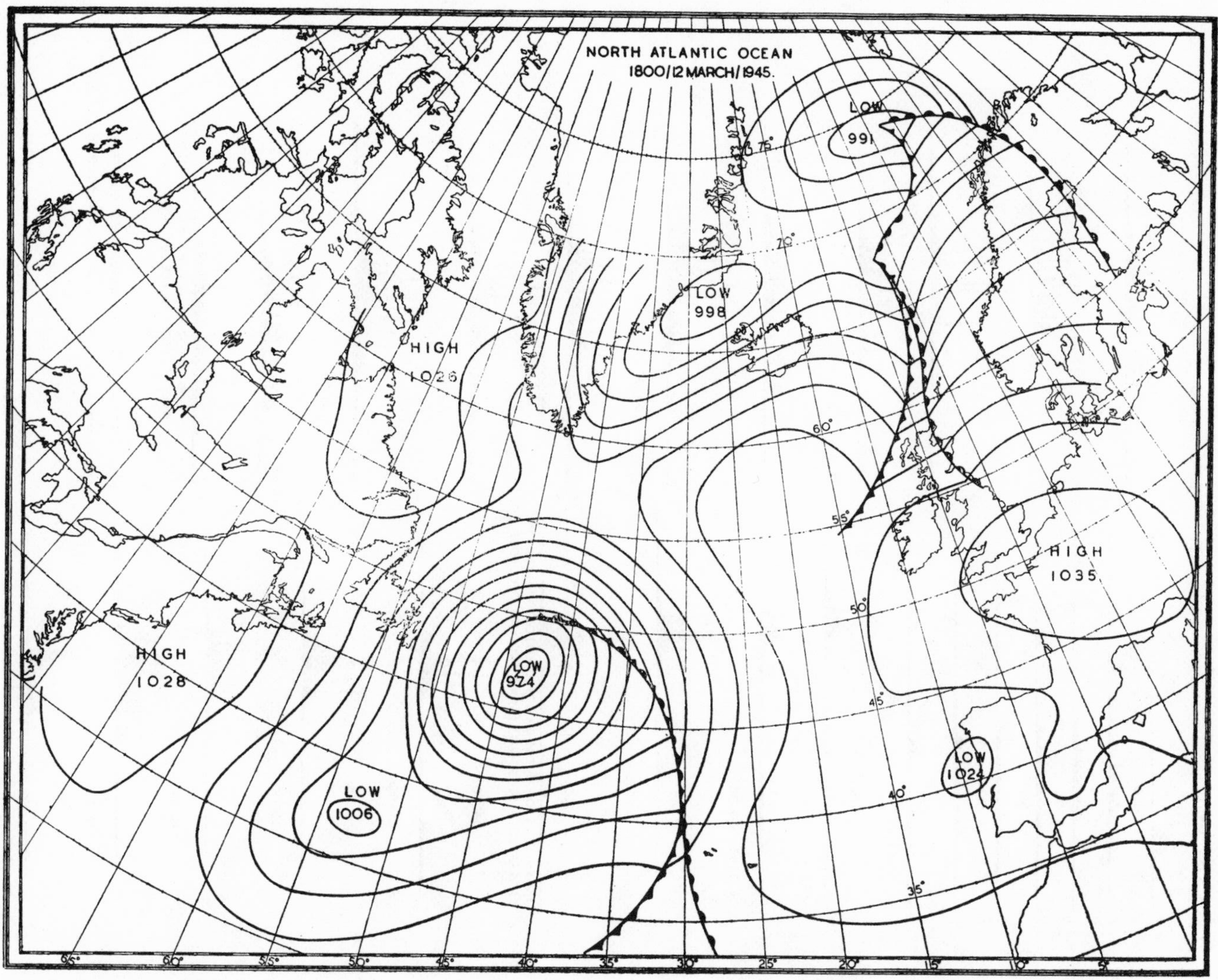

Fig. 3.—Meteorological chart for North Atlantic Ocean at 18.00 hrs. 12 March 1945. The isobar spacing is 4 millibars.

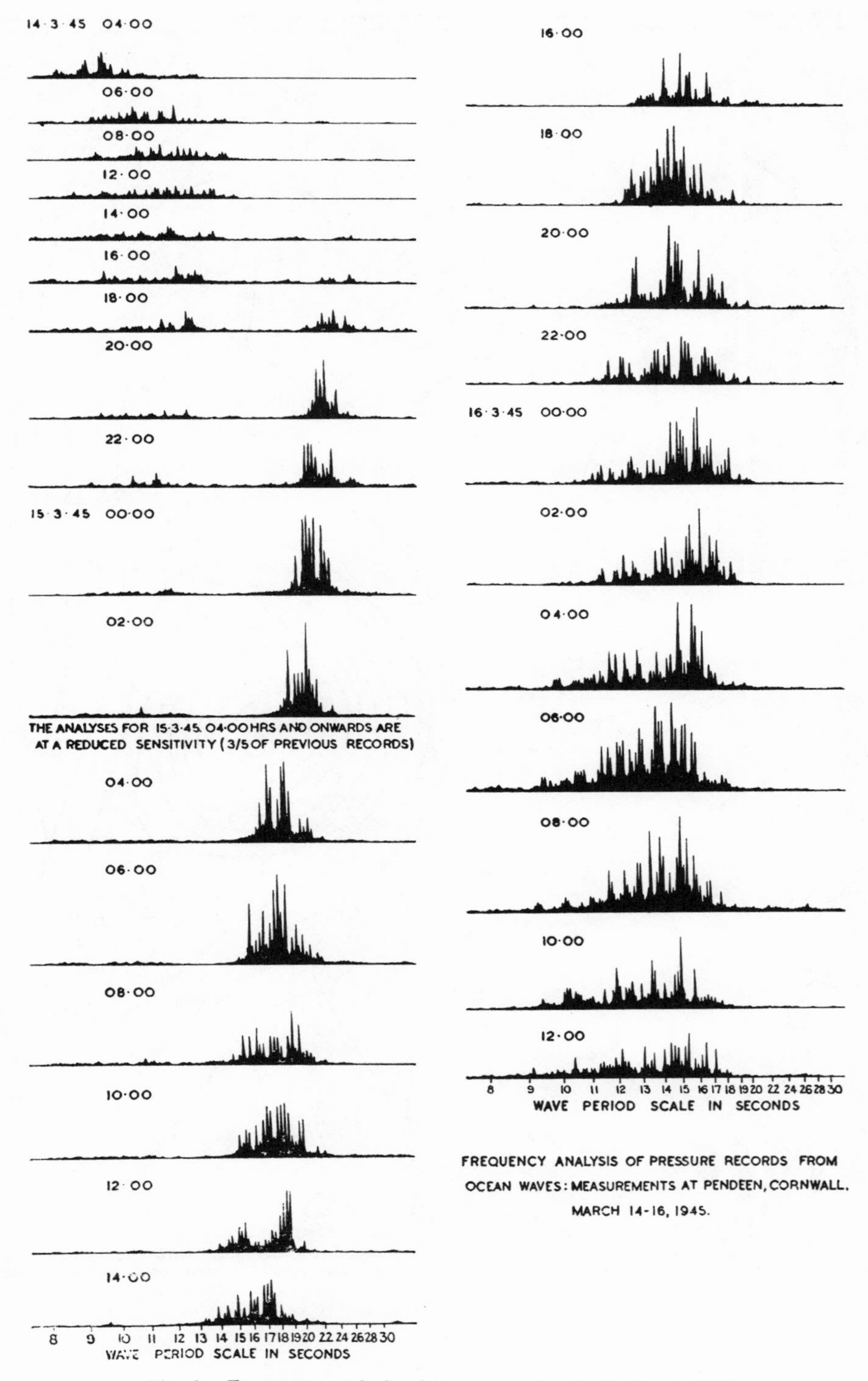

Fig. 2.—Frequency analysis of wave records, 14-16 March 1945.

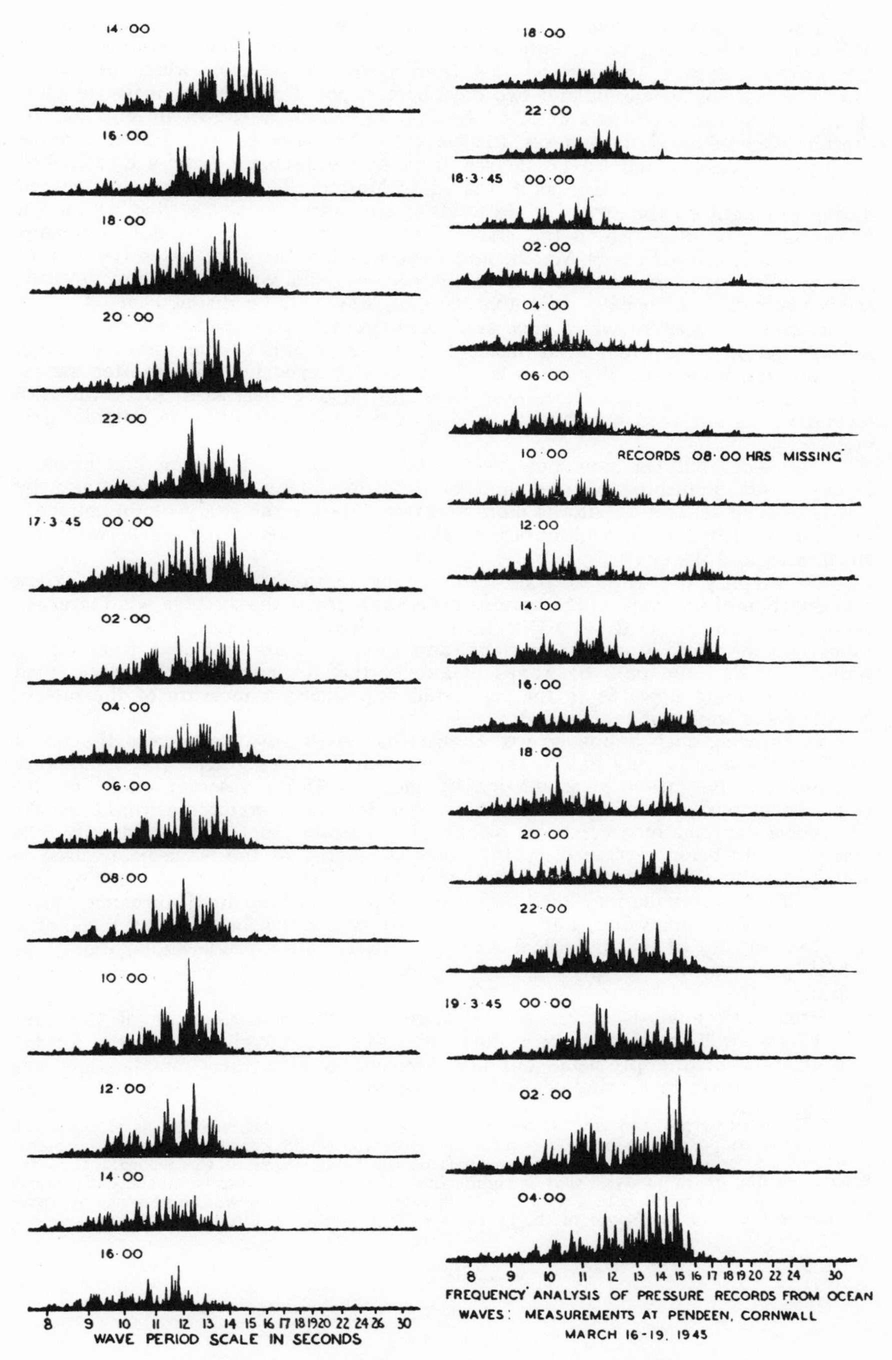

Fig. 4.—Frequency analysis of wave records, 17-19 March 1945.

longer swell began to arrive and the analysis showed periods of 20 to 24 seconds. This was the low long swell, only a few inches high when first measured, which has travelled fastest and arrived first from a deep depression which developed 500 miles S.E. of Newfoundland two days before, on 12 March. A meteorological chart showing the conditions at 18.00 hrs. on 12 March is shown in Fig. 3. As shorter swell followed the longest, the wave band broadened and moved towards the shorter periods, and by 17 March (Fig. 4) the sea near Land's End looked as if it might become as calm as it was on 14 March. This trend was interrupted during the night by the arrival of 18 to 20 second waves from a second depression following 2,000 miles behind the first. The new wave band was not as narrow as that which arrived on 14 March, and there was not the same extensive area of calm weather over the eastern half of the ocean. The same broadening towards shorter periods and increase in amplitude can, however, be distinguished.*

During the year for which such analyses have now been made, waves of 20-24 seconds period have often been detected, but the periods of the waves breaking on the beach were generally much less because of interference with shorter waves. The longest waves to have been seen breaking on our coast seem to be the 22.5 second waves observed by Cornish (p. 14) at Bournemouth after an Atlantic gale in 1899.

Sufficient analyses have now been made to show that the method offers a means of distinguishing between the waves from different storm areas, and by virtue of such separation affords more accurate data for the study of the influence of wind velocity, fetch and duration, and the subsequent changes between the storm area and the coast.

To simplify the work diagrams were prepared to show the intensity, extent and duration of the wind in the various generating areas, the average wind strength being plotted at intervals of 100 miles every 6 hours. Distance from 0 to 2,800 miles was measured from left to right, and time in hours and days from top to bottom, so that the path of waves of any particular period can be represented by a line sloping upwards to the right, the slope being a measure of the rate of travel of the waves of that period.

In drawing such a line it was known that swell advances across the ocean at a speed which is only half that of its individual waves. The rate of advance depends on the rate of transmission of energy. If an observer could fix his attention on waves passing from a storm area into calm water, he would see the first wave continually decrease in height till it became too small to identify, the loss of height being explained by the potential energy of the wave being used to supply kinetic energy to the undisturbed water. The second wave, moving into water already in oscillation, loses height less rapidly so that the disturbance moves ahead into the calm water, but with a velocity which has been shown to be only half the velocity of the individual waves. This is only a crude explanation ; the actual problem, in the presence of waves of different periods, appears very difficult.

Frequency analysis which allows more accurate measurement of the time taken by a swell to travel from a storm area to the coast gives new data for the investigation of the problem, but also emphasizes a further complication, the probable increase in length, velocity and period of swell after it leaves the storm

* The separate peaks in any of the frequency analyses must be regarded as an approach to a Fourier Amplitude analysis, in which the spectrum would consist of a set of isolated ordinates at periodicities which are harmonics of the total length of the record. It is the general outline of the analysis that is significant ; a group which stands substantially above the background, or appears in a number of consecutive analyses, shows the presence of waves of that period, and the height of the group is related to the amplitude of the waves.

area. The theoretical reasons for assuming that there is such an increase appear to be questionable, but there is accumulating evidence in support of it. Thorade (p. 65) shows that wave-periods reported from storm areas are consistently shorter than those reported from distant coasts on which the swell breaks, and although this may be due to greater interference with short waves in the storm area it cannot be ignored. He also quotes an example, based on reports from ships, of the rate of advance of swell increasing with distance from a storm area. The analyses at Pendeen have also given a strong indication that longer wave-lengths are recorded in swell from distant storms than from local storms of comparable intensity. The measurements of the time taken by swell to travel from distant storm areas are not conclusive since it is difficult to decide the starting time within 6 hours or so, but generally there is better correlation between the wave analyses and wind charts when a small increase in velocity with distance is assumed. The increase is an essential part of the United States forecasting technique.

One of the outstanding requirements is the frequency analysis of waves in a storm area and frequency analysis of the resulting swell when it arrives at the distant recording station, and it is hoped to make such analyses in mid-Atlantic and in Cornwall before the summer. Eventually it may be possible to have a network of stations, extending as far as Ascension and St. Helena. Observers in such places exposed to heavy swell could be very useful in making swell predictions. Another outstanding requirement is equipment to measure the direction of swell, to study the interference between waves from different directions; this also is being developed. Most of the apparatus in use or in prospect so far is meant for research purposes, and to obtain all the necessary detail it is elaborate, but efforts are being made to find simpler methods that will encourage more widespread observation, perhaps with some sacrifice of detail. For example it may be found that sufficient information can be obtained about wave periods in different parts of the ocean by recording the movements of certain types of vessels. Preliminary experiments with a trawler have shown that useful information about wave periods can be obtained from the frequency analysis of the pitching of the ship.

This account of waves and swell has dealt mainly with the more or less fundamental aspects to which the attention of the oceanographical group at the Admiralty Research Laboratory has been directed. It leaves unmentioned much work on forecasting, breakwaters, beaches, waves in tidal streams, and other problems. A fair summary of the present position is that wartime requirements have attracted much attention to the subject, and methods of recording and analysis have been developed, or are within reach, which should allow considerable advances to be made within the next few years.

In conclusion, sincere acknowledgment is made to earlier and contemporary workers on the subject, to whom it is difficult to do justice in a short paper, especially at a time like this when there is so much work that is not published. We have had very useful discussions with Commander C. T. Suthons of the Naval Meteorological Service, Professor H. U. Sverdrup of the Scripps Institution of Oceanography, and Dr. T. F. Gaskell, once secretary of the Beach Reconnaissance Committee. The extension of wave recording and analysis could not have been done without the close support given to oceanographical research by those responsible for research in the Admiralty, particularly Dr. C. S. Wright, Dr. J. E. Keyston and Dr. A. B. Wood. I have said little about the individual efforts of members of the oceanographical group, but they will soon publish their work in detail.

REFERENCES.

Cornish, V., 1934. Ocean Waves. Cambridge.

Jeffreys, H., 1934. Additional notes (in Cornish, Ocean Waves). Cambridge.

Krümmel, O., 1911. Handbuch der Ozeanographie. Stuttgart.

Thorade, H., 1931. Probleme der Wasserwellen. Hamburg.

Sverdrup, H. U., Johnson, M. W., Fleming, R. H., 1942. The Oceans. New York.

21

Reprinted from *Nature* **158**(4010):329–332 (Sept. 7, 1946)

A FREQUENCY ANALYSER USED IN THE STUDY OF OCEAN WAVES

By N. F. BARBER, F. URSELL, J. DARBYSHIRE
AND
M. J. TUCKER
Admiralty Research Laboratory, Teddington

A WAVE-ANALYSER was developed at the Admiralty Research Laboratory, Teddington, in 1944 in order to analyse ocean waves and swell and ship movement. The apparatus has been in regular use since February 1945 drawing the frequency spectra of records of wave motion taken near Lands End.

These records of water pressure or depth are taken continuously for 20 minutes, and appear in the form of a black trace of variable width on white photographic paper. Fig. 1 shows a short length of record. On one side of the record is a time trace consisting of a black strip interrupted every 20 sec. By attaching the paper record to the outside of the rotating wheel in Fig. 2, photocells, illuminated by the reflected light from a narrow light beam falling on the record, give a fluctuating electrical output which is a repetition at high speed of the fluctuating trace on the record.

This electrical output is amplified and made to drive a vibration galvanometer. By allowing the speed of the rotating wheel to decrease slowly, the vibration galvanometer is caused to resonate in turn with the various component wave-lengths on the original record. Thus the vibration galvanometer of natural frequency 120 c. per. sec. resonates with the output of a wave-length 1/30 of the periphery of the wheel when the wheel is turning at 4 rev. per sec., but resonates with the output from a wave-length 1/40 of the periphery of the wheel when the speed of the wheel has fallen to 3 rev. per sec. Regarding the record as being compounded of its Fourier harmonics, each having a whole number of wave-lengths on the periphery of the wheel, one can see that provided the vibration galvanometer is sharply tuned and that the speed of the wheel decreases very slowly, the vibration galvanometer will show individual resonances to each Fourier component.

The motion of the galvanometer is detected photo-electrically and the output is amplified, rectified and made to drive a pen recorder the deflexion of which at

Fig. 1. A WAVE-PRESSURE RECORD

Fig. 2. THE FREQUENCY ANALYSER

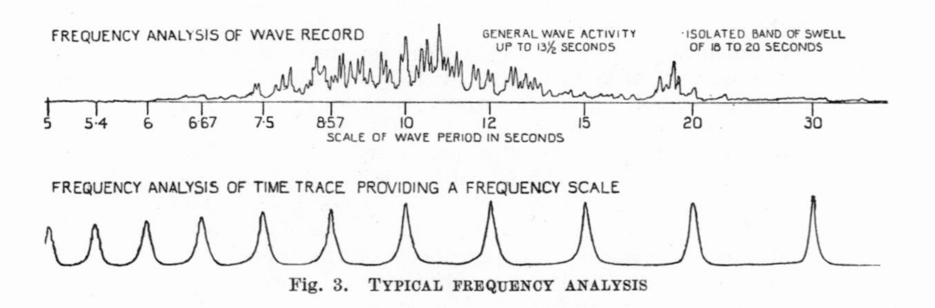

Fig. 3. TYPICAL FREQUENCY ANALYSIS

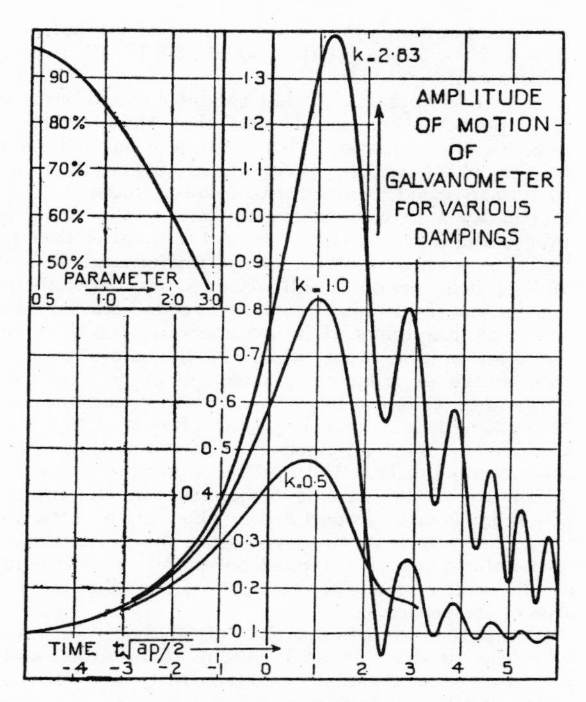

Fig. 4. AMPLITUDE OF OSCILLATION OF GALVANOMETER FOR VARIOUS DAMPINGS

any instant is, therefore, a measure of the amplitude of vibration of the galvanometer. The most recent model uses a vibration galvanometer with an electrical output, developed by G. Collins. The sample pen record shown in Fig. 3 is the analysis of the record of which Fig. 1 is a portion.

A frequency scale is drawn automatically by a second pen giving the lower trace in Fig. 3. This pen is actuated by a second channel working photo-electrically from the time trace and using a resonant filter tuned to 360 cycles. The second pen, therefore, draws a frequency analysis of the time trace, and this consists of a series of isolated peaks which can be used to interpolate a scale of frequency for the trace of the first pen. Because of the 3 : 1 ratio in the resonant frequencies of the two channels, these

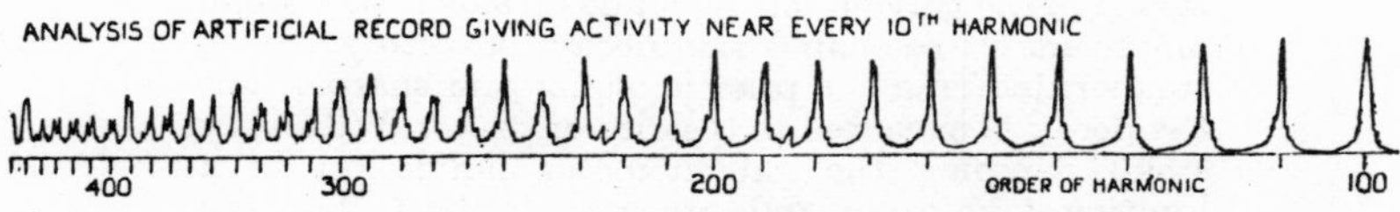

Fig. 5. ANALYSIS OF ARTIFICIAL RECORD SHOWING RESOLUTION OF FREQUENCY BANDS

peaks are equivalent to wave-periods of submultiples of 3 × 20 or 60 sec., that is, 60, 30, 20, 15, 12, 10 sec., and so on.

It will be appreciated that the mechanical parts of the apparatus are simple and that the process of analysis is automatic. There is no mechanical drive to the wheel, which, having been turned by hand to its top speed, continues to revolve under its own inertia at a slowly decreasing speed, the analysis proceeding automatically. In the apparatus already built, the wheel is 30 in. in diameter and weighs 70 lb.; it is carried in ball bearings, and takes about 4 minutes to decrease to half speed. As for the electronic amplifiers, it is not necessary for them to have an amplification which is the same over a wide range of frequency, since the only electrical frequencies that are important are in a narrow belt near 120 c. per sec. It is important, however, that the amplifiers should be linear in the sense that they produce no spurious 120 cycles coming from sum or difference of the various frequencies in the input. Linearity is also important in the optical pick-up from the record, in the sense that the illumination of the photo-electric cell must be strictly proportional to the width of the white part of the illuminated area of the record.

In practice, it is found that an analysis covering four octaves takes place in about 16 minutes, and that an operator can deal conveniently with about fifteen analyses each day.

The analysis approximates to a Fourier amplitude analysis, and it has been found possible to determine theoretically the optimum characteristics of the apparatus. Each of the Fourier components of the record produces an electrical component the frequency of which is slowly gliding as the wheel decreases in speed. If the speed of the wheel decreases at a rate $\exp - at$ and the vibration galvanometer has a natural frequency $p/2\pi$ and a natural rate of decay of free oscillations of $\exp - pbt$, then it can be shown that the manner in which the oscillations of the galvanometer build up and decay in amplitude as the gliding tone passes through resonance is determined by a parameter k, where $k = \sqrt{a/2pb^2}$.

Fig. 4 shows the response curves of the galvanometer for various values of k. They illustrate in particular the effect of changing the damping of the galvanometer without changing the rate of decay in speed of the wheel. With fairly large damping, $k = 0 \cdot 5$, the galvanometer builds up slowly to a small amplitude of resonance and decays smoothly. With smaller damping the peak is higher and sharper, but the decay is executed in a series of beats. With very small damping the galvanometer builds up to a limiting amplitude and proceeds to beat, but the time of decay of the motion is very long. Taking

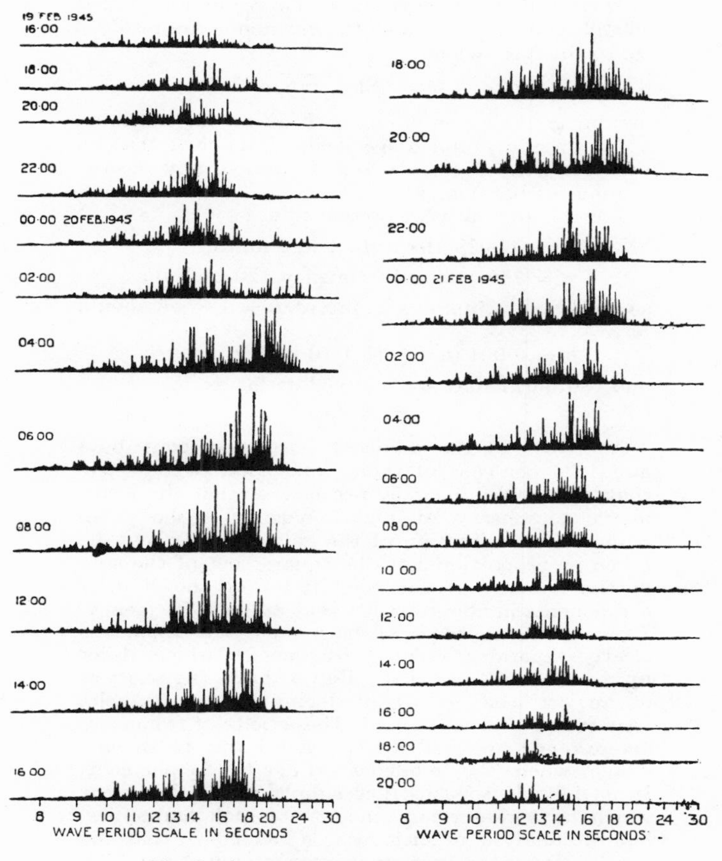

Fig. 6. A SERIES OF WAVE-PRESSURE SPECTRA

the effective width of the response as the interval in which the response exceeds 1/10 of its maximum value, it is clear that there is an optimum damping at which the width of the response curve is least. This is approximately at

$$k \text{ (optimum)} = 1{\cdot}8.$$

This optimum value of k gives the greatest resolution of the Fourier components. When the galvanometer is giving its peak response to one Fourier component, it is being slightly affected by adjacent components the gliding tones of which have either not yet reached the natural frequency of the galvanometer or have passed through it. If we consider the components to be adequately resolved when the contribution from each adjacent harmonic is less than 10 per cent of the peak response to that harmonic, it is possible to show from the curves of Fig. 4 that in any given apparatus all the harmonics are resolved up to the Nth, where

$$N = 0{\cdot}12\sqrt{p/a},$$

assuming that the damping b is at its optimum value for the p and a specified. It is clear that an analyser can be constructed to resolve any desired number of harmonics.

For the apparatus at present in use

$$a = 0{\cdot}0028 \text{ (decay to } \tfrac{1}{2} \text{ in 4 min.)}$$

$$p = 750 \text{ (natural frequency 120 cycles) ;}$$

so that for optimum working at $k = 1{\cdot}8$ we should have

$$b = 0{\cdot}001 \text{ (decay to 1/10 in 3 sec.),}$$

and the harmonics are resolved as far as

$$N = 60.$$

At $N = 120$ the adjacent harmonics contribute about 25 per cent of their peaks, and at $N = 240$ they contribute about 50 per cent, so that the peaks merge together. At higher harmonics the mean amplitude of vibration of the galvanometer may be taken as proportional to the square root of the sum of the squares of the amplitudes of the Fourier components in about a 1 per cent range of frequency. Even at high orders of harmonics the apparatus clearly separates isolated frequencies which differ by more than 3 per cent. Fig. 5 shows the analysis of an artificial record producing frequency belts near every 10th harmonic. These belts of frequency are resolved up to about the 400th and 410th harmonic, where the frequencies differ by 2½ per cent. It is difficult to construct simple artificial records which have prescribed amounts of high harmonics, but the analysis of such records has shown that the amplitudes of the Fourier components up to the 60th are correct to 5 per cent ; this error might be increased to 10 per cent, when a number of adjoining frequencies are present.

A wheel with mechanical drive and variable, controlled, exponential rate of decay, designed by F. E. Pierce, is being constructed in the workshops at the Admiralty Research Laboratory. With this wheel, which can be rotated up to 10 revolutions a second, more favourable characteristics can be chosen for damping and natural frequency, to allow greater resolution in an analysis taking the same time. Complete instruments are being made by Messrs. H. Tinsley & Co. Ltd.

The propagation of waves away from storm areas has been investigated with this analysis. Rules have been found which will allow improvement of methods of forecasting swell, a subject of interest to harbour and shipping authorities.

Fig. 6 shows a series of Fourier amplitude spectra of pressure at the bottom of the sea at a point off the Cornish coast. It is clear that there is a general trend in these analyses ; it will be shown elsewhere that this is consistent with classical hydrodynamicla theory. It is expected that rapid progress will continue to be made, particularly after a network of recording stations has been established. A general account of the problem has been published by Deacon in "Ocean Waves and Swell", Occasional Publications of the Challenger Society, No. 1, April, 1946, pages 1–13 (see *Nature*, 157, 165 ; 1946).

We are indebted to the Board of Admiralty for permission to publish this article.

22

Reprinted from *Nature* **159**:205 (Feb. 8, 1947)

Study of Ocean Swell

THE apparatus for the analysis of waves already described in *Nature*[1] has been used to form spectra of a long series of wave observations made at Pendeen and Perranporth on the north-west Cornish coast.

Comparison of the spectra with the meteorological charts of the North Atlantic shows that the swell arriving from a distant storm shows a more or less narrow band of periods, the minimum period of which decreases with time, in a manner which is consistent with the assumption that the swell travels from the generating area to the coast with a velocity equal to the classical group velocity appropriate to its period on arrival. In one example it is possible to show that the velocity of propagation does not differ by more than five per cent from the theoretical group velocity.

The observations in general suggest that the longest period exhibited by swell due to a distant storm is proportional to the maximum wind velocity in the storm, and does not depend upon the fetch or duration of the wind. This rule may have to be modified in studying generating areas of small size.

The swell is observed to travel great distances. Low swell has been received from a tropical storm off the coast of Florida, 2,800 miles from Cornwall, and from a storm near Cape Horn between 6,000 and 7,000 miles from Cornwall.

It is hoped to publish these results shortly, on behalf of the members of this research group.

N. BARBER
F. URSELL

Admiralty Research Laboratory,
Teddington.

Jan. 25.

[1] *Nature* **158**, 329 (1946).

23

Reprinted from *Physics Today* **2**(8):20–29 (1949)

DANGEROUS CURRENTS IN THE SURF

by Francis P. Shepard

Ewing Galloway.

Rip currents, not undertow, can sweep the unwary swimmer seaward. Knowing how to spot these rips, and what to do if caught in a strong one, may help a swimmer save his life.

Surf bathing leads to the drowning of thousands of people each summer. "Caught in the undertow" is the usual explanation for these disasters and yet there may be no such thing as undertow. In dictionaries it is defined as "the current beneath the surface which sets seaward when waves are breaking on the shore." Many geological textbooks contain brief discussions of undertow, ascribing it to the seaward return of landward driven surface water.

No actual measurements have been made con-

Francis P. Shepard, professor of submarine geology at the University of California's Scripps Institution of Oecanography at La Jolla, California, began his marine research in 1923. A graduate of Harvard, he received his Ph.D. from the University of Chicago in 1922, and from 1922 until 1942 was a member of the University of Illinois geology department. He has published numerous papers in the field and two books on submarine geology. This contribution, from the Scripps Institution of Oceanography, New Series No. 426, represents in part the results of research carried out by the Office of Naval Research under contract with the University of California.

Ewing Galloway.

firming the existence of dangerous subsurface currents of this nature. Our extensive investigations at Scripps Institution, both in the surf zone and outside the breakers, have to date shown no appreciable seaward movement of water along the bottom in areas where the surface current was setting landward. In fact, the net movement at the surface is ordinarily in approximately the same direction as it is at intermediate depths and even near the bottom.

Rip Currents Described

There is another type of dangerous seaward movement for which there is abundant confirmation and which is well known to lifeguards on beaches where there is large surf. This has been called rip tides or sea pusses, but rip currents or rips are more appropriate names because the currents have nothing to do with tides nor, of course, with seagoing pussy cats. These rip currents differ from the hypothetical undertow in that they extend from the surface of the water to the sea bottom both inside the breakers and for some distance outside. Farther out they are confined to the surface and intermediate depths and thus ride over the bottom water quite the opposite of the hypothetical undertow.

These currents move out in narrow bands through the breakers and spread slowly into fan shapes beyond. Between the seaward flowing masses there are broader areas where the water moves towards the land more slowly than the rips move seaward. The shoreward drift also extends from top to bottom of the water column in most places. The result of this shoreward drift is to build up water along the shore with a resultant flow parallel to the shore. These longshore currents are deflected seaward into rip currents.

Rip currents are easy to spot from the air. They can also be seen by a practiced eye from the beach. Whenever there are large waves approaching more or less directly towards a relatively straight beach, one can detect their presence. On clear days it is a rare occasion when we are unable to see them from the upstairs windows of the Scripps Institution which commands a view of the open Pacific. From the air or from any position where you can look down on the surf, rips can be distinguished by the turbulent brownish or greenish masses of water which are moving out from the breaker zone, having somewhat the form of a cauliflower cloud rising from a volcano, or of a thunderhead. The flow is intermittent so that one mass follows another somewhat like puffs of smoke from an engine. The edge of each advancing rip is likely to be outlined by foam which has been carried out from the surf zone.

Viewed from the beach, rips can be detected by the turbulence of the water along the lanes of flow or by a change in color of the water in the surf zone where the rip current or its feeding longshore current has eroded the bottom and thus varied the absorption of the light due to the suspended sediment and to the depth. This same channeling effect may prevent the breaking of the wave as it moves shoreward leaving a gap in the breaker line. If the wave does break in the rip zone, the typical brownish color of the water can be seen in the breaking wave.

The rates of flow in rip currents have not been

Rip currents from the air. Note widening of surf zone in lower portion of picture. Variation is related to submarine canyon which comes in close to coast in the zone of inconspicuous surf. Rip currents are pronounced near margins of wide surf zones. *Scripps Institution photo.*

Rip currents, seen from Scripps Institution pier, are indicated by foam streaks extending out in two gaps in the breaker line.
Scripps Institution photo.

measured under extremely large breaker conditions, but it can be safely assumed that they increase with increase in the size of the waves. Numerous observations when the waves were small (less than about three feet high) indicate that in general rips are so gentle in flow that they are scarcely felt by swimmers. Where the waves attain heights of four feet or more, the currents are likely to be troublesome. With somewhat larger waves, currents of two or three miles an hour have been measured.

In the surf zone, rip currents have a different effect on floating objects which extend partly above the surface than on objects which are slightly submerged or on patches of sea-marker dye. Objects floating on the surface are ordinarily carried seaward as far as the line of large breakers, but then are often thrown shoreward again when they rise into the crest of the breakers, so that they move around in eddies near the breaker line and have little net transport. Sea-marker dye or submerged floats, on the other hand, will move readily through the breakers with only temporary pauses accompanying the passage of large waves.

The reason that a floating object is often stopped at the breaker line can be explained by considering the Hawaiian sport of surfboarding. The transverse profile of an advancing wave crest approaching the breaker line becomes asymmetrical, steepening towards the shore. As a result floating objects, such as surfboards, rise on the steep side of the advancing wave crest and slide down the steep forward slope under the influence of gravity. Thus, in the absence of outflowing currents the net movement of the object is shoreward. In a rip current the shoreward sliding of a floating object tends to offset the seaward movement of the current. It is because of this surfboard action that even a weak swimmer may not be carried through the breakers although he may have extreme difficulty in getting out of the influence of the rip current. On the other hand a drowning man who has become submerged will go out through the breakers without interference.

Black Star.

Rip Currents Explained

The explanation of these near-shore currents appears to be the same as the explanation which was formerly offered for the hypothetical undertow. They are the result of the net shoreward movement of particles of water in waves. The effect of this motion is to pile up water along the shore. As a result there must be net return flow seaward which equalizes the shoreward mass transport. This return could take place underneath the surface or it could take place in restricted lanes both at and below the surface. The evidence now available, as we have already indicated, favors the second of these alternatives.

There are several factors which tend to localize the return flow (that is, the rip currents). Probably the most important of these influences is what is known as wave convergence. Where an even wave crest, moving towards shore from deep water, encounters a sea floor irregularity, even of a very minor nature, the crest is bent. If the irregularity is a shoal or a ridge, the crest is held back and the wave becomes concave towards the shore. If the irregularity is a deep, such as a submarine valley, the crest is bent in the opposite direction since it moves faster along the deep valley than on the shoaler sides. The effect of the ridge is to converge the wave energy and thus cause the waves to increase in height, whereas the effect of the valley is to spread the energy and decrease the wave height.

Inside a ridge or any small bottom projection, the shoreward transport of the water particles is intensified. Near the breaker line this shoreward movement extends from top to bottom although the velocity decreases downward. The next effect of the convergence is to build up water to a greater height shoreward of convergences than elsewhere. This de-

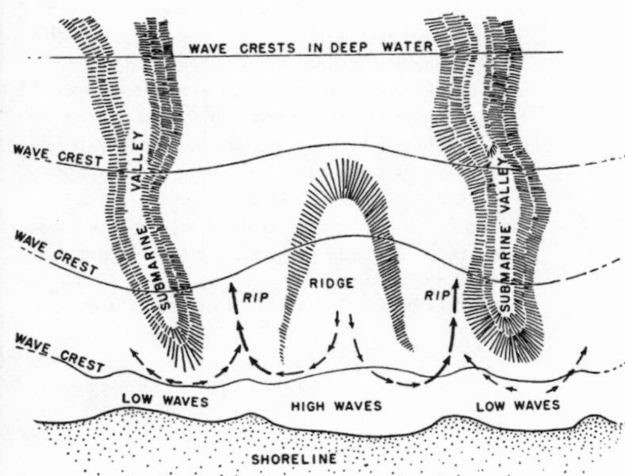

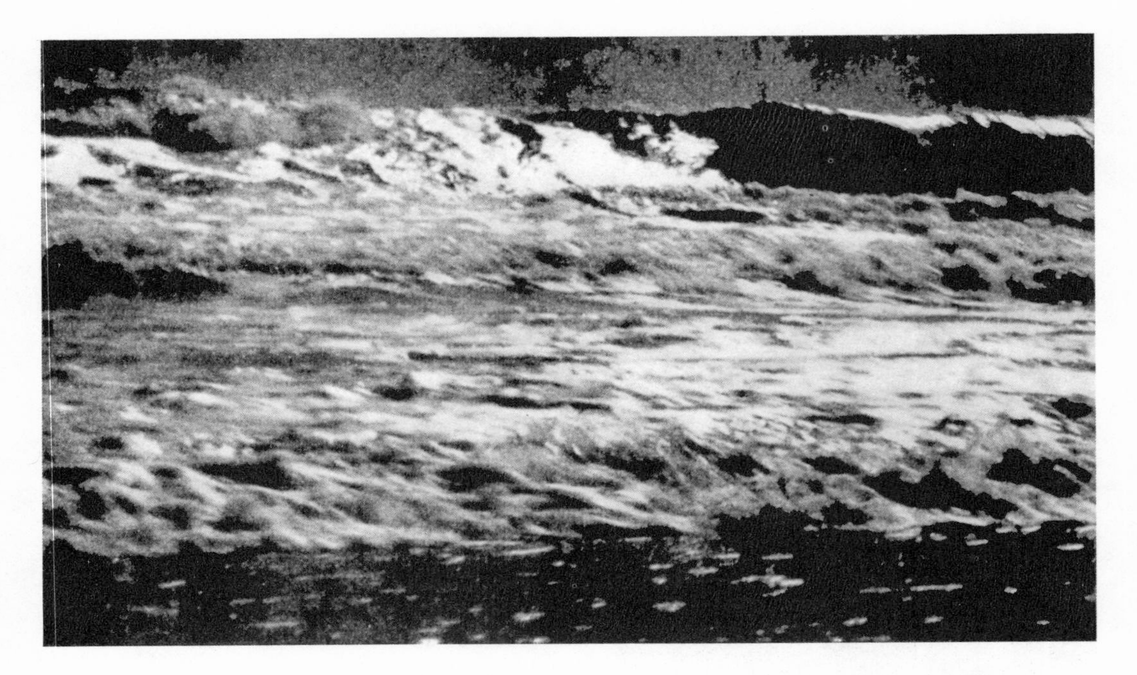

velops a gradient which produces longshore flow away from the zone of convergence on either side. These longshore currents in turn become rip currents where they can break their way seaward. The outflow may occur spasmodically near the point of convergence, but a much more regular outflow occurs outside the area of the highest waves where there is less of the shoreward mass transport to interfere with the outflow.

Measurement of the velocity of rip currents has shown a distinct relation to groups of high waves which ordinarily alternate with groups of low waves. Contrary to a popular concept, the large waves do not come as the seventh wave in a series but at very irregular intervals depending on the times when two or more series of waves of different period and source happen to have coinciding crests. Directly after the large waves break, the water is piled up along the shore resulting in either a strong longshore flow and a resulting strong rip, or an outflow near the point of raised sea level. Because of the lag in developing the return flow, the maximum currents in the rips generally occur during the times of the smaller waves.

Another location in which rip currents are common is on the up-current side of an obstruction (for example, the north side of a pier during south flowing currents). Diagonally approaching waves develop a longshore current in the direction in which they are advancing (except in the case of wave convergence already discussed). These longshore currents turn seaward at obstructions such as projecting points of land. As a result very large rips are

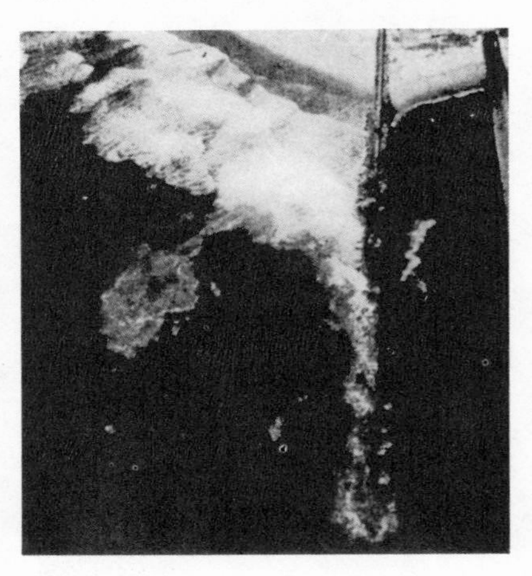

Development of rip current on the up-current side of a jetty.
Scripps Institution photo.

Ewing Galloway.

often seen extending seaward for thousands of feet on one side of a prominent point. Jetties cause similar rips, and even piers are likely to have a rip on one side although they do not offer as much interference to the longshore flow as do the solid structures.

At the head of a submarine canyon, where the waves are generally much reduced, the convex approach of the wave crests results in an impinging of waves on the shore so as to push the water away from the canyon head. The feeble longshore currents are deflected seaward as rip currents when they approach the much stronger rip currents flowing away from the wave convergence area at some distance from the canyon head. These two rip currents join on the outside of a triangular area of quiet water. Inside these triangles the shoreline generally projects seaward due to the deposition of a sand point.

Advice to Surf Bathers

It is not always easy to apply advice on methods of escaping from dangerous rip currents, particularly when a poor swimmer has been carried out beyond his depth. His first consideration should be to keep away from rips. But one has little to fear from dangerous currents along the open coast, where tidal currents are unimportant if the breaker heights are low—that is, less than about three feet. (To test this height one can stand knee deep in water, during upsurges, and see if one can observe the horizon over the top of the advancing breakers. The method will not work if the wave breaks far out from shore.)

If the breakers are large the poor swimmer should keep in shallow water, never getting deeper than waist height, even during the largest waves. He should also avoid bottom irregularities, which indicate the existence of channels cut by the feeder currents of rips. Even if the water in the channels appears to have little current a series of large waves may send a great concentrated surge along the channel, sweeping the bather off his feet and out into the zone of large breakers.

For the swimmer of moderate ability who does get caught in a rip current, the most important advice is to conserve his strength and not to try to force his way shoreward against a powerful surge. Swimming at right angles to the current will almost always allow the swimmer to get out of the rip channel so that he can touch bottom in the shoal water on either side. If the swimmer finds it difficult to get out on one side because large feeder currents are coming in on that side, he should turn around, and swim parallel to shore in the other direction.

It is important also to remember that the advancing breakers can prove very helpful unless they threaten to plunge down directly on top of you. As such a breaker approaches, the swimmer should swim under it in order to avoid being plunged to the bottom with great violence and then driven seaward during the return flow. A good swimmer can ride in on the front of the advancing breakers, even in a rip current, but this is not wise for most bathers. Unless the breaker is of threatening proportions, one should try to get high enough on the advancing front to receive at least a temporary shove towards shore.

If a swimmer is carried out through the breakers by a powerful rip, as happens not infrequently, he should swim parallel to the shore far enough to be able to come in on the side of the rip where he will generally find a favorable current. However, he should guard against being caught in the feeder currents which flow parallel to the shore and may sweep him back into the rip. A feeder current like a rip current varies in intensity. Therefore the swimmer can wait in the comparatively shallow water outside the feeder channel until the current slackens before making the crossing.

It is often possible for a moderately good swimmer to rescue someone in distress in the outer breaker zone by swimming out to him and finding a sufficiently shallow spot nearby so the man can stand up at least during wave troughs. The rescuer may then help the victim or encourage him to swim to the shallow area where he can rest his feet on bottom between wave crests.

There is no doubt but what most drownings take place simply because the bather loses his head and struggles with such violence that he becomes exhausted or has a heart attack. Relaxation and natural buoyancy will get even poor swimmers out of many serious predicaments. One should never forget that the rip currents are spasmodic and that the strong flows are generally of short duration, a few minutes at the most; the belts in the breakers are narrow so that a few strokes may take one clear of the danger area.

Black Star

24

Reprinted from *Contemp. Physics* **8**(2):171–183 (1967)

Modern Studies of Wind-Generated Ocean Waves

D. E. CARTWRIGHT

National Institute of Oceanography, Wormley, Surrey

SUMMARY. A survey is made of some of the major researches on ocean waves of the last 25 years. Starting with the introduction of the wave spectrum by Barber and Ursell and their observations of the decrease in period of swell waves arriving on the Cornish coast, which established the predominant linearity of the propagation of storm-generated waves across the ocean, we proceed to the more precise and ambitious experiments in the Pacific by Munk, Snodgrass and others. The latter work identified the arrival in California of waves generated in the Indian Ocean, nearly half-way round the world, and developed into a detailed study of the attenuation of swell as it travels along a great circle path from beyond New Zealand to Alaska. A summary is given of some basic aspects of the new theory of nonlinear scattering in water waves, and finally, instrumental techniques and theory involved in the study of waves in the North Atlantic Ocean for forecasting purposes are reviewed in simple terms.

1. Swell propagation in the Atlantic

Only a quarter of a century ago, when physicists had learnt how to split the atom, and astronomers had successfully predicted the position of an unknown planet 3 600 million miles away, there existed practically no quantitative knowledge of the physical processes by which waves are generated by wind or of how they are propagated across the ocean. There was a wide gulf between the classical studies of ideal wave forms by mathematicians such as Stokes and Lamb, and the largely visual observations of mariners and of devoted amateurs such as Vaughan Cornish. A few geophysicists, notably Harold Jeffreys, had attempted to bridge the gulf, but the absence of any precise measurements taken at sea made it difficult to assess the applicability of any theory.

The first serious attempt to apply modern techniques of measurement and analysis to waves as they actually occur at sea was by a small group of physicists and mathematicians headed by Dr. G. E. R. Deacon at the Admiralty Research Laboratory, Teddington, in the early 1940's.† The most important piece of work of that period is described in the now classic paper by Barber and Ursell (1948). They recorded by electrical means the changes in pressure at an instrument placed on the sea bed in about 30 metres of water off an exposed coast (near Pendeen) in North Cornwall. Except for wavelengths much less than the depth, the pressure under water waves is very nearly proportional to the instantaneous height of the surface, and so the recorded fluctuations in pressure give a measure of the major waves passing over the instrument.

The first fact Barber and Ursell found was that their wave records bore little resemblance to the periodic waves of constant height discussed in classical works on hydrodynamics. To assign a mean height and period to a 20 min

† Most of this group later formed the nucleus of the National Institute of Oceanography.

record was an over-simplification which gave almost meaningless results. However, a randomly fluctuating quantity can be resolved by spectral analysis into the sum of a number of quasiperiodic oscillations with various amplitudes, which individually may be expected to fit into the framework of classical theory. Accordingly, they devised a spectrum analyser which, given a wave record of about 20 min duration, produced a pen-tracing of amplitude against a scale of wave period.

Figure 1 shows one of the series of wave spectra so derived from pressure records taken at 2-hour intervals during 14th–16th March 1945. It shows the initial development of swell waves arriving on the Cornish coast. The swell first appears as a small hump in the spectrum at a period of about 24 sec; it then grows in amplitude while the period decreases and spreads in range. A continuation of the series (not reproduced here) through to 19th March shows a continuing decrease in period to about 11 sec, then fading in amplitude, and the eventual arrival of another group of swell at about 21 sec period.

This pattern, which was repeated over and over again with minor variation, agrees well with what one should expect from linear hydrodynamic theory. Water waves are unusual among the common wave motions of physics in being ‘dispersive’; that is, their speed of propagation varies with, and in fact is proportional to, their period. A storm in mid-ocean generates a heterogeneous mixture of waves whose energy travels outwards at various speeds according to its position in the spectrum. The spectrum, only somewhat modified by the path of travel, arrives at the distant shore piecemeal, the long-period end first. By the time the 12 sec waves have arrived, the 24 sec group, travelling at twice their speed, have long ceased transmission. The Cornish swell is thus a complete (though somewhat attenuated) sample of the original storm waves, spread out in time.

It is shown in textbooks on hydrodynamics that the speed with which energy is propagated over deep water by waves of period $1/f$ is $g/4\pi f$. This is the so-called ‘group-velocity’, which is half the ‘phase-velocity’ of individual crests. For example, 24 sec waves propagate at about 19 m/s, or 14·5° of latitude per day. By assigning a propagation speed to each part of the spectrum by this simple rule, one can trace back the leading and terminating edges of different observed spectra in distance and time to their common geographical zone of origin. In this case, the waves appeared to originate from a zone 20° to 30° distant from Cornwall between 11th and 13th March. Inspection of the weather maps in fact showed an intense depression south of Newfoundland which satisfied these requirements exactly.

The situation is complicated by the extent and variability of the storms in space and time, but all of the major irregularities were shown to be recognizable in the coastal spectra. The authors studied many similar swell histories, but their most spectacular achievement was the identification of simultaneous arrivals at Cornwall of a swell from the eastern North Atlantic and another, some 12 in. high, which clearly originated in the neighbourhood of Cape Horn, some 100° of latitude and 9 days' travel time away. They also demonstrated a modulation of about 5 per cent in period, due to the Doppler effect of local tidal streams.

By this work, Barber and Ursell for the first time demonstrated the validity of the laws of linear hydrodynamics to the irregular motions of the sea surface.

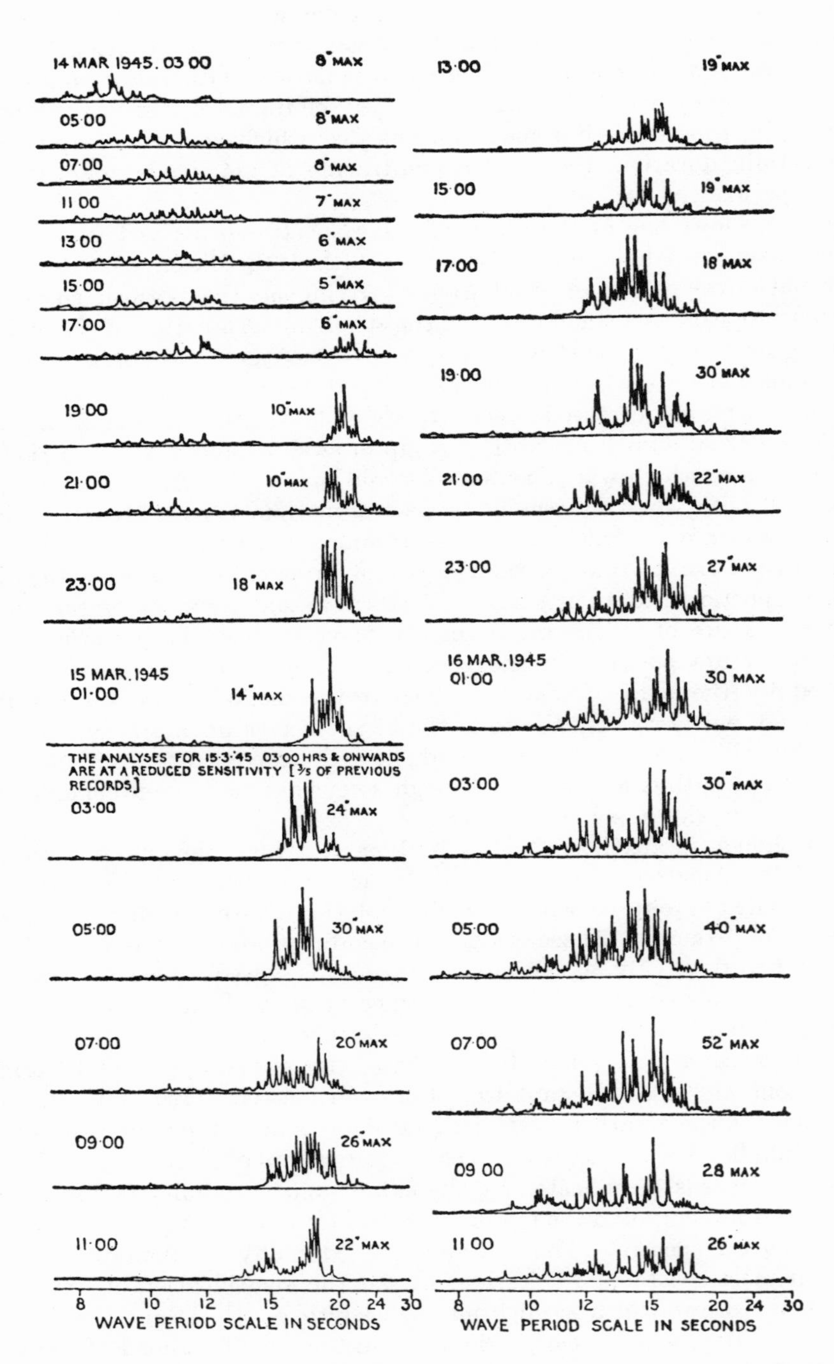

Fig. 1. Spectra of waves arriving at Pendeen, Cornwall, 14th–16th March 1945 (reproduced from Barber and Ursell 1948 by permission of the Royal Society).

2. Studies in the Pacific Ocean

During the 1950's research on measurements of sea waves (as opposed to purely theoretical studies) was mainly concentrated on the problems of generation, to which we shall return later, and of wave statistics. We now consider some more recent investigations which follow more directly on the lines laid down by Barber and Ursell. Munk, Miller, Snodgrass and Barber (1963) applied the methods described above to study the propagation of swell across the Pacific Ocean. The situation in the Pacific is more interesting in that waves can approach Southern California, for example, from a wider range of directions, and have in general much longer travel times. Whereas swell from the southern Ocean is detectable in Cornwall only in special circumstances, because limited by the rather narrow 'viewing angle' subtended between West Africa and Brazil, in the northern Pacific summer it is the general rule, responsible for the impressive breakers and surf of Californian beaches. The storms are so distant that they can be regarded as approximating to 'point sources', which permit easier and therefore more precise analysis.

Munk *et al.* also had the advantage of greatly improved techniques which had developed in the intervening 15 years. In place of the crude hydrophone used by Barber in 1945, they measured bottom pressure by means of the Vibrotron, essentially an air-tight capsule containing a vibrating wire whose frequency varies with the elastic yielding of the capsule under pressure, (a temperature coefficient is allowed for). Pressure is thus recorded as a series of counts of vibrations, in a form immediately available for a digital computer, for precise evaluation of wave spectra. In their first series of experiments, three Vibrotron sensors were placed at the corners of a roughly equilateral triangle of side about 300 m, in 100 m of water off San Clemente Island, near San Diego. This triangle formed a receiving array, which could sense directions of waves in terms of the time differences as they pass over the sensors. (More exactly, one computes the relative phase lags at different spectral frequencies f; two sensors are sufficient for long-crested waves travelling in a unique direction, but three allow one to separate two or three distinct directions, such as those due to reflection at the beach.)

The spectra of the waves were computed at regular intervals from long records, and their high precision enabled contours of equal energy density to be plotted against frequency and time. An example covering most of the month of October 1959 is shown in fig. 2. This is really a refined and extended version of the type of representation in fig. 1, but some superficial differences deserve comment. First, cross-sections of fig. 2 at given times are smooth curves, whereas each component diagram of fig. 1 is spiky and irregular. Fig. 2 is the correct representation; the appearance of fig. 1 is due to the analogue instrument and poor statistical sampling used in the earlier work. Second, fig. 1 refers to wave periods, while fig. 2 refers to frequency in cycles per kilosecond, which is the reciprocal. This is again the result of better understanding of spectrum statistics, whose properties are now known to be expressed more naturally in terms of frequency. The contours are expressed in units of cm^2 per frequency band of 1 c/ks, as in almost all modern usage. They are presented for convenience in logarithmic steps.

Another advantage of using frequency as a base is that wave energy at frequency f from a storm at distance Δ and time t_0 should, from the previous

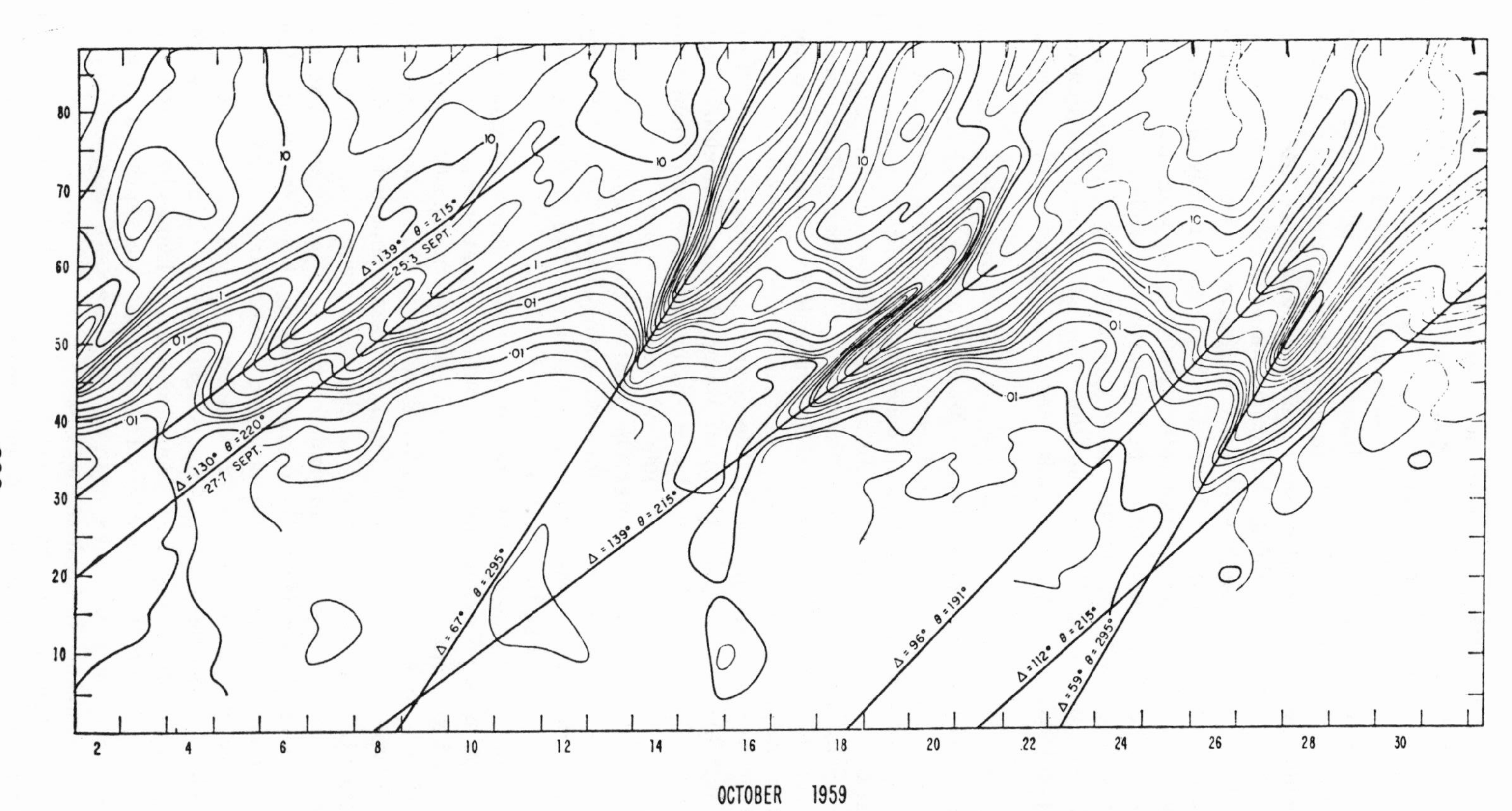

Fig. 2. Contours of spectral energy density plotted against time, for waves at San Clemente Island during October 1959 (reproduced from Munk *et al.* 1963 by permission of the Royal Society).

discussion, reach the recording station at time $t = t_0 + 4\pi f\Delta/g$. That is, the frequency of the leading edge of the swell spectrum should increase linearly with time, at a rate proportional to Δ. The ridges in fig. 2 are in fact seen to be very nearly straight, and the lines drawn designate the individual storm events. Their intercept at $f = 0$ gives the times of origin t_0, and their gradients are proportional to Δ, whose numerical values in degrees of latitude are shown. Also shown are the angles of approach relative to local North, deduced from the relative phases of the triad of sensors. All effects of refraction and other modification in the shallow water of the recording area have been corrected.

The exact position and time of origin of each linear event are thus well determined. Those with t_0 at the end of the 8th and on the 23rd have the same bearing, 295°, and hence originate in the North Pacific, at distances 67° and 59° respectively. The rest originate in the winter storm belt of the southern Ocean, rather near the edge of the pack ice. The greater distance (gentlest slope) in fig. 2 is 139°, with origin some 30° south of western Australia. Three earlier events were recorded with distances between 165° and 175°, nearly half-way round the world. At first sight, an ' antipodal swell ' seems geographically impossible, but close inspection of the globe reveals a narrow great circle path through a ' window ' south of New Zealand, reaching out to a stormy region between Kerguelen Island and Madagascar, which is indicated as the origin. The measured direction of approach to San Clemente, around 220°, confirmed that this was the route taken.

At this point it is worth reminding readers of some basic properties of wave propagation on a spherical ocean. It is rather like light waves imagined to be constrained to propagate along great circles instead of straight lines. Land masses, even small oceanic islands, are usually many times larger than the wavelength, and so cast shadows which can be treated as approximately geometrical. Fig. 3 sketches the land-bounded great circles enclosing waves

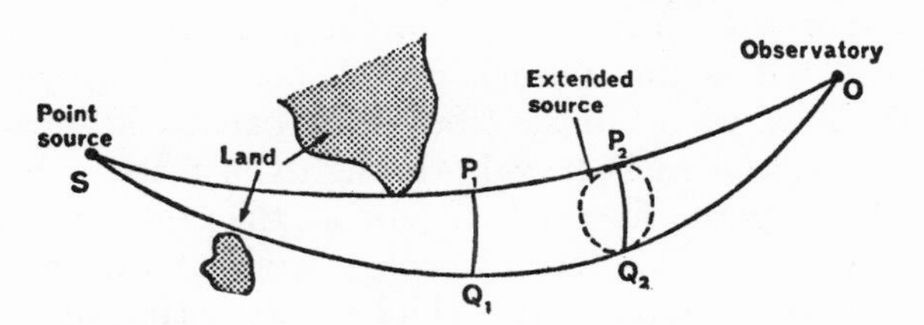

Fig. 3. Geometry of wave propagation on a sphere.

propagated from a point source at S. The amount of wave energy crossing various sections P_1Q_1, P_2Q_2 at distances Δ_1 and Δ_2 from S at the appropriate frequency and times is constant, so the energy density per unit frequency per unit of crest length (which is what is normally measured) is inversely proportional to $\sin \Delta$, the angle S being supposed fairly small. The observed energy density therefore decreases with distance up to $\Delta = 90°$, then begins to increase until it becomes very large near the antipode, O. Thus, in spite of their small visible aperture, an observatory at O can expect to record measurable intensities from storms near its antipode, as well as from closer storms which may subtend large angles.

If we keep O as the observing station, but suppose P_2Q_2 to define the extent of a closer storm, then the waves arrive at O spread over the angle POQ.

But for constant PQ this angle, too, is inversely proportional to $\sin \Delta_2'$, where $\Delta_2' = OQ_2$. Therefore the energy density per frequency band per crest length *per unit angle* is approximately constant along any unobstructed path from a given storm. Only the angles themselves (relative to local North) change, in accordance with the path geometry. The quantity just defined, usually expressed in the units $cm^2/(c/ks)/radian$ (omitting the constant factors g and water density necessary to express it as physical energy), is known as the ' directional wave spectrum ' and is fundamental to modern wave studies. We shall refer to it again in § 4. Several aspects of the spherical propagation geometry just described were confirmed by the measurements of Munk and his colleagues.

The above description neglects possible attenuation along the wave path by means other than geometrical spreading. In fact, comparison of swell spectra such as in fig. 2 with spectra derived by people working on waves recorded in the North Atlantic storm areas (e.g. Darbyshire, *O.W.S.*) showed fair agreement only below about 50 c/ks. Above 50, the swell spectra fall rapidly below the generation spectra, being typically two orders of magnitude low at 80 c/ks (12·5 sec period). This is in agreement with common observation. All common storms in deep water generate a great deal of energy at frequencies of 100 c/ks and above, which is off the range of fig. 2. A typical r.m.s. wave height in a storm is 4 m, whereas the corresponding height of swell waves in fig. 2 (estimated from $2\sqrt{2} \times$ the square root of the integral of the spectrum) is about 40 cm. Evidently, the higher frequencies have been heavily attenuated somewhere along their travel path.

It is impossible to account for this attenuation by molecular viscosity. The calculation of viscous dissipation is classic, and yields a result several orders of magnitude too low for waves in the frequency range considered. Conversion of wave energy into small-scale turbulence by the process of breaking is an important source of loss in the generating area, and of course on the final beach, but is rare elsewhere. A possibility suggested by O. M. Phillips is that the local waves in the trade-wind belt could break preferentially on the crests of the swell as it travels through the tropics, because opposing waves are steepened by the forward current at the crest and correspondingly flattened at the trough. This loss of energy would be drawn partially from the swell itself. Another possibility, which is attracting increasing attention since its first suggestion by Phillips (*O.W.S.*), is the scattering of waves by nonlinear interactions between different parts of the swell spectrum itself. We shall consider this mechanism in more detail in the next section.

In order to study the attenuation of the Pacific swell more closely, Munk and his colleagues undertook one of the most ambitious series of measurements in the history of oceanography (Snodgrass *et al.* 1966). They set up six simultaneous wave-recording stations along a great circle from New Zealand to Alaska during the summer of 1963, in order to track the southern swell throughout its journey across the Pacific Ocean. Besides Cape Palliser, N.Z., and Yakutat, Alaska, Vibrotron gauges were laid offshore from Tutuila (Samoa), Honolulu and the uninhabited atoll Palmyra. To break the long gap between Honolulu and Yakutat, a special ship known as ' Flip ', which is capable of floating with its length vertical to form a stable platform, was equipped with two Vibrotrons and posted in an appropriate position in the North Pacific.

Swells from twelve major storms were tracked across the whole route by spectral plots similar to that in fig. 2. Source distances deduced from all stations and directions computed from a two-sensor array at Honolulu agreed well with each other and with the weather maps. Compensations were made for geometrical spreading and land shadows as discussed above, to give in most cases six comparable sets of spectra for each storm event.

The results were variable and difficult to summarize adequately in a single diagram. Above 70 c/ks there was a general tendency for the spectrum to decrease at a fairly uniform rate of about 0·1 dB per degree along the whole path. At 70 c/ks and below there was no such tendency, and in some cases the energy actually rose with distance, possibly for the reasons discussed in the next section. There was no special reduction across the trade wind belt, and so the hypothesis of interaction with the trade wind systems is unconfirmed or at any rate unimportant. However, the major result of the experiments was that the attenuation above 50 c/ks in the first 20° or so from the generating area was much greater than that observed over the whole of the rest of the path to Alaska. There is good reason to suppose that this rapid attenuation is due to scattering by nonlinear interactions in the swell itself.

3. Nonlinear scattering of water waves

The outstanding feature of all these swell propagation studies is the predominant linearity of the wave motion; wave groups from different parts of the spectrum travel independently at the velocity predicted by linear theory. The basic equations of motion are actually nonlinear, since they contain second-order terms such as the well-known Bernoulli pressure effect, proportional to the square of the velocity; but the error in ignoring such terms is of order ka, where k = wavenumber = 1/wavelength and a is the primary wave amplitude, and for sea waves ka is usually rather small. The usual method of treating the nonlinear terms is first to obtain the linear (primary) wave solution by setting the nonlinear terms to zero, then to substitute the primary solution into the nonlinear terms and obtain a secondary solution containing terms of order (ka). One may then continue in a smaller manner to solve for tertiary corrections of order $(ka)^2$ and so on, to produce a series which can usually be assumed convergent.

If the primary wave is a pure sinusoid, the secondary wave involves a small 'second harmonic' and a slight change in mean surface level, which in general makes the wave crests sharper and the troughs flatter, as in the classical 'Stokes wave'. In the general case of a mixture of primary waves of various amplitudes, frequencies and wavenumbers, one gets terms involving products of pairs of primary waves, which in turn can be expressed in terms of secondary waves with sum and difference frequencies and wavenumbers. Now, by the linear dispersion relation, any pair of primary waves must satisfy

$$k_1 = (2\pi/g)f_1^2, \quad k_2 = (2\pi/g)f_2^2,$$

but it can be shown that no secondary combination

$$\boldsymbol{k}_3 = \boldsymbol{k}_1 \pm \boldsymbol{k}_2, \quad f_3 = f_1 \pm f_2,$$

can satisfy such a relation. (We here use $\boldsymbol{k}$ to denote a vector wavenumber whose direction is that of wave propagation and whose magnitude is $k = |\boldsymbol{k}| = 1$/wavelength.) The secondary waves are therefore always forced waves, inseparable from their primaries, and strictly bounded in magnitude.

It was first noticed by O. M. Phillips (*O.W.S.*) that if one considers tertiary waves, formed by the product of three primary wave components,

$$\boldsymbol{k}_4 = \boldsymbol{k}_1 \pm \boldsymbol{k}_2 \pm \boldsymbol{k}_3, \quad f_4 = f_1 \pm f_2 \pm f_3,$$

then certain combinations *can* be found such that $k_4 = (2\pi/g)f_4^2$. This means that the tertiary wave is a free wave in its own right, which can travel independently of its primaries. We therefore get a resonance condition, whereby energy is transferred from the primary waves to the new wave $\boldsymbol{k}_4$, whose energy grows linearly with time.

A special case of this tertiary interaction occurs when a primary wave k_1 interacts with the secondary component of another wave k_0 to generate the tertiary wave $\boldsymbol{k}_4 = 2\boldsymbol{k}_0 - \boldsymbol{k}_1$. In order that k_4 should resonate, we must have

$$|2\boldsymbol{k}_0 - \boldsymbol{k}_1| = (2\pi/g)(2f_0 - f_1)^2 = (2\sqrt{k_0} - \sqrt{k_1})^2.$$

To satisfy this equation, given $\boldsymbol{k}_0$, $\boldsymbol{k}_1$ must lie somewhere on a certain figure-of-eight locus drawn in fig. 4. In a given directional spectrum, therefore, all pairs of wavenumbers in the configuration of fig. 4 (and in general all triplets $\boldsymbol{k}_1$, $\boldsymbol{k}_2$, $\boldsymbol{k}_3$, satisfying more complicated relationships) will leak energy at a steady rate into the corresponding wavenumber $\boldsymbol{k}_4$. In other words, the primary wave system is *scattered* in direction and frequency. Resonant interactions of the type shown in fig. 4 have been confirmed experimentally by two independent groups (see Longuet-Higgins and Smith 1966 and following paper in the same publication).

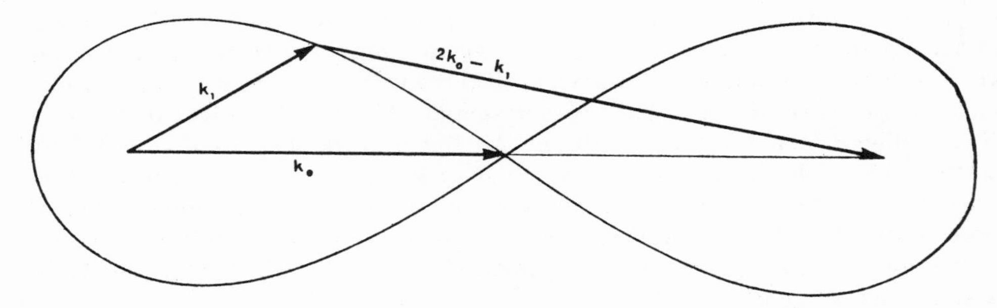

Fig. 4. Locus of wave numbers k_1 which can interact with the secondary component associated with a primary wave k_0 to transfer energy to the wavenumber $2k_0 - k_1$ (after Phillips).

Hasselmann (see Snodgrass *et al.* 1966, and other references therein) has shown that this scattering process is analogous to the collision of particles and anti-particles in high-energy physics, and can be similarly represented by 'Feynman Diagrams'. Wavenumbers are analogous to momentum, and frequency to particle energy. A 'collision cross-section' can be computed from the hydrodynamic equations to determine the actual rate of wave energy transfer. Applying the theory to waves leaving the storm area to form the swell systems discussed above, Hasselmann computed that waves travelling along the chosen great circle would lose energy by scattering at a very high rate above 90 c ks, but would be almost unaffected, or even gain energy by the same process at lower frequencies. The rate, being proportional to the third power of amplitudes, will also decrease as time proceeds, on account of spreading of the energy in space by the linear dispersion process. As a result, the scattering would be reduced to rather small proportions after the first 20° or so. These properties are qualitatively close to the observed facts, even if they do not agree down to the last detail.

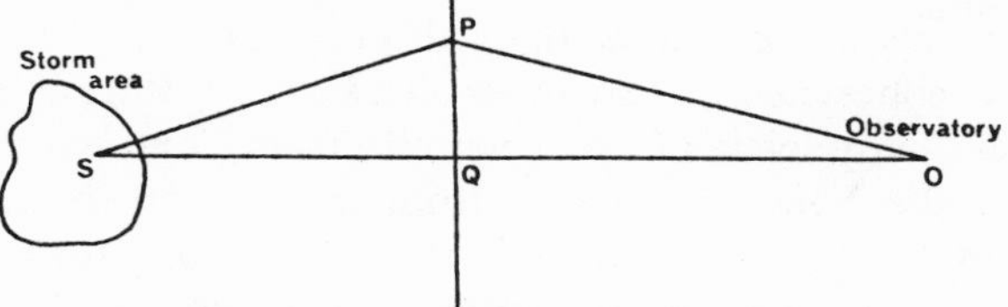

Fig. 5. Direct and scattered wave propagation.

Beyond the initial region of rapid scattering, when the swell has settled down into a fairly narrow beam, the situation is as sketched in fig. 5, where for simplicity great circles are represented by straight lines. The observatory at O receives most of its energy from within a narrow angle of the direct path SO. On the whole, what is lost along the path is gained by receiving energy scattered from other wave beams scattered from the storm; for example, from the area P. The net result is merely to broaden the beam angle slightly at O. However, if there are islands at places such as P, O will receive less than its full share of scattered wave energy, and a reduction in total spectral energy will result. Similarly, if there is an island in the main beam at Q, but not so broad as to block it entirely, and the spectrum at O is compensated numerically for the direct shadowing effect, the compensated energy at O will be too great, because O still receives the same amount of scattered energy as if the island were absent. Hasselmann calculated that this effect could cause fluctuations of ± 6 dB along the path, and these are in fact observed.

Another interesting result of tertiary interactions is the production of very low frequency waves known as 'surf-beat'. Because of the narrowness of their spectra in frequency and wavenumber, swell waves tend to show marked beating, or alternation of groups of high and low waves at intervals of 2 or 3 minutes in time and a few kilometres in space. These are of course due to differences of close pairs of primary waves, which also produce secondary waves with (f_1-f_2) and (k_1-k_2)—the same frequencies and wavenumbers as suggested by the beats. The secondary wave is below the mean level under the large beat amplitudes, and above it under the small amplitudes. Its dynamics have been thoroughly analysed by Longuet-Higgins and Stewart (1964), who show an analogy to radiation stress in other physical wave motions. As previously mentioned, it is a forced wave, because (k_1-k_2) cannot equal $(2\pi/g)(f_1-f_2)^2$. However, the action of an irregularly shaped sea bed (such as a shelving beach), can be described as that of a 'bottom wave' of zero frequency and a spectrum of wavenumbers k_0. This bottom wave can convert the forced wave into a free wave by modifying its wavenumber to $(k_1-k_2-k_0)$ without altering its frequency. While the primary waves are eventually destroyed on the beach, this long free wave is released into the ocean, where it echoes back and forth for days, having a low reflection coefficient. It is called surf-beat because it has the same frequency as the groups of high surf activity observed on the beach. It accounts for the very uniform background energy level below 30 c/ks in fig. 2, and is also responsible for severe oscillations known as 'range action' in certain resonating harbours.

4. Some wave forecasting problems

We now return to the North Atlantic Ocean, from where much of the practical impetus for wave research has originated. It will be readily appreciated that the general understanding of waves, in say, the Bay of Biscay, is a good deal more difficult than the studies of swell propagation with which this paper has been mainly concerned. The cases studied by Barber and Ursell were selected for their simplicity of weather situation, but even storms centred near Newfoundland are not too distant from the eastern shores compared with their widths, and the 'point sources' from the Cape Horn area are trivial, except as scientific curiosities, compared with the great waves generated by the local North Atlantic gales. In order to study these waves, therefore, we have to contend with wide-angle swell sources, possibly from more than one locality, as well as local generation and strong scattering processes.

Soon after the experiments of Barber and Ursell it was realized that the Cornish coast was not a very suitable recording area because of the wide shallow continental shelf. There being no oceanic islands suitably placed, one had to record waves from a ship in deep water. One instrument which has been widely used in weather ships and elsewhere is the N.I.O. shipborne wave recorder. Signals from pressure sensors at a point low in the hull are combined with twice-integrated signals from accelerometers to give a continuous electrical output approximating to the height of the wave surface. It is the only deep-sea wave recorder yet devised which can be used in all weather conditions. However, the instrument suffers from being affected by the interference of the ship itself with the waves, and it gives no information about wave direction, except by rather tedious measurements estimating the Doppler shift in recorded wave frequency as the ship steams on a series of courses.

For more precise research work the National Institute of Oceanography has developed a series of wave recording buoys, used with a ship in attendance. An accelerometer, held vertical by a gyroscope, registers the heaving of the buoy, which is virtually that of the water surface; its signal can be integrated to give true displacement. The angular motions from the vertical, referred to North and East axes, give direction sensitive signals. To be precise. comparison of the spectrum of the northerly slope with that of vertical acceleration gives a mean value of the directional wave spectrum weighted by the function $\cos\theta$, where θ is direction of wave propagation relative to north. Similarly, easterly slope gives $\sin\theta$ weighting, and comparisons between the slope signals themselves give $\cos 2\theta$ and $\sin 2\theta$. Readers familiar with Fourier analysis will recognize these weighted means as the leading coefficients of the Fourier series expansion of the directional spectrum referred to in § 2. They can be interpreted to yield mean direction and spread in direction of a wave system, or to separate two wave systems propagating in different directions (Longuet-Higgins, Cartwright, and Smith (*O.W.S.*)). A later development measures the three components of wave curvature as well as slopes, allowing a continuation of the Fourier series to include arguments 3θ and 4θ. Measurements in the North Atlantic have confirmed wide directional spread of order $\pm 45^\circ$ in all common wave systems.

Studies of wave generation by wind in the past for purposes of forecasting have tended to avoid the complexity of the physical processes involved by searching for rough empirical relationships. Some, for example Darbyshire (*O.W.S.*), are based on a great deal of data and have proved very serviceable. However, in the last few years, theoreticians such as Phillips, Miles and Hasselmann have considerably improved our understanding of the physical processes, and a few workers are beginning to apply them to actual wave measurements. In brief, a wind blowing over the water surface generates waves by two mechanisms. First, the turbulent pressure spectrum travelling over the surface with the mean wind velocity w sets up a resonance with those wave components with the same velocity which satisfy the free wave condition (that is, waves travelling in the direction ϕ relative to the wind which have frequency $f = (g/2\pi w)\cos\phi$). This process produces a growth of wave energy linear in time.

The second process, generally considered to be more important under normal conditions, is an unstable interaction between air and sea motions. The water waves force corresponding waves in the air flow with consequent pressure variations. If the air were a perfectly non-viscous fluid, these pressure variations could easily be shown to be 90° out of phase with the velocity normal to the water surface, and so would do no net work. The turbulent air motion due to viscosity, however, retards the phase of the pressure variations causing a tranfer of energy to the water waves from the wind. The effect being proportional to the existing wave amplitude, the energy growth rate is exponential.

At the same time there is a loss of energy due to waves breaking at their crests (strictly, the energy is converted into heat). Little is known about this very complex process, but empirical studies, backed up by certain theoretical considerations, have shown that the energy in any part of the spectrum grows until a certain density is reached, which is proportional to $1/f^5$. At this density it is supposed that a balance exists between energy input and energy loss due to white-capping and non-linear scattering to other spectral zones. The spectrum is said to be ' saturated '. Obviously, the high frequencies reach saturation more quickly than the low frequencies, so the spectrum grows in time from the high-frequency end downwards as far as the lowest frequency which can be generated by the ambient wind speed (roughly $g/2\pi w$), as sketched in fig. 6. Pierson and Moskowitz (1964) of New York University have recently proposed a realistic empirical formula for the final ' saturation spectrum ' in any given wind speed.

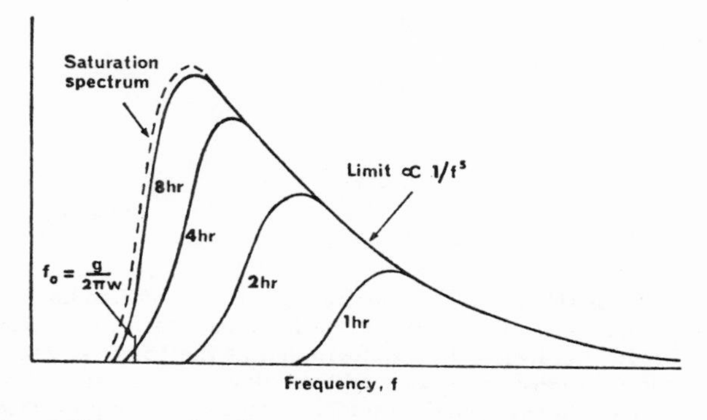

Fig. 6. Schematic growth of a wave spectrum towards its saturation limit.

A full wave-forecasting system has to compute the energy input and loss for each segment of the directional wave spectrum at each of a network of areas covering the whole ocean. Of course, wide coverage of meteorological data has first to be provided. In addition to local generation and loss, the propagation and scattering from one area to another must also be taken into account, as considered in detail in the earlier sections of this article. Several factors, such as the influence of air stability on the wave growth rates, and the spectrum of wind turbulence responsible for the initial wave growth, are still rather unknown and difficult to deduce directly from measurements, but it is hoped as always to make up for lack of knowledge by means of empirical constants which fit average conditions. In fact, progress in the field largely consists in the replacing of empirical terms by mathematical expressions which have been shown by theory and experiment to model the physical processes of fluid mechanics. In any case, this article should have made it clear that the science of sea waves has advanced greatly in the twenty-odd years since Barber and his colleagues first installed their hydrophones off the shores of Cornwall.

References

References marked *O.W.S.* are papers presented at the Conference on ' Ocean Wave Spectra ', Easton, Maryland, 1961. (Prentice-Hall, 1963), 357 pp.

Barber, N. F., and Ursell, F., 1948 *Phil. Trans., Roy. Soc.* A, **240,** 527–560.

Longuet-Higgins, M. S., and Stewart, R. W., 1964, *Deep-Sea Res.*, **11,** 529–562.

Longuet-Higgins, M. S., and Smith, N. D., 1966, *J. Fluid Mech.*, **253,** 417–436. (Followed by a paper describing similar experiments by other authors.)

Munk, W. H., Miller, G. R., Snodgrass, F. E., and Barber, N. F., 1963, *Phil. Trans. Roy. Soc.* A, **255,** 505–584.

Pierson, W. J., and Moskowitz, L., 1964, *J. Geophys. Res.*, **69,** 5181–5190.

Snodgrass, F. E., Groves, G. W., Haselmann, K. F., Miller, G. R., Munk, W. H., and Powers, W. H., 1966, *Phil. Trans. Roy. Soc.* A, **259,** 431–497.

The Author:

David Cartwright, B.Sc. (Hons.) in Mathematics, graduated at London University in 1951. He is at present a Principal Scientific Officer at the National Institute of Oceanography, where he has worked since 1954; was Research Oceanographer at the University of California, San Diego, 1964–65. His research interests and activity have developed from oscillatory motions of ships to ocean waves of various classes, including tidal motions. He has published papers on statistical theory of ship motions and sea waves, internal waves, sand waves, tides, and mean sea level.

Part V

CHEMISTRY OF SEA WATER

Editor's Comments on Papers 25 Through 29

25 LAVOISIER
Mémoire sur l'usage de l'esprit-de-vin dans l'analyse des eaux minérales

26 MARCET
Excerpt from *On the Specific Gravity, and Temperature of Sea Waters, in Different Parts of the Ocean, and in Particular Seas; with Some Account of Their Saline Contents*

27 DITTMAR
Excerpt from *Report on Researches into the Composition of Ocean-Water Collected by HMS* Challenger *During the Years 1873–76*

28 MILL
The Pettersson-Nansen Insulating Water-Bottle

29 KNUDSEN
Preface to *Practical Manual of the Analysis of Sea Water. I. Chlorinity by Knudsen's Method*

Ancient scientists and medieval philosophers alike were interested in the cause of the saltness of the sea (Sarton 1927–1948) and by the seventeenth century there were many different suggestions to choose from. One explanation attributed saltness to the action of the sun's rays on the sea and from this it was argued that all but the surface water might be fresh. Some travelers' accounts seemed to support this idea. This was one of the reasons why the Royal Society in their *Directions* (Paper 1) were so interested in sampling the water of the ocean in the deeper layers and why Hooke (Paper 2) and others took such trouble to devise water-sampling apparatus (Deacon 1971, ch. 4).

Robert Boyle's "Observations and experiments about the saltness of the sea" (1673) was the most successful of his four essays dealing with marine science (Deacon 1971, ch. 6). He had samples of sea water from different depths collected for him and compared their specific gravities. This satisfied him that the sea was salt throughout, though isolated patches of fresh water might occur around submarine springs.

Boyle was also interested in learning whether the amount of salt in the sea varied from place to place. He lent a hydrometer to a traveler going to the West Indies who reported that the specific gravity increased until they approached the tropics and then stayed the same. Other travelers reported that, instead of, as might have been expected, being the saltiest part of the ocean, the water at the equator had the same specific gravity as the sea in the latitude of the Cape of Good Hope.

Boyle was less successful in his attempts to find what constitutes the saltness in sea water. It was not until the second half of the eighteenth century that analytical chemistry developed far enough for the investigation of sea and mineral waters to be successfully attempted. The first full-scale analyses of sea water were made by the French scientist Antoine Lavoisier (Paper 25) and the Swedish scientist Torbern Bergman (1779, vol. 1, pp. 179–183).

Lavoisier, 1743–1794, who was guillotined during the French Revolution, announced his findings in 1772. He used sea water collected off Dieppe. Its saline content of 1.7 percent was considerably lower than average sea water, so presumably it was somewhat dilute. Lavoisier heated the sea water. The first solid matter to appear he identified as a mixture of *"terre calcaire"* (calcium carbonate) and *"sélénite"* (calcium sulphate). He evaporated the remainder to dryness and washed the residue with alcohol. A small portion of the residue was redissolved and was later found to be *"sel marin, à base terreuse ordinaire"* (calcium and magnesium chlorides). Lavoisier dissolved the rest of the residue by heating it in a mixture of alcohol and water. When it cooled, a power was precipitated which he found to be Glauber's salt and Epsom salt (sodium and magnesium sulphate). He then evaporated the remainder and obtained *"sel marin à base d'alkali fixe de la soude"* (sodium chloride) and *"sel marin à base de sel d'Epsom"* (magnesium chloride) (Riley and Skirrow 1965).

Lavoisier recognized that the salt in the sea had been accumulated over a long period as the result of the solvent action of rain and rivers on the earth. This idea was only one of the many explanations of salinity still being debated in the eighteenth century. Halley (1715) had suggested that if salinity were measured over a period of time it would be found to increase, and that from this figure it would be possible to calculate the age of the earth.

Further developments in chemistry enabled scientists to make individual measurements of the elements present in a solution by using reagents to separate the acids and bases. This was done by John Murray (1818a), lecturer in materia medica at Edinburgh University, not to be confused with the John Murray of the *Challenger*, and also by Alexander Marcet (1819). Marcet, 1770–1822, was Swiss but studied medicine at Edinburgh and worked in London. With Smithson Tennant he began collecting samples of sea water from all over the world in order to compare their salinity. Tennant died but Marcet eventually completed the work, incorporating material provided by the British Arctic expeditions of 1818 (Deacon 1971, ch. 11). The result was an extremely thoughtful piece of work, based both on his own work and on the ideas and observations of previous writers, and went far beyond a simple chemical description (Paper 26).

Marcet first described the distribution of salinity, as well as he was able, bearing in mind the unequal distribution of his samples. In order to compare specific gravities, he had also to take account of temperature and this led him on to consider the idea already in the air (see Editor's Comments for Papers 3 through 12) that inequalities in density distribution might contribute to ocean currents, temperature measurements in the depths of the sea and the behavior of sea water at low temperatures. It had been widely assumed, on slight authority, that sea water, like fresh water, reached its temperature of maximum density some degrees above freezing. Marcet showed that in sea water the temperature of maximum density was actually below its freezing point. This meant that, other things being equal, one would expect to find the coldest water at the bottom of the sea, unlike lakes where the dense bottom water would be some degrees above freezing even in the coldest weather. Marcet's result was improved by later workers such as Erman (1828) and Neumann (1861); but up to the late 1860s it was still widely believed that sea water was no different from fresh water in this respect. Marcet himself regarded it as strengthening the case for a thermal circulation in the ocean along the lines proposed by Rumford (1800) and Humboldt (1814).

Marcet's analyses of sea-water samples from different parts of the world led him to formulate the important "conservative" principle that although the total concentration of salts in a sample may vary from place to place, the proportions of the individual salts to each other remain the same within very narrow limits. He also announced the discovery of potassium in sea water, the last of the major constituents to be found, by W. H. Wollaston.

During the nineteenth century, as methods of chemical analysis improved, chemists gradually enlarged the number of trace elements identified in sea water. By 1865 there were altogether twenty-seven elements known. The position was summarized by the Danish mineralogist Georg Forchhammer (1865) who had been working on sea water for over twenty years. He also confirmed Marcet's finding that the principal constituents of sea salt maintain a very nearly constant proportion to each other and suggested that the chloride content might be used as an indicator of salinity.

William Dittmar, 1833–1892, in reporting on the chemical work of the *Challenger* expedition greatly improved on Forchhammer's determination of the proportions of the major constituents (Riley and Skirrow 1965) and stated that any of them might be used to compute the total salinity. He showed that the principle of the almost constant proportions of the constituents was valid for all the major oceans and for all depths. Paper 27 is an extract from his report, which was published in 1884.

As Dittmar had recommended, the determination of salinity on the basis of a measurement of the chloride content was adopted and this, made by titration, replaced the old method, employed on the *Challenger,* which had been to convert a measurement of specific gravity obtained with a hydrometer. The Danish physicist Martin Knudsen, 1871–1949, improved the method of determining chlorinity by titration and initiated the manufacture of Standard Sea Water, first made by I.C.E.S. and now by the Institute of Oceanographic Sciences, Wormley, England, in order to prevent the discrepancies that had been occurring between the work of individual observers (Paper 29). This method of determining salinity was only replaced by measurements of electrical conductivity or other physical properties within the last few years.

Paper 28 is a description by the British geographer H. R. Mill, 1861–1950, of one of the water samplers devised by Otto Pettersson and Fridtjof Nansen. In deep water, this device was replaced by the Pettersson-Nansen reversing water bottle which, with its thermometer attached, is still a basic piece of the oceanographer's equipment today. The 1890s and early 1900s in Scandinavia were notable for the development and testing of much new apparatus as well as for the emergence of new ideas and observations. This was particularly the role of the Central Laboratory of I.C.E.S., run by Nansen at Christiania from 1902 to 1908 (Went 1972a and 1972b).

During the present century, marine chemists have done much to assist biologists, geologists, and physical oceanographers. They have helped to show for example, how nutrients, trace elements, and gases are distributed in the sea, how they help support marine plant and therefore animal life, and how, when the cycle is completed, the elements are returned to the water. They have helped to trace movements in the ocean and to explain anomalies in the distribution of animal populations to which they may give rise. Dr. Leslie Cooper (1967), formerly chemist at the Plymouth Laboratory of the Marine Biological Association, recently gave a new look to temperature and salinity profiling by making observations more closely spaced in the vertical than had been customary. Such close observations showed much sharp microstratification instead of smooth, continuous changes. It is seen very plainly in modern, continuous, STD records.

25

Reprinted from *Histoire Acad. Royale Sci. 1772*, pt. 2, 555-563 (1776)

MÉMOIRE SUR L'USAGE DE L'ESPRIT-DE-VIN DANS L'ANALYSE DES EAUX MINÉRALES.

Par M. LAVOISIER.

LA partie de la Chimie qui porte le nom de *Halotechnie*, celle qui traite des Sels, eſt une des dernières qui ſemble avoir fixé l'attention des anciens Chimiſtes; l'analyſe des Eaux minérales, qui appartient eſſentiellement à cette partie, s'eſt reſſentie de ce retard; à peine y a-t-il cinquante ans que les Chimiſtes commencent à acquérir des idées nettes ſur les différentes ſubſtances qui entrent dans leur compoſition, encore eſt-ce de nos jours que ces progrès ont été les plus rapides.

Ceux qui ſe ſont occupés particulièrement de cet objet, ſavent qu'il reſte encore beaucoup à faire, & les différences énormes qui ſe trouvent dans les analyſes d'une même eau, faites par différens Chimiſtes, prouvent combien cet Art peut encore prêter à l'arbitraire, ou au moins combien eſt grande l'extenſion des erreurs qu'on peut commettre: j'avoue que c'eſt quelquefois plutôt à l'Artiſte qu'à l'Art qu'il faut imputer ce défaut de ſuccès; mais il n'en eſt pas moins vrai qu'en ſimplifiant l'Art, on le mettra à portée d'un plus grand nombre d'Artiſtes.

La difficulté de l'analyſe des Eaux minérales, conſiſte principalement à ſéparer les différentes ſubſtances qui s'y rencontrent, à purifier les ſels qui ſouvent ſont imprégnés d'eau-mère, de matières extractives, ou de parties bitumineuſes.

C'eſt pour faire cette ſéparation dans la plus ſcrupuleuſe exactitude que je propoſe aujourd'hui une méthode, non pas peut-être abſolument neuve, puiſqu'elle exiſte entre les mains des Chimiſtes, mais dont ils ne paroiſſent pas avoir ſenti tout le mérite & toute l'importance, & dont je ne ſache pas qu'il ait encore été fait aucune application ſuivie.

M. Macquer eſt le premier qui ait entrepris une ſuite d'expériences ſur la ſolubilité des ſels dans l'eſprit-de-vin, & qui ait déterminé juſqu'à quel point alloit cette ſolubilité; les connoiſſances qu'on avoit acquiſes avant lui ſur cet objet n'étoient pas très-étendues; elles ſe trouvoient d'ailleurs éparſes dans un grand nombre d'Auteurs. Le travail de M. Macquer les a raſſemblées, y a infiniment ajouté, & a complété en quelque façon toute cette partie de la Chimie.

Le Mémoire de M. Macquer, qui ſe trouve dans la Collection académique de Turin, eſt donc la baſe de ce que je donne aujourd'hui; c'eſt le point d'où je ſuis parti, & par d'autres expériences, j'ai reconnu qu'indépendamment des ſels qui ſe diſſolvent dans l'eſprit-de-vin le plus déflegmé, la plupart des autres devenoient également ſolubles dans ce menſtrue, en y mélangeant une certaine portion d'eau, & je me ſuis même aſſuré qu'il étoit poſſible, dans pluſieurs circonſtances, de tellement proportionner les doſes, que le mélange pût diſſoudre un ſel, ſans en attaquer un autre.

Les bornes que les circonſtances me preſcrivent *, ne me permettent pas d'expoſer ici comment j'ai été conduit à cette découverte, ni de préſenter dans tout leur détail les expériences nombreuſes que j'ai été obligé de faire pour en tirer parti.

Je me contenterai donc de dire qu'après avoir mélangé de l'eſprit-de-vin & de l'eau diſtillée dans huit proportions différentes, j'ai examiné, ſoit à chaud, ſoit à froid, quelle étoit l'action de ces mélanges ſur différentes eſpèces de ſels, & que j'ai reconnu:

---

* Ce Mémoire étoit deſtiné pour une Séance publique.

1.° Que le sel marin & le nitre à base terreuse, se dissolvoient dans l'esprit-de-vin avec beaucoup de facilité.

2.° Que le même esprit-de-vin seul ne dissolvoit ni le sel marin, ni le sel de Glauber, ni l'alkali de la soude, ni le sel d'Epsom, ni le sel marin à base de sel d'Epsom; mais qu'il enlevoit seulement au sel de Glauber son eau de cristallisation, & le réduisoit en une poudre fine.

3.° Qu'un mélange de deux parties d'esprit-de-vin & une d'eau, dissolvoit à chaud une quantité considérable de sel marin, sans qu'il se fît aucune cristallisation par le refroidissement.

4.° Que le sel de Glauber ne se dissolvoit point à froid dans tout mélange, où il entroit plus d'esprit-de-vin que d'eau; qu'il se dissolvoit au contraire en quantité notable par l'ébullition, mais que la totalité cristallisoit par le refroidissement, sur-tout si l'on avoit employé un mélange de deux parties d'esprit-de-vin, contre une de sel de Glauber.

5.° Que le sel d'Epsom donnoit dans sa solution, par un mélange d'eau & d'esprit-de-vin, à peu-près les mêmes résultats que le sel de Glauber, à l'exception qu'il étoit un peu moins soluble; de sorte que, par exemple, si ces deux sels avoient été dissous par un même mélange à chaud, le sel d'Epsom cristallisoit ou se déposoit le premier.

Il auroit été intéressant, sans doute, d'étendre ces expériences aux différentes espèces de sels que nous connoissons, & de compléter, s'il avoit été possible, cette partie de la Chimie; le temps ne m'a pas encore permis d'exécuter ce travail; mais en attendant j'ai cru devoir publier les expériences qui ont un rapport plus immédiat avec l'analyse des eaux.

Il ne sera pas inutile, à cette occasion, de donner ici une idée générale des substances salines qui se rencontrent dans les eaux, ou qui peuvent s'y rencontrer; le nombre de ces substances est moins considérable qu'on ne le croiroit au premier coup-d'œil; il se réduit à peu-près aux suivantes: *la terre calcaire, la sélénite, l'alkali fixe de la soude, le sel marin à base saline & terreuse; le sel de Glauber, le sel d'Epsom & l'alun.* Je ne parlerai pas ici du fer & du cuivre, l'alkali phlogistique est un spécifique sûr pour reconnoître la présence de ces métaux, & pour en évaluer la quantité.

Toutes les analyses d'eaux minérales, données jusqu'ici, prouvent que ces substances sont à peu-près les seules qui se rencontrent dans les eaux; mais quand il seroit possible de former quelques doutes à cet égard, ils seront facilement détruits par les réflexions suivantes.

Les Chimistes & les Naturalistes conviennent la plupart qu'il n'existe que deux acides dans le règne minéral, l'acide vitriolique & l'acide marin; ils ne sont pas tous d'accord, il est vrai, sur l'origine de celui de nitre, les uns pensent qu'il appartient au règne végétal, les autres qu'il est le produit de la putréfaction des matières, soit animales, soit végétales; mais tous conviennent au moins qu'il est étranger au règne minéral, & qu'il ne s'y trouve que par accident.

Je sais qu'un Auteur moderne a cru devoir introduire un nouvel acide dans le règne minéral, sous le nom d'*acide phosphorique;* mais quelqu'ingénieuse que soit la théorie qu'il adopte, comme elle ne paroît pas encore suffisamment appuyée par l'expérience, je crois qu'on peut généralement réduire à deux le nombre des acides vraiment minéraux. D'ailleurs, quand il en existeroit d'autres, il est probable qu'ils forment des sels absolument insolubles, & que c'est par cette raison qu'on ne les trouve pas dans les eaux. On peut donc regarder comme constant, d'après l'expérience & d'après la théorie, qu'il ne peut se trouver dans les eaux minérales que des sels vitrioliques ou marins; or, il est aisé de faire voir que le nombre des sels de cette classe qui peuvent être chariés par les eaux n'est pas très-considérable.

Premièrement, la plupart des demi-métaux ne sont point susceptibles de dissolution dans l'acide vitriolique, ils exigent du moins un acide vitriolique concentré & bouillant: or, ces deux circonstances réunies, ne se rencontrent jamais dans la Nature, puisque même l'acide vitriolique ne s'y trouve que très-rarement à nu. On ne doit donc pas s'attendre à trouver dans les eaux, ni mercure, ni antimoine, ni cobalt, ni bismuth dissous par l'acide vitriolique, il en est à-peu-près de même des métaux, sur-tout des métaux blancs, auxquels les Chimistes ont coutume de donner le nom de *métaux lunaires;* ils sont de même indissolubles dans l'acide vitriolique, à moins qu'il ne soit bouillant & concentré. On ne doit donc pas s'étonner, s'il ne se trouve dans la Nature, & particulièrement dans les eaux, ni vitriol d'or, ni d'argent, ni de plomb, ni d'étain; le fer & le cuivre sont les seuls que l'acide vitriolique puisse dissoudre aisément, & c'est ce qui fait que ces deux métaux, sur-tout le premier, se trouvent si communément dans les eaux. On a indiqué plus haut les moyens de reconnoître la présence de ces métaux, & d'en évaluer la quantité. Tout ce qu'on vient de dire des dissolutions métalliques par l'acide vitriolique, peut également s'appliquer à celles par l'acide marin; toutes ces dissolutions, à l'exception de celles du fer & du cuivre, se font avec beaucoup de difficulté, elles exigent même la plupart des manœuvres particulières que la Nature ne peut employer; tous les sels de cette section doivent donc être mis au nombre de ceux qui ne peuvent se rencontrer dans les eaux.

On ne connoît jusqu'à présent que trois alkalis dans le règne minéral, celui de la soude, l'alkali terreux, ou la terre calcaire, & la base du sel d'Epsom; je ne parle pas de la base de l'alun parce que sa nature n'est pas encore suffisamment déterminée, & qu'on ne voit pas d'ailleurs qu'elle se trouve bien communément dans les eaux. Ces trois alkalis combinés avec les deux acides proprement appelés *minéraux*, ne peuvent former que six espèces de sel, la sélénite, le sel de Glauber, le sel d'Epsom, le sel marin, le sel marin à base terreuse, le sel marin à base de sel d'Epsom: ces six sels sont ceux qui se trouvent communément dans les eaux, & c'est pour cette raison, comme je l'ai déjà dit, que j'ai examiné de préférence l'action que l'esprit-de-vin pouvoit avoir sur eux.

L'eau de mer est le résultat du lavage de toute la surface du globe; ce sont en quelque façon les rinçures du grand laboratoire de la Nature, on doit donc s'attendre à trouver réunis dans cette eau, tous les sels qui peuvent se rencontrer dans le règne minéral, & c'est ce qui arrive en effet: comme cette eau est la plus compliquée de toutes celles que j'ai eu occasion d'examiner, je l'ai choisie pour donner un exemple de l'application de l'esprit-de-vin à l'analyse des eaux minérales.

J'ai pris quarante livres d'eau de mer, qui avoit été puisée à la côte de Dieppe, à quatre lieues en mer, par un temps calme, je les ai fait évaporer lentement au bain-marie & au feu de lampe, dans une capsule de verre, que j'avois soin de remplir à mesure que l'eau s'évaporoit: jusqu'à plus de moitié de l'opération, il ne s'est montré ni terre, ni sélénite, ni sel; mais enfin, vers cette époque, il a commencé à se former une pellicule qu'il étoit aisé de reconnoître pour de la terre calcaire & de la sélénite; j'ai continué d'évaporer jusqu'à ce que les premiers vestiges de sel marin, commençassent à paroître; alors j'ai décanté, j'ai changé de capsule, j'ai mis soigneusement à part la terre & la sélénite, je l'ai lavée avec un peu d'eau distillée, pour la dépouiller de toutes parties salines; enfin lorsqu'elle a été bien sèche, je l'ai portée à la balance, & j'ai trouvé qu'elle pesoit 4 gros 56 grains.

Je placerai ici une observation qui m'a conduit à séparer d'une façon mécanique, la terre calcaire d'avec la sélénite. La première se dépose sous forme pulvérulente, tandis qu'au contraire la sélénite cristallise en petites aiguilles à six pans, presque imperceptibles, qui se réunissent & se confondent, il arrive de cette différence de configuration, que si l'on lave avec de l'esprit-de-vin, un mélange de terre & de sélénite, qu'on agite un peu rapidement la liqueur, & qu'après l'avoir laissé reposer pendant quelques minutes, on la décante encore trouble, toute la sélénite reste dans le fond du vase, tandis que la terre plus divisée reste nageante dans l'esprit-de-vin, & ne se dépose que dans un intervalle de temps beaucoup plus long; cette première ſéparation ne doit pas être regardée comme ſcrupuleuſement exacte, & il reſte preſque toujours une portion de ſélénite mêlée avec la terre; mais il eſt aiſé de la ſéparer par une ſeconde opération, ainſi qu'on le verra dans un moment.

Lorſque la terre calcaire & la ſélénite ont été ſéparées, ainſi que je viens de l'expoſer, j'ai continué d'évaporer; d'abord j'ai obtenu de beaux criſtaux de ſel marin, mais ſur la fin de l'opération, la criſtalliſation eſt devenue confuſe, & les ſels ſe ſont trouvés imprégnés d'une eau-mère épaiſſe & viſqueuſe, & ce n'eſt qu'avec peine que j'ai pu évaporer juſqu'à ſiccité; j'y ſuis cependant parvenu, & le réſidu que j'ai obtenu, s'eſt trouvé peſer un peu plus de douze onces; j'ai pris toute cette maſſe ſaline & je l'ai miſe dans un matras, j'ai paſſé deſſus de bon eſprit-de-vin froid; ce menſtrue a acquis une couleur jaunâtre aſſez marquée, il a diſſout toute la ſubſtance viſqueuſe, & il n'eſt reſté qu'une maſſe ſaline d'une très-grande blancheur: j'ai reconnu depuis que cet eſprit-de-vin n'avoit attaqué que l'eau-mère du ſel marin, autrement dit le ſel marin à baſe terreuſe.

La maſſe ſaline reſſéchée enſuite de nouveau, ne peſoit plus que dix onces deux gros.

Cette première ſéparation faite, j'ai pris un mélange de deux parties d'eſprit de-vin & d'une d'eau, je l'ai verſé ſur la ſubſtance ſaline, & j'ai fait chauffer fortement, preſque tout s'eſt diſſout, mais ayant laiſſé refroidir, il s'eſt précipité une poudre blanche, qui n'étoit autre choſe que du ſel de Glauber & du ſel d'Epſom: cette poudre peſoit quatre gros vingt-ſix grains.

Il s'agiſſoit de ſavoir ſi la ſéparation des ſels, ainſi faite, étoit rigoureuſement exacte, & s'il n'en reſtoit pas encore quelques-uns de mélangés les uns avec les autres; pour cet effet, j'ai examiné d'abord la ſélénite que j'avois ſéparée mécaniquement de la terre, & je l'ai trouvée abſolument pure, ſans mélange & parfaitement analogue à la pierre à plâtre & à toutes les autres ſélénites qui ſe rencontrent dans la Nature ; quant à la portion pulvérulente que j'avois obtenue par décantation, j'y ai verſé de l'eſprit-de-vin, rendu acidule par le moyen d'une petite portion d'acide nitreux légèrement fumant ; cette liqueur a formé un nitre à baſe terreuſe ordinaire, qui s'eſt diſſout dans l'eſprit-de-vin ſurnageant, & il m'eſt reſté en outre quelques portions de matière inſoluble que j'ai reconnue pour être encore de la ſélénite.

Il eſt aiſé de voir que l'uſage de l'eſprit-de-vin dans cette opération, eſt préférable à celui de l'eau diſtillée. On ſait, en effet, que la ſélénite eſt ſoluble dans l'eau, ſur-tout dans l'eau acidule, tandis qu'elle ne l'eſt point dans l'eſprit-de-vin déflegmé.

J'ai enſuite remis en évaporation, à un feu très-lent, toute la portion que j'avois miſe en diſſolution par un mélange de deux parties d'eau & d'une d'eſprit-de-vin. On a vu plus haut que cette diſſolution ne devoit contenir que du ſel marin ; cependant, comme j'avois obſervé qu'en verſant ſur cette ſolution quelques gouttes d'alkali fixe en *deliquium*, il ſe faiſoit un précipité terreux blanc ; j'ai cru devoir rechercher la cauſe de cet effet, & m'aſſurer s'il tenoit à l'eſſence même du ſel marin, ou à un ſel à baſe terreuſe mélangé avec lui. Cet examen me paroiſſoit d'autant plus intéreſſant, qu'il pouvoit jeter quelque lumière ſur une queſtion qui a diviſé deux Savans célèbres, M. du Hamel & M. Pott. En conſéquence, j'ai ſéparé en douze fractions le ſel qui s'eſt formé par la criſtalliſation, je les ai miſes chacune à part dans des flacons différens, & j'ai remarqué que les premières portions qui avoient été criſtalliſées étoient d'une ſalure agréable, mais qu'à meſure que l'évaporation s'avançoit, le ſel devenoit de plus en plus âcre & amer ; à la fin j'ai obtenu un ſel qui ne criſtalliſoit plus en cubes, mais d'une façon aſſez irrégulière ; il peſoit une once juſte. J'ai fait diſſoudre dans douze verres une égale portion de chacune de ces fractions de ſel dans de l'eau diſtillée ; après quoi j'ai verſé dans chaque verre quelques gouttes d'alkali fixe purifié ; à peine y a-t-il eu de précipitation ſenſible dans le premier numéro ; mais à meſure qu'on approchoit des derniers, la précipitation devenoit plus abondante; de ſorte qu'il a été démontré à mes yeux que l'amertume & l'âcreté ne venoient uniquement que d'une portion de ſel marin à baſe terreuſe, qui ſe combinoit avec les criſtaux de ſel marin. Je ferai voir dans un autre Mémoire, que cette terre n'eſt pas la terre calcaire ordinaire; auſſi ce ſel marin à baſe terreuſe, diffère-t-il eſſentiellement du ſel marin à baſe terreuſe, notamment par la propriété de criſtalliſer aiſément & d'être indiſſoluble dans l'eſprit-de-vin.

Il me reſtoit enſuite à examiner la portion pulvérulente de ſel qui s'étoit dépoſé au fond du mélange de deux parties d'eſprit-de-vin & d'une d'eau, à meſure que la liqueur s'étoit refroidie; j'ai reconnu que ce n'étoit autre choſe que du ſel de Glauber, que j'ai obtenu en très-beaux criſtaux, & un peu de ſel d'Epſom; le tout peſoit 4 gros 26 grains.

Enfin, j'ai verſé, dans un alambic de verre d'une ſeule pièce, l'eſprit-de-vin qui avoit ſervi à diſſoudre les ſels à baſe terreuſe, je l'ai bouché avec un bouchon de criſtal, & j'ai diſtillé avec un appareil de vaiſſeaux enfilés à la façon de Glauber; l'eſprit-de-vin eſt paſſé pur ſans huile ni bitume.

Il m'eſt reſté au fond de l'alambic une eau-mère, qui miſe dans une capſule, m'a donné, par la ſeule évaporation au bain-marie, de beau ſel marin à baſe terreuſe ordinaire, en criſtaux confus & bien ſecs; il peſoit *une once cinq gros dix grains.*

On trouvera, en rapprochant les réſultats rapportés ci-deſſus, que l'eau-de-mer contient:

	Pour 40 livres d'eau de mer.			*Pour chaque liv. d'eau de mer.*	
1.° Terre calcaire ſoluble dans les acides, & qui paroît ne pas différer de la terre calcaire commune. 2.° Sélénite ou ſel gypſeux.		4.	56		$8\frac{1}{6}$
	onces	*gros*	*grains.*	*gros.*	*grains.*
Sel marin à baſe d'alkali fixe de la ſoude....	8.	6.	32	1.	$54\frac{4}{5}$
Sel de Glauber & ſel d'Epſom............	"	4.	26	"	$7\frac{17}{20}$
Sel marin à baſe de ſel d'Epſom..........	1	"	"	"	$14\frac{3}{4}$
Sel marin, à baſe terreuſe ordinaire, mêlé de ſel marin à baſe de ſel d'Epſom..........	1.	5.	10	"	$23\frac{13}{23}$

26

Reprinted from pp. 185–194 of *Royal Soc. London Philos. Trans.*
109:161–208 (1819)

ON THE SPECIFIC GRAVITY, AND TEMPERATURE OF SEA WATERS, IN DIFFERENT PARTS OF THE OCEAN, AND IN PARTICULAR SEAS; WITH SOME ACCOUNT OF THEIR SALINE CONTENTS

A. Marcet

[*Editor's Note:* In the original, material precedes and follows this excerpt. Plate 12 and the table referred to on p. 263 have not been reproduced because of limitations of space.]

On the other hand, Lieut. PARRY, who had the command of the ship Alexander, in Captain Ross's expedition (and is now appointed commander of the second expedition to Baffin's Bay), fully confirms the observations made by Captain Ross, and also by Captain SABINE,* on board Captain Ross's ship; so as to place beyond all doubt the fact of Baffin's Bay being colder at the bottom than it is at the surface.†

But although these points may be considered as satisfactorily established, it must be admitted that the various modes

* Captain SABINE has been so obliging as to furnish me with a table containing some of his observations on the subject, which will be found in the Appendix.

† Captain PHIPPS also states in his Journal (Appendix, page 142), that he found the temperature in Baffin's Bay, at the depth of 680 fathoms, as low as 40°, the surface being 55°, and the air 66½°.

Other observers have obtained, in other seas, analogous results. Thus, DE SAUSSURE having examined with great care the temperature of the Mediterranean at various depths, found it in two different places to be 10,6° of REAUMUR's scale, or about 56° FAHR. at the depth of 900 and 1800 feet, the surface being about 71°; and he was induced to conclude that the temperature of the Mediterranean at great depths is uniform, and not likely to be affected by the vicissitudes of the atmospheric temperature, or by changes of season (Voyage dans les Alpes, III. § 1351 and § 1391).

M. de HUMBOLT, whose attention was often directed to this subject, makes the following curious observation. "In the seas of the tropics we find that at great depths the thermometer mark 7 or 8 centesimal degrees (or about 45° FAHR.). Such is the result of the numerous experiments of Commodore ELLIS and of M. PERON. The temperature of the air in those latitudes being never below 19° or 20° (or about 56° FAHR.), it is not at the surface that the waters can have acquired a degree of cold so near the point of congelation, and of the maximum of the density of water. The existence of this cold stratum in the low latitudes is an evident proof of the existence of an under-current, which runs from the poles towards the equator: it also proves that the saline substances, which alter the specific gravity of the water, are distributed in the ocean, so as not to annihilate the effects produced by the differences of temperature." ("*Personal Narrative of Travels*," English edition, Vol. I. page 63.)

in which the experiments were made, could not be relied upon as to perfect accuracy.*

It is obvious that these defects in the methods employed, though affecting the precision of the results, and rather tending to render them less striking, could not in the least degree invalidate the general conclusion, that in Davis's Straits, and in Baffin's Bay, the sea, at great depths, is considerably colder than at the surface; while to the east of Greenland, and in rather higher latitudes, the temperature of the ocean follows precisely the opposite law.

These various facts having an obvious and immediate connexion with the density of water under different temperatures, my attention was naturally directed to that circumstance in respect to sea water, which had not yet, I believe, been the subject of direct investigation. It had been long suspected, but was first established by Deluc, and afterwards correctly ascertained by Sir Charles Blagden, that water, in cooling towards the freezing point, ceases to contract when its temperature reaches about the 40th degree; but that, on the con-

* Captain Ross, who generally used a register-thermometer, might easily have detected, by a comparative observation, any material error made in ascertaining the temperature of the mud which he brought up by his apparatus; and as he appears to have occasionally availed himself of that mode of checking his observations, we may presume that his results were free from any considerable error. Lieut. Franklin, on the other hand, when he could not reach the bottom, and was therefore unable to make use of my machine, employed that used by Dr. Irving, consisting of a leaden cylindrical vessel with two valves; a convenient apparatus, but which, as I before observed, is liable to some inaccuracy. He sometimes also used a corked bottle, which he sunk to a great distance from the surface, and by means of which he obtained, doubtless, water from considerable depths; but it was obviously impossible to estimate with exactness the precise depth from which this water was procured, or the change of temperature which it had undergone in traversing the upper strata.

trary, it begins to expand, and continues to do so till it becomes solid, at which moment it undergoes a farther and much more considerable expansion.* The question which I was desirous of ascertaining was, whether the same, or any analogous law, prevailed in regard to sea water.

The mode in which I first attempted to decide this point, was simply by cooling sea water, by means of cooling mixtures, till it reached the freezing point, and ascertaining its specific gravity, at each degree of temperature, as it approached congelation. Researches of this description are liable to a variety of practical difficulties, which I could not altogether overcome by this method, and the results which I obtained, offered slight inconsistencies, which prevented my relying upon their strict accuracy.† Still however they uniformly led me to the conclusion, that the law of greatest specific density at 40°, did not prevail in the case of sea water; but that, on the contray, sea water gradually increased in weight down to the freezing point, until it actually congealed.

Soon afterwards I used another method, which afforded more precise, and, as far as I am able to judge, decisive results. Instead of weighing the water, I measured its bulk, under various temperatures, by means of an appropriate apparatus. A sketch of this instrument (which was executed by Mr. NEWMAN) is given in Plate XII., and an explanation is annexed, which supersedes the necessity of any farther de-

* Philosophical Transactions for 1788, page 143.

† In experiments of this kind it is always necessary to make an allowance for the contraction of the glass vessel, the effect of which is to produce an apparent expansion of the fluid contained in it. There are formulæ for this purpose, and in particular that derived from ROY's experiments, which was adopted in GILPIN's Tables. According to ROY, a vessel of glass of the capacity of 10.000.000, would enlarge, by 1 degree, to the capacity of 10.000.129.

scription. The general conclusion drawn from four experiments, the results of which did not essentially differ from each other, was, that if a vessel filled with sea water of the specific gravity of about 1027, and of any temperature above the freezing point, be gradually and slowly cooled, the water contracts in bulk; and that this contraction continues to proceed, though in a diminishing ratio, till the temperature has reached 22° of FAHRENEIT'S scale. At this point the water appears* to expand a little, and continues to do so till its temperature is reduced to between 19° and 18°, at which point the fluid suddenly expands to a very considerable degree, shooting up with great rapidity, and forcing itself out at the open end of the tube. At the same moment the thermometer rises to 28°, and remains at that point. The liquid is now

* I say *appears*, because the rise of the column, occasioned by the contraction of the glass, may in part account for this effect. It would have been extremely difficult to have estimated this circumstance with precision in the above experiment, because the tube belonging to my apparatus was not perfectly uniform in its bore. But by ascertaining the capacity of a given portion of the tube, as well as that of the bulb of the apparatus, and calculating the contraction produced in the glass by a reduction of four degrees of temperature, I have been able to satisfy myself that the effect arising from this contraction could only produce about one half of the rise of the column observed in this experiment. So that it can hardly be doubled but that some expansion, however small in its amount, takes place in sea water when cooled from 22° to 18°. But I hope to be soon able to repeat the experiment in a more perfect manner, by a method similar to that employed, for an analogous object, by MM. DULONG and PETIT, and described in their excellent paper on the "Mesures des Temperatures, &c. 1818."

It may also be objected to this experiment, that the bulb has not its interior cooled uniformly, since the surface must be acted upon by the application of cold before the central parts. This is true to a certain extent. But from the great slowness of the experiment (which lasted about three hours at each time), this source of error is in a great degree avoided; and, that the greatest degree of cold actually reached the centre of the vessel, was proved by a nucleus of ice being formed in it, which closely invested the bulb of the thermometer.

found frozen, and in a few minutes the maximum of expansion is obtained. During this congelation the apparatus was never broken, and I satisfied myself by various trials with other vessels, that if a vent, however small, be allowed to sea water at the moment of freezing, the vessel is preserved entire, which, it is well known, scarcely ever happens in the case of common water.*

A singular consequence to be drawn from these experiments seems to be, that, since sea water does not begin to expand till it has been cooled below the point at which it usually freezes, if its congelation were not retarded, it would become solid without undergoing any previous expansion, and the law in question would altogether cease to exist in the case of sea water.

With regard to the singular anomalies of temperature in the Arctic seas, which have given rise to this digression, though some of the facts in question may now be more easily understood, it would be premature, until the observations have been multiplied, and the facts themselves more accurately investigated, to attempt to bring them under any

* The ice thus produced, it should be remembered, is very different from that which forms on the surface of the sea, since the latter parts with its salt in the act of freezing, a separation which can but very imperfectly take place in confined vessels. Accordingly I found the ice produced in this experiment soft and compressible like the water-ice of confectioners.

With regard to the quantity of expansion which sea water undergoes, in confined vessels, at the moment of freezing, I have been able to estimate it with ease, and with sufficient accuracy, by freezing a known weight of water in a phial, connected with an open tube, and ascertaining exactly the proportion of water forced into the tube during congelation. The result of two experiments which agreed perfectly with each other, was, that the expansion of sea water, when passing to the state of ice, is equal to 7,1 per cent of it bulk.

general law, or to explain the phenomena by particular theories.* Why, for instance, two neighbouring and almost contiguous portions of the ocean, placed nearly alike in regard to solar influence, should differ so widely in the temperature of their waters, the warmer strata being, in one case, found lying above the colder, while in the other that order is reversed, appears perfectly unaccountable. Whether, also, this singular circumstance may lead to inferences bearing upon the question now at issue respecting a north-west passage, I shall not presume to decide. But I may be allowed to indulge a hope, that the facts collected in this paper, may assist future inquirers in forming more accurate views of those grand phenomena of nature, in which the navigation of certain seas, the vicissitudes of seasons, and the geological history of the globe are so essentially concerned ; or that they may at least be the means of inducing other and abler observers to turn their attention to this interesting subject.

Count Rumford, in one of his Philosophical Essays (Vol. II. Essay VII.), in endeavouring to trace this class of natural phenomena to final causes, was led to some speculations and generalizations on the comparative temperature of the seas, and of large lakes, at their surface and at different depths, and on the relation which these temperatures bear to climate and to human comfort, which, however hypothetical, possess considerable plausibility and interest. Count Rumford's general idea was that the uniform temperature of large lakes at great depths, which De Saussure found in the Swiss lakes to be constantly between 41° and 42°, was naturally explained from the circumstance since discovered, of water possessing its greatest density at about that temperature ; and he conceived that the object of this law of nature was to preserve in winter a store of warmth at the bottom of these lakes, by which their freezing was retarded at the surface, and altogether prevented at a great depth, thus affording a check to the effects of severe winters. With regard to salt water, however, he took it for granted that the law which fixes the greatest density at about 40°, did not prevail ; but that, on the contrary, sea water being denser in proportion is at is colder, the coldest strata must occupy the bottom of the sea, while the warmest arising to the surface, serve to moderate the effects of the Arctic cold. He then

§ II. *On the Saline contents of the Waters of different Seas.*

I confined my remarks, in the first part of this paper, to the subject of the specific gravity and temperature of sea water, in various seas and in different latitudes. It remains for me to offer a few observations on the saline contents of these waters.

An accurate analysis of all the specimens which I have noticed in this paper would have been a most laborious, and indeed almost interminable undertaking, which would not have afforded any adequate object of curiosity or interest. All that I aimed at, therefore, was to operate upon a few of the specimens, so selected as to afford a general comparison between the waters of the ocean in distant latitudes and in both hemispheres, and to enable me also to ascertain whether particular seas differed materially in the composition of their waters.

For this purpose, availing myself of the experience I had obtained, in former inquiries of this kind,* respecting the

supposed that the colder and heavier strata would form sub-marine currents, constantly moving from the vicinity of the poles towards the equator, and occasioning upper and warmer currents precisely in an opposite direction. It is obvious that this theory, though capable of explaining some of the phenomena above mentioned, cannot apply to those of an opposite nature, also related in this paper. Yet these may possibly depend upon peculiar and local causes; and I cannot omit to observe, that M. De Humboldt, in the work already quoted, entertains notions of an exchange constantly going forward between the waters of the Polar regions and those of the Equatorial seas, which bear considerable analogy to those of Count Rumford, and cannot fail to give them additional weight.

* See an 'Analysis of the Brighton Chalybeate,' published in Dr. Saunders's Treatise on Mineral Waters, 1805. Also ' An Analysis of the Waters of the Dead Sea and

difficulty, and indeed the impossibility, of analyzing complex solutions of saline substances with a view to obtain a precise and certain knowledge of the state of combination in which the salts exist in these solutions, I contented myself with ascertaining, first, the proportions of saline matter yielded by a given quantity of each water, and afterwards, the proportions of acids and earths contained in these respective waters; thus presenting data which are quite divested of theoretical views, and from which the composition of those waters may at any time be inferred in the way which may be deemed most eligible.

It has been long known that the principal salts contained in sea water are muriate of soda and muriate of magnesia, and that it contains also sulphuric acid and lime. But whether these ingredients existed in the form of sulphate of soda, or of sulphate of lime, or muriate of lime, or sulphate of magnesia, was more or less a matter of conjecture, as the different states of binary combination which they assume, are modified during evaporation by the different degrees of solubility which the salts possess, and are liable to be influenced by heat and concentration, the very processes which are used in attempting to resolve the question. These difficulties have been ably discussed by Dr. MURRAY,* whose reasonings and experiments on the subject have given great plausibility to the doctrine which he has proposed, according to which the salts contained in sea water are supposed to be:

River Jordan;' Philos. Trans. 1807. And 'An Analysis of an Aluminous Chalybeate Spring in the Isle of Wight;' Geolog. Trans. Vol. I. 1811.

* See 'An Analysis of Sea Water' read in 1816, and published in the Edinburgh Transactions, Vol. VIII. and also a 'Formula on the Analysis of Mineral Waters,' printed in the same volume.

Muriate of soda,
Muriate of magnesia,
Muriate of lime,
Sulphate of soda.

Still however, it must be admitted that a degree of doubt remains respecting the mode in which the sulphuric acid is combined, and that we can only pronounce with certainty upon the proportions of acid and base taken singly, as I have explained above. My experiments, therefore, were confined to the following points.*

1st. To ascertain the quantity of saline matter contained in a known weight of the water under examination, desiccated in a uniform and well defined mode; and to compare it with the specific gravity of the water.

2ndly. To precipitate the muriatic acid from a known weight of the water, by nitrate of silver.

3dly. To precipitate the sulphuric acid by nitrate of barytes, from another similar portion of water.

4thly. To precipitate the lime from another portion of water, by oxalate of ammonia.

5thly. To precipitate the magnesia from the clear liquor remaining after the separation of the lime, which is best effected by phosphate of ammonia, or of soda, with the addition of carbonate of ammonia.

The soda, by this method, is the only ingredient which is not precipitated, and which therefore, can only be inferred

* It is but just to mention that I received, in this part of the inquiry, much valuable aid from Mr. Wilson, who has many years acted as assistant, to my colleagues and myself, in the Chemical Theatre of Guy's Hospital.

by calculation. But if the processes are conducted with sufficient care, this mode of estimating the proportion of alkaline muriates is susceptible of great accuracy, as I had an opportunity of ascertaining by some comparative experiments which I related at full length in the analysis of the waters of the Dead Sea.*

The whole of the results obtained by this mode of investigation, has, for the sake of brevity, been condensed into a table which is annexed to this paper, and upon which it is unnecessary to detain the Society by any farther comment. It will be seen by this table that, with the exception of the Dead Sea, and of the Lake Ourmia,† which are mere salt ponds, perfectly unconnected with the ocean, all the specimens of sea water which I have examined, however different in their strength, contain the same ingredients all over the world, these bearing very nearly the same proportions to each other; so that they differ only as to the total amount of their saline contents.

* In devising the above method, I followed, step by step, the plan which I had myself pointed out, and actually used, in various analyses, and particularly in that of the Dead Sea, and of an aluminous chalybeate, in the Isle of Wight, as may be seen by a reference to these papers. It is satisfactory to observe that Dr. MURRAY adopted, several years afterwards, from considerations of the same kind, a mode of proceeding precisely similar, and indeed that he proposed in a subsequent paper, a general formula for the analysis of mineral waters, in which this method is pointed out as likely to lead to the most accurate results. And this coincidence is the more remarkable, as it would appear, from Dr. MURRAY not mentioning my labours, that they had not at that time come to his knowledge.

† I had only between 2 and 300 grs. of water from this curious lake, which is so nearly saturated, that it begins to deposit crystals the moment that heat is applied to it. Though it contains no lime, it yields about 20 times as much sulphuric acid, and six times as much muriatic acid as sea water does, as may be seen by the annexed table. Dr. WOLLASTON has also detected traces of potash in this water.

27

Reprinted from pp. 199–209, 227–230 of the *Report on the Scientific Results of the Voyage of HMS* Challenger. *Physics and Chemistry,* vol. 1, pt. 1, Her Majesty's Stationery Office, London, 1884, 251 pp.

REPORT ON RESEARCHES INTO THE COMPOSITION OF OCEAN-WATER COLLECTED BY HMS *CHALLENGER* DURING THE YEARS 1873–76

William Dittmar

[*Editor's Note:* In the original, material precedes this excerpt.]

SUMMARY OF RESULTS

The foregoing memoir, though ostensibly only a report on a series of investigations into the composition of ocean-water, which it has been my privilege to carry out under the auspices of the Director of the Expedition, includes also the final elaboration of all Mr. Buchanan's work connected therewith, and is, consequently, a complete record of what the Challenger Expedition has added to our knowledge on the subject. But the greater part of my paper consists of more or less lengthy discussions of chemical methods and matters of calculation, which, as mere means to an end, are of little interest to the general reader, and may prove tiresome even to the scientific critic, if he do not happen to be a professional chemist.

This is my reason for drawing up the following summary, in which I have endeavoured to collect the general results of the whole investigation, and, for the benefit of non-professional readers, to explain their oceanographic significance.

The configuration of the ocean, broadly speaking, must have been the same as it is now for thousands of years. Hence its bed may be regarded by this time as having been almost deprived of all the more soluble components. No mineral, it is true, is absolutely insoluble in water; the ocean, consequently, must still be presumed to continue taking up soluble matter from the volcanic and other minerals with which it is in contact on the floor of the ocean, and what it thus gains is probably in excess over what it contributes towards the matter of new deposits. It must be granted also that it is continuously taking in large masses of dissolved mineral matter from rivers, and that what it receives from these two sources in a single year, if measured by ordinary standards, amounts to an immense quantity. But the gain even in a century is a mere trifle in comparison with what it already contains,—far less no doubt than the relative errors in our most exact methods of measurement.

The ocean, of course, takes in gases from the air as well as solids from the earth's crust; but this is a case of mutual exchange, in which the gain and loss, on either side, must long since have arrived at a state of equilibrium.

Hence the absolute composition of the ocean as a whole, meaning the total number of kilograms of water, chloride of sodium, &c., &c., present in it, though subject probably to an extremely slow increase in the dissolved saline matter, is practically constant and invariable. The percentage composition of a given sample of ocean-water is, of course, liable to variation according to the place where and the time when it was collected. This holds true more especially of the volatile components, viz., for the dissolved nitrogen and oxygen, the merely dissolved part of the carbonic acid, and last, and not least, of the water which forms the bulk (some 96 per cent. or more) of the whole.

Water, even at the lowest temperatures occurring on the surface of the globe, is appreciably volatile, and its volatility, as part of the sea, is not very materially diminished by the salts dissolved in it. Hence, from the whole of the area of the ocean, myriads of molecules of vapour of water are continuously being given out into the atmosphere; at the same time molecules previously given out are returning whence they came, the tendency, however, in every portion of atmosphere touching the ocean being to establish the maximum vapour-tension corresponding to the prevailing temperature. This vapour-tension is the greater the higher the temperature, and it increases more rapidly than does the temperature. Hence the rate at which the air takes up water from the sea is very great in the tropics, less in our latitudes, and far less in the circumpolar regions. On the basis of some law of distribution of temperatures, it would be a matter of calculation to inquire what this would lead to if the atmosphere were in a state of stagnation. But the atmosphere is not in such a state, and cannot be. The moist air in the equatorial regions, being relatively warm and consequently light, ascends, while relatively cold and dry air streams into its place from the north and south: a corresponding part of the uppermost stratum of the aerial ocean wells over and flows towards the poles. The consequence is that the greater part of the moisture taken up by the warm air of the tropics is not recondensed there, but is deposited as rain in the colder latitudes. Hence the sea must be less saline there than in the lower latitudes; but the permanence of a great excess of salinity anywhere is precluded by the oceanic currents.

To map these currents accurately and determine their velocities is the most important problem of general oceanography, and the solution of this problem would obviously be greatly facilitated if we had a correct and complete representation of the contour surfaces of equal salinity. As a means towards this end, Mr. Buchanan, in the course of the Expedition, collected thousands of samples of ocean-water from a great variety of places and depths, and defined their salinity by determining their specific gravity at known temperatures. His results are detailed and discussed by himself in his Report on the Specific Gravity of Ocean-Water.* My own connection with this part of his work is but slight. All I did was,—firstly, to work out experimentally the mathematical relation between salinity and temperature on the one hand, and specific gravity on the other, so that Mr. Buchanan's numbers might be reduced to a standard temperature, and be translated into salinities; and secondly, to determine the salinities of some 160 of Mr. Buchanan's

*Phys. Chem. Chall. Exp. part ii.

water samples by a more direct (chemical) method, and compare the resulting values with those computed from Buchanan's specific gravities by my formula. This comparison led to the very satisfactory result, that the "probable error" in any one of Mr. Buchanan's specific gravities is rather less than ±0·1; that of pure water being taken as =1000.

A little reflection shows that no number of analyses will enable one to calculate with any degree of exactitude the mean salinity of the ocean as a whole; but even my 160 salinity determinations, since they correspond to a great variety of places, suffice to give an idea of the limits between which the quantity fluctuates. Expressing the salinity in "parts of total salts per 1000 parts of sea-water," I find that (of the 160 values)

The lowest (from the southern part of the Indian Ocean, south of 66° lat.) is . . 33·01
The greatest (from the middle of the North Atlantic, at about 23° lat.) is . . 37·37

(Some few samples from narrow straits or close to certain coasts are omitted, as being probably diluted to an abnormal extent with fresh water.)

So much as to the ratio of the water to the sum total of the salts dissolved in it. Let us now inquire into the percentage composition of the salt mixture itself.

A priori, we should say that this composition cannot be subject to any great variation; because, if there were no chemical changes going on in the ocean, and no gain or loss of dissolved individual salts, this composition would now, after thousands of years' constant intermixture, be absolutely the same everywhere; and what is going on in the shape of reactions and importation or exportation of individual salts, really amounts only to an extremely minute fraction of the whole, even in the course of a century. This conclusion is confirmed by the analyses of several hundred samples of surface-waters, which were carried out by Forchhammer in connection with a great research which he published in 1864.* According to his results, if we confine ourselves to the open ocean, we find that everywhere the ratios to one another of the quantities of chlorine, sulphuric acid, lime, magnesia, and total salts, exhibit practically constant values. With the view chiefly of supplementing Forchhammer's work, I have made exact determinations of the chlorine, sulphuric acid, lime, magnesia, potash, and soda in 77 samples of water collected by the Challenger from very different parts of the ocean:—

12 from the surface.
10 from depths of 25 to 100 fathoms.
21 from depths of over 100 to 1000 fathoms.
34 from greater depths.

* *Phil. Trans.*, 1865, vol. clv. p. 203.

The results, while fairly agreeing with Forchhammer's, were in still closer accordance with one another, and thus showed that Forchhammer's proposition may be extended from surface-waters to ocean-waters obtained from all depths.

The solid matter dissolved in sea-water, though strictly speaking, and we may add necessarily, of a very complex composition, consists substantially of the muriates and sulphates of soda, magnesia, lime, and potash. Forchhammer, after having satisfied himself that all the other constituents taken conjointly amount to only a small fraction of one per cent. of the total solids, in his individual analyses limited himself to exact determinations of the chlorine, sulphuric acid, lime, and magnesia. The potash he determined only in a comparatively small number of samples; and where he reports the soda this component is calculated by difference, on the assumption that the acids and bases present exactly neutralise each other. But this assumption had never been proved to be correct, and *à priori*, is improbable, because it leaves out of reckoning the carbonate of lime which many animals need for forming their shells. I therefore in my analyses made it a special point, in addition to the other bases, to determine also the soda by a method independent of the assumption quoted; and on calculating my first set of (21) Challenger water analyses, had the satisfaction of finding that they had all given a small surplus of base, amounting on an average to 86 equivalents per 10,000 equivalents of acid present, corresponding (if we assume the excess of base to be present as normal carbonate) to about 0·11 gram of carbonic acid, equivalent to 0·25 gram of carbonate of lime per 1000 grams of sea-water analysed. While recognising the importance of this result, I was keenly alive to the possibility of its having been brought about by a constant positive error in my sum total of base determinations, and accordingly sought for an exact direct method for the determination of the surplus base in a given sea-water.

One of the methods I tried was to distil a measured volume of the sea-water with a certain proportion of sulphate of ammonia; then in a strictly comparative manner to repeat the experiment with an exactly neutral artificial sea-water substituted for the natural sea-water, and to determine the ammonia in the two distillates. There was a distinct surplus of ammonia in the distillate from natural as compared with that from the artificial sea-water, proving the existence of surplus fixed base in the former; but the results were not sufficiently constant to pass for quantitative determinations of the "alkalinity." The problem was subsequently solved in a surprisingly simple manner by the chemists of the Norwegian Expedition.* I lost no time in testing their method by synthetical trials, and finding it trustworthy, applied it to 154 samples of Challenger water. They all proved to be alkaline, but the mean value corresponded to only 54·7 milligrams of carbonic acid (CO_2 present as R_2CO_3) per kilogram of sea-water, which showed that my complete analyses had considerably over-estimated the surplus base. I refer to my memoir in regard to the manner in which I utilised my

* The Norwegian North Atlantic Expedition, 1876 to 1878; Chemistry by Tornøe, Christiania, page 31.

alkalinity determinations in the final adjustment of my numbers for the average composition of ocean-water salts, as deduced originally from the 77 complete analyses; but I must not omit to state that in this final calculation the values for the lime were taken, not from those analyses, but from the results of a later series of more exact determinations made with three mixtures of Challenger waters representative of certain ranges of depth; namely, a mixture (I.) for depths from 0 to 50 fathoms; a mixture (II.) for depths from 300 to 1000 fathoms; a mixture (III.) for depths from 1500 fathoms and more. As shallow shore waters do not occur in the series of Challenger samples, I also analysed, as No. IV., a water which had been collected for me near Port Louis in Arran, Scotland, at a shallow place where there is abundance of marine vegetation.

The same set of four waters had before served for an elaborate research on the relative quantity of bromine in ocean-water salts.

From the 77 complete analyses, as thus corrected and supplemented, I calculated the following numbers for the average composition of ocean-water salts. The numbers under Forchhammer are transcribed from his memoir.

Average Composition of Ocean-Water Salts.

	Per 100 parts of Total Salts.	Per 100 of Halogen calculated as Chlorine.	
	Dittmar.	Dittmar.	Forchhammer.
Chlorine,	55·292 }*	99·848	Not determined.
Bromine,	0·1884 }	0·3402	Not determined.
Sulphuric Acid, SO_3,	6·410	11·576	11·88
Carbonic Acid, CO_2,	0·152	0·2742	Not determined.
Lime, CaO,	1·676	3·026	2·93
Magnesia, MgO,	6·209	11·212	11·03
Potash, KnO,	1·332	2·405	1·93
Soda, Na_2O,	41·234	74·462	Not determined.
(Basic Oxygen equivalent to the Halogens),	(−12·493)	...	...
Total Salts,	100·000	180·584	181·1

* Equal conjointly to 55·376 parts of chlorine, which accordingly is the percentage of "halogen reckoned as chlorine" in the *real* total solids. Compare second footnote on page 138.

Combining Acids and Bases in the (arbitrary) mode shown, we have from my Numbers—*

Chloride of Sodium,	77·758
Chloride of Magnesium,	10·878
Sulphate of Magnesium,	4·737
Sulphate of Lime,	3·600
Sulphate of Potash,	2·465
Bromide of Magnesium,	0·217
Carbonate of Lime,	0·345
Total Salts,	100·000

As a general result of Forchhammer's and my own analyses, *the above numbers may be taken as holding approximately for any sample of ocean-water.* Of the degree of approximation we can form an idea by comparing my numbers for the percentages of chlorine, sulphuric acid, magnesia, and potash, with the corresponding entries in the 77 reports tabulated on pages 23 to 25, and the numbers for the lime there with one another. The percentages of soda being too largely affected by the cumulative error of the other determinations, had better be left out of consideration. But even if we do so, we often meet with fluctuations which are too great to be taken as arising from analytical errors, and consequently must correspond to differences in the actual composition. I have taken great pains in trying to explain these differences by natural causes, but have not been very successful. The final results of my inquiries may be summed up as follows :—

From my analyses (which I do not pretend exhaust the subject), it would appear that the composition of sea-water salt is independent of the latitude and longitude whence the sample is taken. Nor can we trace any influence of the depth from which the sample comes, if we confine ourselves to the ratio to one another of chlorine, sulphuric acid, magnesia, potash, and *bromine.* I emphasise the bromine because, while present in very small proportion, it is taken up preferably by sea-plants, and consequently must be presumed to be more liable than any of the major components to at least temporary local diminution. And yet my analyses of the three mixtures of Challenger waters, and of the Arran water referred to, gave identical values for the bromine present per 100 of chlorine. But the determinations of the lime in the same set of waters make it most highly probable that the proportion of this component increases with the depth. Referring to 100 parts of halogen calculated as chlorine, we find for the quantity of lime :—

* The mode here chosen differs somewhat from the one I adopted on page 138, which represents the carbonic acid as carbonate of magnesia. Neither can claim to be the true mode ; but I now think the one chosen here is the more accurate.

In deep-sea waters—	
Mixture III., ...	3·0307
In surface waters—	
Mixture I., ...	3·0175
Difference, ...	0·0132
In medium depth waters—	
Mixture II., ...	3·0300
In surface waters, ...	3·0175
Difference, ...	0·0125

and either of the two differences is five to six times as great as even the absolute sum of the probable errors of the respective two terms. A discussion of the quantities of lime brought out by the 77 analyses had given a similar result, but exaggerated the difference between deep-sea on the one hand and shallow or medium depth on the other.

But there can be no doubt that, if I had applied even as exact a method in the 77 analyses as I did subsequently in the special investigation on the lime, I should have arrived at a greater difference than 0·013 between certain individual samples.

The result under discussion received a valuable confirmation from the alkalinity determinations to which I had occasion to refer above. Following the example of the Norwegian chemists, I measured the surplus base (*i.e.*, the base left unsaturated by the sulphuric and hydrochloric acid) by the weight of carbonic acid (CO_2) which it would need to convert it into normal carbonate, and referred it to 1 litre of water analysed. But it struck me that in discussing any series of such determinations, they must be referred to a constant salinity, and I accordingly reduced all my numbers to 100 parts of total salts or 55·42 of halogen counted as chlorine; so that with me "alkalinity," as designating a quantity, means "the weight of carbonic acid (CO_2) present as normal carbonate (*i.e.*, in forms similar to carbonate of lime) in every 100 parts of total salts," which, on an average of 130 cases, and if the number of parts by weight of carbonic acid be taken in grams, corresponds to 2·78 litres. Omitting a number of abnormally high or low values, and a few suspected analyses, which left 130 cases for discussion, I found the alkalinity in the whole set to range substantially from 0·140 to 0·164, and then, confining myself to "surface" waters (meaning waters from depths not exceeding 100 fathoms) and bottom waters, and referring on both sides to 100 samples, I found that alkalinities from 0·140 to 0·148 occur preferably in surface waters, while from 0·148 to 0·160 the bottom waters were in the majority. From a graphic representation * showing the frequency of occurrence of certain narrow ranges of alkalinity, I concluded that the most frequently occurring value is

For surface waters, ...	0·146 ± 0·002
For bottom waters, ...	0·152 ± 0·003

* See diagram on p. 130.

which values may be adopted, *provisionally*, for the two kinds of ocean-water. In fifteen cases I was in a position to compare with one another the alkalinity of a surface water and the bottom water at the same Station. In two cases the balance was in favour of the surface water, the numbers being 0·015 and 0·010 respectively; in one case the difference was *nil;* in the remaining twelve cases it was in favour of the bottom water, the differences ranging from 0·002 to 0·019. According to the above two averages the alkalinity of bottom water exceeds that of surface water by 0·006, meaning of course 0·006 grams of carbonic acid per 100 grams of total salts, or 0·014 grams of lime CaO per 100 of chlorine, if we assume the increase in alkalinity to be owing to additional lime. My determinations of the lime, as stated, had shown the presence of 0·013 grams of extra lime in deep-sea as compared with shallow waters. The *closeness* of the agreement is of course accidental. That the surplus base in a sea-water is not owing entirely to carbonate of lime is too obvious to be specially pointed out. In sea-water (as in any mixed salt solution) each base is combined with each acid, and as there are four acids and four bases there must be sixteen salts, the individual percentages of which we have no means of determining. But there are reasons for assuming that the carbonic acid being a feeble acid, is combined chiefly with the weakest bases, and consequently chiefly with the magnesia, and in the second instance with the lime. So we should say, if the arrangement of the bases and acids into salts were a mere matter of tendency to form simple salts. But magnesium has a characteristic tendency to form double chlorides with potassium and sodium, and there is superabundance of chloride of sodium in sea-water. Hence, probably, most of the magnesium is not there as carbonate but as double sodio-chloride, and the lime takes the greater share of the carbonic acid. The alkalinity in any case represents the potential, and may fairly be presumed to measure approximately the actual, carbonate of lime. This is the only answer to that often raised question about the presence of ready-formed carbonate of lime in sea-water, which some chemists, who at the time must have deliberately shut their eyes to the established propositions of chemistry, have endeavoured to solve by direct experiment. Supposing actual carbonate of lime could be extracted from sea water without the co-operation of external matter (I greatly doubt whether this has ever been done), the weight of such extracted carbonate of lime could not reasonably be assumed to be equal to that which was originally present in the water. Sea-water is alkaline, all the alkalinity must be owing to carbonates, and of these carbonate of lime must be one. This is, and for a time is likely to be, the sum total of our knowledge on this point.

When I said that the alkalinities in the samples considered ranged on the whole from 0·140 to 0·164, I meant to hint that these limits do not embrace even all the 130 cases admitted for the general discussion. Less values than 0·140, it is true, do not occur; but there are seven cases in which the alkalinity was decidedly greater than 0·164, as shown in the following table, in which the first column gives the Challenger number of the respective sample, the fourth names the Station, and the fifth gives further geographical notes. "B" in Column II. means that the sample came from the bottom.

I. No.	II. Depth if not B.	III. Alkalinity per 100 of Solids.	IV. Station.	V.
586	B.	0·1707	191A	All these waters came from that archipelago north of Australia, and the Stations lie within the latitudes 5° S. and 18° N., and the longitudes 117° E. and 135° E.
596	B.	0·1693	193	
616	50 fathoms	0·2079	198	
656	B.	0·1704	206	
205	B.	0·1647	97	Atlantic, 11° N., 5° west of coast of Africa.
378	B.	0·1731	152	Southern Indian Ocean, latitude 61° S., longitude of Madras.
878	300 fathoms.	0·1888	240	North Pacific, latitude 35° N., 14° east of Yeddo, Japan.

The very alkaline water No. 616 came from a point close to Celebes, in the Molucca Passage. Of the few anomalous alkalinities which were excluded from the general discussion, those in which the abnormal results could be accounted for by an abnormal condition of the samples (generally the presence in the bottle of mud or other kind of ocean-deposit) may well be passed over. If we do so there remain only two cases, which however are very interesting; I refer to the two following samples: No. 5, a surface water from Station 2, North Atlantic, near the Canary Islands; and No. 31, a surface water from Station 12, about the middle of the ship's track from the Canaries to the West Indies. The latter is one of the waters which had been completely analysed for chlorine, sulphuric acid, lime, magnesia, potash, and soda, long before the alkalinity was determined. Both samples had deposited in their respective bottles large quantities of crystalline matter, which exhibited the reactions of a mixture of carbonates of lime and magnesia, and *may* have included sulphate of lime; but I unfortunately neglected to test for sulphuric acid.

In No. 5 the alkalinity per 100 of salts amounted to only 0·0756; but adding in that corresponding to the lime and magnesia in the deposit, on the supposition of its being all carbonate, I calculated that the original alkalinity must have had the high value 0·291, and the original lime must have amounted to 3·300 per 100 of chlorine instead of the 3·026 brought out as a general mean by the 77 analyses which have been so frequently referred to.

No. 31. There was not enough of this water left for a satisfactory determination of the alkalinity by Tornøe's method, and the calculation of the alkalinity from the complete analysis would be of no value. From the analyses of the deposit and the water-remnant left, and the complete analysis previously made, I calculated that the original water should have contained, per 100 of chlorine, the quantities of lime and magnesia given below and contrasted with those present in average surface water, *according to the* 43 *analyses quoted on pp.* 30 *and* 31—

	Lime.	Magnesia.
Water (No. 31) in its original condition by calculation, . .	3·496	11·163
Average surface or small-depth water, ...	3·018	11·203

Hence, taking 0·146 as corresponding to ordinary surface-waters, I calculate for the surplus alkalinity 0·1779, which, together with the normal value 0·146, gives 0·324 per 100 of salts for the original water. The calculations which led to these high values for Nos. 5 and 31, do not, it is true, rest upon a perfectly secure basis; but I believe the results are approximately correct, and besides the high number 0·2079, which was found quite directly for the normal water, No. 616 is beyond suspicion. I have no doubt that far higher alkalinities than even 0·33 occur locally in many parts of the ocean, wherever there is abundance of carbonic acid and of carbonate of lime or magnesia at the same time. To obtain some insight into the possible extreme limit, I took a sea-water whose alkalinity was 50·2 milligrams per litre, the surplus base being present substantially as bicarbonate, and after having saturated it with carbonic acid, I digested it, in one case with carbonate of lime, in another with carbonate of magnesia. The filtered liquors showed immense alkalinities, the increase being

	Lime.	Magnesia.
Increase of alkalinity, . .	314·2 mgrms. per litre,	1234·0 mgrms. per litre.

A similar set of trials with the *natural* water gave, in the case of magnesia, an increase in alkalinity of 10·6; in the case of lime there was a decrease of 3·2 milligrams per litre.* The decrease of alkalinity caused by the addition of carbonate of lime is difficult to explain. Perhaps it is only the outcome of an observational error; but in any case my experiments show that it would be quite possible for the alkalinity to increase beyond the maxima that occurred in the 130 samples. It is very curious that in my experiments with sea-water saturated with carbonic acid, carbonate of magnesia proved far more abundantly soluble than carbonate of lime. My explanation is that a considerable portion of the magnesia, immediately after having been dissolved by the carbonic acid, suffered double decomposition by the large mass of chloride of sodium present, with formation of carbonate of soda and a double chloride of sodium and magnesium; so that for this part of the process a very small proportion of free or loosely combined carbonic acid would have sufficed, as it always comes back in the reaction, which I suppose to go on, as an equivalent quantity of bicarbonate of soda. The tendency of magnesium to form such double salts explains how those Challenger waters deposited

* Compare page 131.

their large excess of surplus base, as carbonate of lime and not as carbonate of magnesia, although the latter is the weaker and more abundant base. Reference may here be made to certain observations of Sterry Hunt's, which he made in the course of his experiments on the formation of dolomites. He found that a litre of water, containing 3 to 4 grams of sulphate of magnesia, can dissolve 1·2 grams of carbonate of lime, and in addition thereto 1 gram of carbonate of magnesia, forming a strongly alkaline solution, which, on long standing, deposits the whole of its lime as crystals of hydrated carbonate ($CaCO_3$ $5H_2O$).

Our hypothesis, that a small quantity of free carbonic acid in sea-water enables its chloride of sodium to dissolve carbonate of magnesia as a double chloride of magnesium and sodium, would explain the local prevalence in the sea of excessive alkalinity, without the assumption of the presence of any exceptional proportion of carbonic acid. But the presence of excess of carbonic acid, whether the sea be in contact with carbonate of lime or with carbonate of magnesia, strongly adds to its tendency towards an increase in the alkalinity. The question of alkalinity is, in fact, inseparable from that of the carbonic acid in ocean-water, to which we now turn.

[*Editor's Note:* Material dealing with the work on carbonic acid, nitrogen, and oxygen in sea water has been omitted at this point.]

Suggestions for Future Work.

In conclusion, I may be permitted to offer a few suggestions in regard to the manner in which these researches on the composition of ocean-water should be continued.

That they ought to be continued, and extended, and that it is the special vocation of this country to take the matter in hand, will be admitted.

The work involved may be arranged under two heads, one of which would comprise the various kinds of observations and experiments which might easily be carried on by any intelligent seafaring man, even if he were devoid of all professional knowledge of chemistry. I here refer chiefly to,—(1) Salinity determinations by means of the hydrometer and a good thermometer. A set of handy and relatively small hydrometers, graduated so as to give the specific gravity for say 60° F. quite directly and without the aid of attached weights, would easily be supplied by a good mechanician for a few pounds sterling; and any intelligent man would soon learn to use these. Each of Her Majesty's ships should be provided with such a set, and a number of good thermometers, both verified by a scientific chemist or physicist. (2) Observations on the behaviour of sea-water on standing in a bottle of clear hard glass and provided with a good glass stopper, to see whether any deposit of carbonate of lime is formed, and in order to identify the places where the water has already come up to the state of saturation in regard to this component. (3) Alkalinity determinations by the method of Tornøe, as described on pages 106 and 124. The standard solution of hydrochloric acid could easily be provided in large quantities and at a low price, and even when used by itself, *i.e.*, without an auxiliary solution of caustic alkali (which probably only a chemist could manipulate correctly on board a ship), would give valuable approximations in the hands of any intelligent man who had been taught to use it in a laboratory. (4) Rough determinations of the carbonic acid by means of aurine as an indicator, and the normal hydrochloric acid for the alkalinities. I found by experiments made a short time ago that a sea-water becomes neutral to aurine, when, by addition of hydrochloric acid, the ratio of surplus base to carbonic acid has come down to the value 1 [NaOH] to 1·36 or 1·46 times [CO_2]. Hence, supposing 1 litre of a sample of sea-water to contain surplus base equal to 50 mgrms. of carbonic acid as normal carbonate, and 1 litre of the same water, after adding aurine and then hydrochloric acid in the cold, required hydrochloric acid equal to 10 mgrms. of carbonic acid before the violet aurine colour gives way to the yellow tint, then the total carbonic acid present would amount to $(50-10) \times 1{\cdot}4 \times 2$ mgrms. And a sea-water which does not become violet on addition of aurine, but yellow, is sure to contain at least $0{\cdot}41 \times 44$ mgrms. of free carbonic acid for every 1 mgrm. equivalent of base (meaning $\frac{1}{2}$ Na_2O or $\frac{1}{2}$ CaO, etc.) present as bicarbonate, *i.e.*, for every one molecule of bicarbonate CO_3 R′H. Hence, if Tornøe and Buchanan assure us that all sea-water becomes violet on addition of aurine, this *in itself* is quite compatible with the assumption that all the samples which they thus examined contained *free carbonic acid gas* in addition to fully saturated bicarbonate. The free carbonic acid must rise to 0·8 mgrms. for every one mgrm. of CO_2 in the R_2CO_3 part of the bicarbonate (*i.e.*, for every one mgrm. of "alkalinity per litre") before the aurine reaction ceases. Of all the 195 samples of sea-water which Mr. Buchanan analysed for carbonic acid, only two (Nos. 532 and 383) came even approximately up to this limit.

Let seafaring men search for waters which assume a *yellow* colour on addition of aurine. Wherever such water is found a volcanic carbonic acid spring must be close at hand.

Under my second heading fall such kinds of work as demand a skilled chemist for their performance, and it will be convenient to take them up in the order in which they appear in my memoir.

1. *Further researches on the Composition of Ocean Salt.*—By Forchhammer's and my own analyses it is proved that the percentages of the several components are subject to only slight variations. Apart from the one success with the lime, I was not able to trace back the fluctuations to natural causes. Hence new analyses are absolutely useless

unless these are executed with the highest attainable precision. All the components must be determined in the style adopted for the lime (in the supplementary work) and for the bromine. I could not possibly have determined all the saline components in my 77 waters by similarly refined methods for sheer want of material, and besides, the large number of analyses required would have rendered the work almost impracticable.

What ought to be done is to collect waters at different times throughout the year at two stations, one might be selected somewhere in the middle of the Pacific, and a second at some place in the middle of the Atlantic Ocean. In each case two large samples should be taken, one from a little *below* the surface (to preclude abnormal dilution with rain-water), another at some 50 fathoms above the bottom to avoid admixture of solid bottom matter, which in the bottle would gradually dissolve.

Supposing we had, from each of the stations, six surface and six bottom samples, or twenty-four samples in all, we should begin by determining the chlorine in each sample *à haute precision*, and then do the same with the lime. The six samples from each place should then be mixed together (in equal volumes), so as to produce four samples, each representative of one of the four places. In each of them the chlorine, lime, sulphuric acid, magnesia, potash, and alkalinity should now be determined by at least triplicate analyses executed with the highest precision. Ships which happen to pass the localities might be instructed to collect samples as indicated, and bring them home.

This would enable us, before trying to find out the difference between Atlantic water on the one hand and Pacific on the other, to inform ourselves as to the extent to which Pacific or Atlantic water at a given place is liable to vary. But before even this can be done successfully we must have sufficiently exact methods for the execution of the analyses. Hence, first and foremost, a chemist should be appointed to work out (by synthetical experiments in the first instance, and repeated analyses of some one sea-water in the second) a series of methods by means of which the sulphuric acid, magnesia, and potash could be determined with at least that degree of precision which I attained in regard to the lime.

Another useful investigation would be the exact determination of the minor components (iodine, silica, fluorine, iron, aluminium, manganese) in a large mass of some one kind of sea-water. If a chemist succeeded in devising easy and yet sufficiently exact routine methods for determining one or other of these components, its comparative determination in different sea-waters might be undertaken.

2. *Alkalinity.*—Tornøe's method is sufficiently exact, and if applied to a very large number of judiciously-selected samples would be sure to give valuable results.

3. *Carbonic Acid.*—In regard to the methods for determining the carbonic acid, there is room for much improvement. For oceanographic purposes, carbonic acid deter-

minations are of little use henceforth, unless carried out with a multitude of freshly drawn samples, and coupled with alkalinity determinations. The difficulty is to discover a method which would combine high precision with sufficient ease and rapidity of execution.

My method, described in the memoir as the "vacuum method," would work as easily as Buchanan's did on board ship, but either is troublesome, and would become very tedious if duplicate or triplicate analyses were demanded, as they ought to be. The most practical plan, perhaps, would be to combine the determination of the carbonic acid with that of the nitrogen and oxygen, as proposed by me on page 105; that is to boil out the gases in Jacobsen's apparatus in the presence of hydrochloric acid, to seal them up, and subsequently analyse them at home.

4. *The Absorbed Oxygen and Nitrogen.*—Jacobsen's method is the only one which would work on board ship, and it certainly is susceptible of a fair degree of exactitude. But in any future expedition it would be desirable to have all gas-extractions done in duplicate or triplicate, in order to supply the one item without which no series of analyses can be properly discussed, namely, the probable error in the single determination.

Meanwhile the best thing that could be done in regard to all the analytical problems referred to would be to work many times on samples of the same kind of water, with a view of improving upon the methods and ascertaining the extent to which that one water fluctuates in its composition.

28

Reprinted from *Geog. J.* 16:469–471 (Oct. 1900)

THE PETTERSSON-NANSEN INSULATING WATER-BOTTLE.*

By HUGH ROBERT MILL, D.Sc., LL.D.

PROF. PETTERSSON has, in conjunction with Prof. Nansen, completed a modification of his well-known apparatus for obtaining samples of sea-water without change of temperature. A specimen of the improved water-bottle constructed by Messrs. L. M. Ericsson & Co., of Stockholm and London, was exhibited in the museum arranged for the illustration of papers read to the British Association at the Bradford meeting. The purpose of this apparatus is to enclose a quantity of sea-water at any desired depth, to hold it securely, and to bring it to the surface without any change of temperature exceeding one-hundredth of a degree Centigrade. The previous form of insulating water-bottle was found by Dr. Nansen in his arctic expedition to be less trustworthy at great depths than in shallow water, hence the suggestions which resulted in the new apparatus.

The insulation, which is the essential feature of the water-bottle, is secured by a series of concentric chambers of non-conducting material, which are simultaneously filled with water, and so protect the portion, measuring about 2 litres, which occupies the large central tube. The cylindrical body is so constructed as not to become heated by compression at the greatest depth. This is effected by using metal, which is heated by compression, and indiarubber or ebonite, which is cooled by compression, in such proportions as to ensure constancy of temperature for the whole structure.

The water-bottle when set (see Fig. 1) is held apart so that the base, cylindrical body, and lid are separated, and the water passes freely through the concentric tubes, which occupy the cylindrical body, as the apparatus descends. When the apparatus is being drawn up, the propellor (which, during the descent, revolves freely) engages with a screw and releases the shackle supporting the lid. A heavy weight hung from the sides of the lid causes it to drop on to the top of the cylinder, which in turn is driven against the base, and the three parts of the

* Read at the British Association meeting at Bradford, September, 1900.

water-bottle are locked rigidly together (see Fig. 2). On closing, the indiarubber discs which cover and project from the lower surface of the lid and the upper

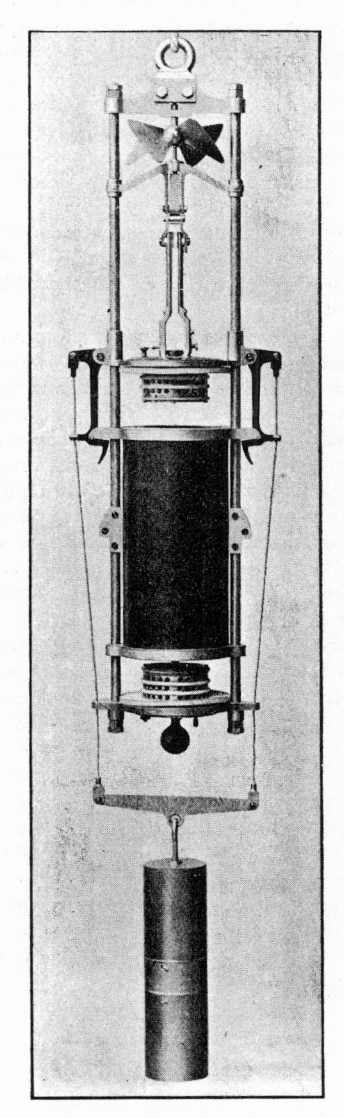

FIG. 1.—PETTERSSON-NANSEN WATER-BOTTLE DESCENDING, SET AND READY FOR USE.

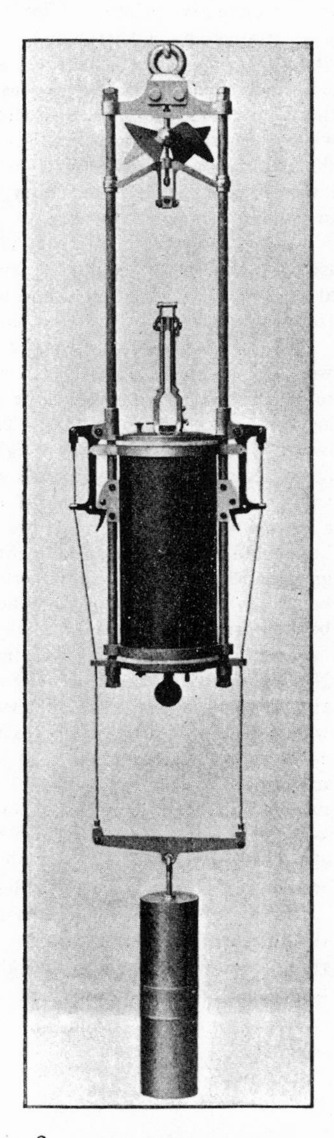

FIG. 2.—DITTO ASCENDING, CLOSED AND LOCKED.

surface of the base-plate completely shut all the concentric tubes, and prevent any movement in the enclosed water. An arrangement is provided for the relief of pressure as the included water expands on being hauled up, the indiarubber ball seen in the photograph below the base-plate being in communication with the central tube. The propellor can be adjusted so as to release the catch after one, two, or more revolutions. After it has done so, it revolves freely. The temperature is ascertained by a thermometer, protected against pressure (the metal case of which is alone shown in the figure), enclosed in the central tube, and projecting sufficiently far to be easily read. If preferred, the aperture for the thermometer may be closed (as shown) by a screw, and the thermometer inserted when the water-bottle is brought up. After the temperature has been observed, the water may be drawn off from the central tube by an ingenious stopcock, the lever of which works horizontally, so as to run no risk of being opened by pressure in moving through the water.

A reversing thermometer to give the temperature of the water independently may be attached to the upper part of the water-bottle by detaching the ring shown at the top of the instrument and bolting on a metal frame with a ring above. The thermometer is set in action at the moment of closing.

The whole apparatus weighs about 50 lbs., and is used on a wire line and worked by a steam-winch. Its framework consists of two solid brass rods parallel to one another, supporting the propellor frame above, the base-plate below, and the locking-blocks in the middle. The cylindrical body and the lid slide on these rods. The lower part of the weight, which serves both as a sinker and the actuator of the locking-gear, is made hollow and detachable, closed below by an indiarubber valve opening inwards, so that if it strikes the bottom it will enclose and bring up a sample of the sediment. However, it appears to me to be inexpedient to risk so heavy and delicate an instrument in such close proximity to the bottom, since at great depths it would almost certainly fall on its side and get clogged with mud before the running out of the line could be checked. It will probably be found better in practice to attach the deposit collector to the deep-sea sounding-lead.

During August, 1900, the improved water-bottle has been tested by Prof. Nansen on board the *Michael Sars* in the sea between Iceland and Spitsbergen, and at the greatest depth met with (3000 metres; 1670 fathoms) the insulation was perfect. On August 11 a sample was taken from 3000 metres, and when it came up the thermometer read $-1°\cdot285$ C.; after 5 minutes, $-1°\cdot283$; after 9 minutes, $-1°\cdot270$; and after 11 minutes, $-1°\cdot210$. On August 13, from 2000 metres (1110 fathoms), the thermometer showed $-1°\cdot135$; after 5 minutes, $-1°\cdot135$; after 6 minutes, $-1°\cdot130$; and after 8 minutes, $-1°\cdot110$. It is considered essential to use an included thermometer to secure exact results when working in polar waters, for which, indeed, the water-bottle was specially designed.

Prof. Nansen has also experimented with an insulating water-bottle in which the insulating material consists, in addition to the concentric water-tubes, of a ring of eighty exhausted glass tubes, similar to the vessels used by Prof. Dewar in his experiments on liquid gases. Its insulation was proved to be perfect.

29

PRACTICAL MANUAL OF THE ANALYSIS OF SEA WATER. I. CHLORINITY BY KNUDSEN'S METHOD

Mieczyslaw Oxner
Assistant at the Musée Océanographique

with a preface by

Martin Knudsen
Professor at the University of Copenhagen
Director of the Hydrographic Laboratory

This preface was translated by Margaret B. Deacon expressly for this Benchmark volume, from "Manuel Pratique de l'analyse de l'eau de mer. I. Chloruration par la méthode de Knudsen," in Comm. Int. Explor. Scient. Mer Méditerr. Bull. 3:1–36 (April 1, 1920)

PREFACE

It is Professor Otto Pettersson to whom the merit is due of having introduced chlorine titration into hydrography, that is to say the establishment by direct titration of all the determinations of salinity obtained by systematic hydrographic researches. Although individual titrations can be made with both precision and speed, the determination of the concentration of a solution requires certain work which must be done very meticulously and of which only highly trained and skillful chemists are capable. It became apparent, in fact, after the introduction of the method of chlorine titration in other countries, that the numerical results found for the same water sample varied from each other in an undesirable way.

A series of analyses made with the view of studying the salinity of the surface water of the North Atlantic Ocean, on the shipping route between Scotland and Iceland, especially attracted my attention to the fact that to give real significance to the salinity determinations made by different people at different times in these waters required safeguards greater than those available up to the present time. It was to this end that I suggested to the First International Conference on the Exploration of the Sea, which took place, at the proposal of Professor Otto Pettersson, at Stockholm in 1899, the establishment of an international institution of which the principal aim would be the production of a standard sea water (*eau normale*) for use in all kinds of titrations of sea water. According to the suggestion I made in print (*Conférence Internationale pour l'exploration de la mer*, Stockholm 1899, *Supplément 4, p. XLII*), the proposed institution would equally concern itself with other researches and examinations of importance in hydrographic studies.

This proposal without doubt contributed to the nomination of a commission composed of MM. John Murray, Knudsen, Pettersson, Nansen, Krümmel, H. N. Dickson and Makaroff, charged with experimentally establishing the constants of sea water at the Polytechnic in Copenhagen. According to the terms of the resolution adopted, the researches were to aim at experimentally determining the relations that exist between the quantity of halogens and the density; it also expressed the urgent need for the revision by experiment of tables by Makaroff, Krümmel and others, for the reduction of the specific gravity of sea water, and for an exact determination of the relations that exist between density and salinity.

Thanks to the cooperation of various scientists, physicists and chemists, the determinations of the relevant constants were made within the space of two years, and I thus found myself able to lay the Hydrographic Tables, established on the basis of the completed researches, before the second Conference for the exploration of the sea held at Christiania in 1901. The rules to be followed in making titrations were drawn up at the same time as the constants were determined, and in such a way that the Hydrographic Tables could be used to evaluate the numerical results of analysis. Several of the water samples whose constants had been determined were enclosed in sealed glass tubes for future use. These tubes, approximately 200 in number, marked the beginning of the standard sea water used at present.

The tubes were distributed among the different countries who had been participating in the research and at the setting up of the Central Laboratory at Christiania in 1903, under the direction of Professor Nansen, several of the tubes were sent to this institution where they served as the basis for the production of standard sea water which was made there until the closure of the Central Laboratory in 1908. I was then invited to take charge of the production and distribution of standard sea water again.

Through the use of standard sea water it becomes possible to make an exact comparison between the titrations made by different people at different times, and the practical execution of titration itself is simplified to the point where it is not necessary to be a chemist to make the analyses and calculations with great accuracy. Detailed instructions are, however, an absolute necessity for those who have no knowledge of chemical experiment, and I have often regretted the lack of a manual of this kind whenever I was asked how to make the titrations. In consequence, I am especially pleased that Dr. Oxner has been so good as to take upon himself the difficulties involved in the preparation of this kind of guide.

[*Editor's Note:* Oxner's description of the chlorine titration method is too long to reproduce here. His list of references is given below.]

INDEX BIBLIOGRAPHIQUE

1. (1901) Thoulet (J.), Échantillons d'eau et de fonds provenant des campagnes de la "*Princesse-Alice*." (Résult. des Camp. Scient. Fasc. XXII. Monaco 1902).
2. (1890) Thoulet (J.), Océanographie statique.
3. (1883) Attlmayr (F.), Handbuch der Oceanographie Bd. I.
4. (1910) Krümmel (O.), Handbuch der Ozeanographie Bd. I., 1907.
5. (1910) Richard (J.), L'Océanographie.
6. (1912) Helland-Hansen (B.), The Ocean waters . . .I general Part. (Internationale Revue der Ges. Hydrobiologie).
7. (1904) Sabrou, (Bull. Mus. Océanogr. No. 22, 10 décembre 1904).
8. (1901) Knudsen, Hydrographical Tables.
9. (1902) Knudsen, Berichte über die Konstantenbestimmung (Mém. Ac. Sc. de Danemark. 6e Serie t. XII. No. 1).
10. (1903) Knudsen, On the Standard-Water used in the Hydrographical Research . . . (Publ. de Circonst. No. 2).
11. (1904) Knudsen, α^t Tabelle (Public. de Circonst. No. 11).

Part VI

DEPTHS OF THE OCEAN AND THE SEA BED

Editor's Comments on Papers 30 Through 36

It is a curious fact, considering how little was known about the element, that Renaissance philosophers, discussing how best to measure the depth of the ocean, frequently expressed the fear that because of the intervention of currents the seaman's lead and line would be less reliable than some kind of mechanical device. This was the reasoning that lay behind apparatus such as the lineless sounder (Papers 1 and 2). Many different kinds of sounding machine were proposed during the eighteenth and nineteenth centuries. Stephen Hales, 1677–1761, who was best known for his work on physiology (Clark-Kennedy 1929), constructed an instrument based on Boyle's law: that the volume of a gas varies inversely with pressure. It consisted primarily of a musket barrel up which the water rose as the depth increased. Red dye floating on the surface of the water inside the barrel would leave a mark at its highest point on a fixed rod. When the apparatus returned to the surface the rod could be removed and the height reached by the water measured. Unfortunately, Hales relied on a wooden float to bring the apparatus up again and, as we would now expect, at the trial related in Paper 30, it never reappeared.

As well as manufacturing increasingly successful mechanical depth and pressure gauges, nineteenth-century scientists discussed the possibility of using sound and other waves to measure depth. William Henry Fox Talbot (1833), the pioneer of photography, suggested that depth could be calculated from the time taken by the sound of a shell exploding on the sea bed to reach the surface. He wrote:

> The method which I would propose, with some hope that it would prove successful, is to let fall from the deck of a ship one of the newly-invented percussion shells, which would explode on striking the ground; and the interval of time before the explosion was heard, would give the depth of water with great accuracy. The experiment should be first tried in a known depth of water, say a hundred fathoms, or whatever lesser depth would be consistent with security. The descent of the shell through the water would after the first few seconds be uniform, as is well known to be the case with all heavy bodies moving in a resisting medium. The time taken by the sound in returning through the water might be neglected, unless great accuracy were required; since it would move at the rate of a mile in half a second.
>
> If it should be objected that the report of the shell might not be audible at great depths, I would remind the reader that in M. Colladon's experiments the sound of a *bell* was distinctly heard through the water of the lake of Geneva for a distance of *nine miles.* (Talbot 1833, p. 82)

Echo sounding was attempted by an American scientist, Charles Bonnycastle in the 1830s (Drubba and Rust 1954). It was not, however, until the development in the early twentieth century of the necessary electronic equipment for receiving and amplifying signals that this method became practicable. Using wave theory, Alexander Dallas Bache (1856) was able to calculate the average depth of the Pacific by working from the time taken by seismic surges generated by an earthquake in Japan to reach California in 1854.

Thus, in spite of all these developments in theory, man's first acquaintance with the ocean bed came at last with the lead and line. Sir James Clark Ross (1847) made several soundings in the region of 4000 fathoms during his Antarctic voyage of 1839–1843. American surveyors took up deep-sea sounding in the 1840s, using very thin line, and reported depths of up to 8000 fathoms.

When the possibility of a submarine cable between Great Britain and the United States arose it was necessary to make precise surveys to select the best route. Matthew Fontaine Maury, 1806–1873, studied the soundings available and found that depths of 5000–7000 fathoms had been reported because the surveyors, when timing the rate at which the rope went out, had missed the first slight check which meant that the rope had hit the bottom (Paper 33). When Commodore Cadwallader Ringgold of the U.S. North Pacific Surveying Expedition reported a great depth, Maury wrote to him:

> I hope your cast off the Cape of Good Hope was taken from a boat and well timed! 8000 fathoms—9 miles—is a very deep hole, and I hope therefore the data and particulars which you furnish will enable us to speak positively on the subject. (12 June 1854)

For deep soundings up to and including the *Challenger* expedition, rope was used; but Sir William Thomson, Lord Kelvin, 1824–1907, had already in 1872 developed an apparatus for sounding with piano wire, Paper 34, which was both strong and offered less resistance to the water than rope—an important consideration when precious instruments had to be hauled up from a depth of several thousand fathoms. Thomson's sounding machine was successfully used by the U.S.S. *Tuscarora* under George E. Belknap while surveying a telegraph cable route between America and Japan in 1874 (Deacon 1971, p. 350) and rapidly became a standard piece of equipment.

The first deep soundings in the North Atlantic showed the sea bed to be made up of foraminiferal ooze (see Editors Comments for Papers 37–39). It was not until the *Challenger* expedition (Deacon 1971, ch. 15; 1972) that the full range of marine sediments was discovered (Murray and Renard 1891). The expedition found, Paper 35, that within a few hundred miles of land the terrigenous sediments were replaced by others purely marine, or pelagic, in character. They found globigerina ooze extensively on the deep-sea bed but below a depth of about 2000 fathoms it disappeared and was replaced by another formation that they named red clay. The reason for the disappearance of the globigerina ooze was found to be the solvent action of the carbonic acid in sea water which attacked the calcium carbonate of which the globigerina shells are composed. At first it was thought that red clay might be the residue of this process but Murray showed that it was formed by the decomposition of volcanic material. The presence of fossilized bones and teeth showed that it accumulated extremely slowly. Associated with it were large collections of manganese nodules which Buchanan analyzed but which defied attempts at explanation. In far southern regions where the surface waters were too cold for globigerina their place was taken by diatoms and their remains predominated on the sea bed. Being made of silica, both diatom and radiolarian oozes were found at all depths since they are not affected by the increased effect of carbonic acid in deep water.

The geological results of the *Challenger* became the special province of Canadian-born Sir John Murray, 1841–1914, one of the naturalists on the expedition. He was assisted in their preparation by the Belgian geologist Alphonse Renard, 1842–1903.

By the early nineteenth century, the nature of geological stratification was well understood. Many of the best-known fossil faunas were marine and on the basis of Lyell's theory of uniformitarianism, or continuity of change through continuously operating causes, it was assumed that land and sea periodically changed places. The *Challenger* results were among new evidence that led some geologists and others to assert a contrary view—the concept of permanence of ocean basins (Mill 1893). Murray was led to adopt this view because of the distinctive character of the deep-sea sediments that were absent from the known geological successions. Geologists such as Suess, on the other hand, pointed out the great extent of marine transgressions over the continents. Suess coined the term "eustatic" to describe these worldwide changes in sea level, though Darwin had mentioned them in connection with his atoll theory, half a century before.

Other evidence seemed to show that the level of the water in the oceans had at times been many hundreds of meters lower. This was deduced by geologists studying the submarine canyons of the continental shelves which they believed could only have been formed by subaerial erosion. Great emergences were also suggested by biologists seeking to explain how, in the light of Darwin's theory of evolution, closely related plants and animals could have reached places like Madagascar, now divided from the main continental masses by water of moderate depth.

Some interesting early work on submarine erosion, though its significance was not fully appreciated until comparatively recently, was done by John Milne, professor of Seismology at the Imperial College of Tokyo in the 1880s and 1890s. He linked cable breaks on the sea bed with submarine earthquakes and pointed out that in many cases it was not the quake itself but the vast landslips generated by it on the continental slope that caused the breakage. He wrote:

> The concentration of detritus derived from continental surfaces along coastlines on tracts which are comparatively small, indicates that beneath the sea the growth by sedimentation is greater per unit area than the similarly estimated loss is by denudation on the land. This rapid submarine growth, largely under the influence of gravity, but modified by hydrodynamic action, leads to the building up of steep contours, the stability of which may be destroyed by the shaking of an earthquake, the escape of water from submarine springs, the change in direction or intensity of an ocean current, or by other causes which have been enumerated. That submarine landslides of great magnitude have had a real existence is proved for certain localities by the fact that after an

> interval of a few years very great differences in depth of water have been found at the same place, whilst sudden changes in depth have taken place at the time of and near to the origin of submarine earthquakes (see p. 272). Large ocean waves unaccompanied by volcanic action indicate that there have been very great and sudden displacements of materials beneath the ocean. The most important evidence of sub-oceanic change is, however, to be found amongst the archives of the cable engineer. The routes chosen for cables are carefully selected as being those where interruptions are least likely to occur, and yet, as it has been shown, something which is often of the nature of a submarine landslip takes place and some miles of cable may be buried. Here we seem to have proof positive, especially along the submerged continental plateaus, of sudden suboceanic dislocation. Because these changes are frequent, it is reasonable to suppose that sedimentation and erosion, and other causes which lead up to the critical conditions, are geologically rapid. (Milne 1897, pp. 281–282)

We now know that it is commonly the turbidity currents triggered off by the slumping of sediments on the continental slope that help scour out the tremendous canyons found there.

A comparable debate took place over the origin of coral reefs. Adelbert von Chamisso (1822), naturalist on Kotzebue's first voyage, had supposed that the coral atolls of the Pacific were formed by reefs growing on the tops of submarine banks. He wrote:

> Their situation with respect to each other, as they often form rows, their union in several places in large groups, and their total absence in other parts of the same seas, induce us to conclude, that the corals have founded their buildings on shoals of the sea; or, to speak more correctly, on the tops of mountains lying under water. On the one side as they increase, they continue to approach the surface of the sea, on the other side they enlarge the extent of their work. The larger species of corals, which form blocks measuring several fathoms in thickness, seem to prefer the more violent surf on the external edge of the reef; this, and the obstacles opposed to the continuation of their life, in the middle of a broad reef, by the amassing of the shells abandoned by the animals, and fragments of corals, are probably the reason that the outer edge of the reef first approaches the surface. (Chamisso 1822, p. 38)

Charles Darwin (1842) believed, on the other hand, that atolls and barrier reefs originated as the fringing reefs growing along the shore of islands or mainland coasts which had then subsided. Darwin's views were endorsed by James Dwight Dana, 1813–1895, naturalist on the United States Exploring Expedition of 1838–1842 (Dana 1853). He added further proofs of subsidence, including the existence of drowned erosional valleys on some of the mountainous island groups of the Pacific. In Papers 31 and 32 Dana examines the influence of temperature and ocean currents on the distribution of reef-building corals and the area over which subsidence had apparently taken place. The Darwin and Dana theory of coral reefs was later attacked by some scientists, including John Murray, on the basis that the mechanism of subsidence was not explained and that other evidence suggested that the ocean basins were permanent. Murray (1880) preferred an explanation of atolls resembling Chamisso's. Recent work has shown that subsidence can be explained in two ways; volcanoes may sink below the surface as the slopes of the cooling mid-ocean ridges subside or, if formed on the ocean bed, sink isostatically as the weight of extruded material displaces the earth's crust below (Menard 1969). Neither explanation is considered completely adequate by Fairbridge (1966, p. 786) who also considers a shift of the poles relative to the crust as causing a geoidal change in sea levels.

The forces behind major earth movements were not at all well understood in the nineteenth century. It was generally supposed that the newly formed earth had contracted as it cooled and that down-faulting of large portions of the crust might have been the cause of the ocean basins. In 1879, G. H. Darwin suggested that the Pacific Ocean had been formed by the moon's breaking away from the earth. The Reverend Osmond Fisher, geologist and mathematician, suggested in his book *Physics of the Earth's Crust* (2nd ed. 1889) that the remainder of the earth's crust had divided up and drifted to its present position. He realized that the low-density continental rocks were of a type distinct from the denser rocks of the ocean floor and thought that the continents had been moved by convection currents in the earth's liquid substratum, rising beneath the oceans and descending beneath the continents. He wrote:

> Let us now consider what would happen beneath the oceans, where the ascending currents impinge. The liquid tending to spread laterally will produce a tensile stress in the central parts, which will become converted into a compressive stress as the continental areas are approached. In the central parts we may therefore expect that the crust will be fissured, and that volcanic eruptions will be the consequence. This may explain why so many volcanic islands are found in mid ocean, and why so many eruptions take place in the bed of the sea, even where no permanent volcanic islands are formed. It is evident that whatever amount of compression is caused by this kind of action in the continental areas, must have its correlative extension in the width of fissures beneath the oceans, which will become dykes of igneous rock in the suboceanic crust. (Fisher 1889, p. 322)

Fisher's ideas did not have much impact in Britain or America. A. Hallam (1973) in his history of the idea of continental drift allies his work more closely to geological thought in Germany. The American scientist F. B. Taylor in 1910 and slightly later, the German meteorologist and geophysicist Alfred Wegener both put forward arguments for continental drift, though neither were able to propose convincing causes and this weakness gave their opponents grounds for rejecting their theories *in toto.* Stronger cases were made by the British geologist Arthur Holmes, who revived the idea of convection currents, and Alexander du Toit who demonstrated the geological relationship between Southern Africa and South America and showed that the continents fitted together much better if one took the dividing line at the edge of the continental shelf, rather than at the coastline.

During the 1950s, the results from several different branches of submarine geology gradually disposed marine geologists to think more favorably of the idea of continental drift (Schlee 1973). Maurice Ewing's seismic refraction work with the *Vema* and other ships belonging to the Lamont Geological Observatory of Columbia University showed that the sediments of the ocean floor were much thinner than anticipated but became thicker as one approached the continents. Furthermore, dredging from sea mounts revealed no rocks older than Cretaceous times. The mid-ocean ridge of the Atlantic, first discovered by the *Challenger* (Deacon 1971, pp. 354–355), was found to extend right round the world. Sir Edward Bullard, the Cambridge geophysicist, discovered a concentration of high heat flow through the ocean bed, precisely through the mid-ocean ridge. It seemed at length that these and other facts (e.g., the absolute dates of the crust) could only be explained by viewing the suboceanic crust as a kind of conveyor belt, formed at the mid-ocean ridge and moving out on either side to the continental margins, where in certain regions it sank below the continental blocks creating the deep offshore trenches, found principally around the Pacific, and associated earthquake and volcanic activity. The theory was ventilated by Harry Hess in 1960 but the first published account of it was by R. S. Dietz in 1961, Paper 36.

New discoveries confirming the idea of continental drift and seafloor spreading soon appeared, for example the recognition of F. J. Vine and D. H. Matthews (1963) of mirror-image patterns of magnetic anomalies on opposite sides of the mid-ocean ridges. Since then, the subject has developed rapidly to the point where many of the old questions seem to have been reopened. Plate tectonics has replaced simple continental drift; and convection currents are no longer considered adequate as the motive force for the movement of the plates which, it is now thought, move as entities rather than the continental blocks and upper mantle moving independently of each other. This revolution in geology stemming from the sea has been compared to Darwin's discovery of evolution. It has had a major impact on all the earth sciences during the last two decades and will apparently continue to do so for some time to come.

30

Reprinted from *Gentleman's Magazine* 24:215–219 (May 1754)

A DESCRIPTION OF A SEA GAGE, TO MEASURE UNFATHOMABLE DEPTHS

Rev. Stephen Hales, D.D. F.R.S.

[*Editor's Note:* Because of the poor quality of this reprint, the text has been re-typed and follows.]

IN my *Vegetable Statics*, under experiment 89, I propoſed a method for finding out the depth of the ſea where it is unfathomable; which method the ingenious Dr *Deſaguliers* put in practice before the Royal Society, with a machine which he contrived, the deſcription of which he has given in the *Philoſophical Tranſactions*, numb. 405. I ſhall here give a more particular deſcription how to prepare and graduate this ſea gage.

Suppoſe A B (*Fig.* I.) to be an iron tube or muſket-barrel of any length, as 50 inches, having its upper end A well cloſed up. If this tube be let down in this poſition about 33 feet into the ſea, a column of water of that height is nearly equal to the middle weight of our atmoſphere, and conſequently from a known property of the air's elaſticity, it will be compreſſed into half the ſpace it took up before; ſo that the water will aſcend half way up the tube; and if the tube be let down 33 feet deeper, the air will be compreſſed into one third of its firſt dimenſions, and ſo on, one fourth, one fifth, one ſixth *&c.* the air being conſtantly compreſſible, in proportion to the incumbent weight; whence, by knowing to what height the water has aſcended in the tube, we may readily know to what depth the tube has deſcended into the ſea.

Now to meaſure the depth of one of theſe columns of ſea-water; firſt, by a line, let the iron tube, with a weight at its bottom, ſink about 33 feet, which depth, in ſalt water, will nearly anſwer to the weight of the air at a mean height of the barometer; then draw up the tube, and obſerve how far the water roſe. If 33 feet of water is equal to one atmoſphere, then will the water riſe ſo high as to fill exactly one half of the tube. But if the water riſes higher or lower than half way, then, by the rule of three, ſay, as the number to which the water riſes, is to one, ſo is thirty-three, to the number of feet meaſuring the depth of the column required. For example, ſuppoſe the water riſes (when the tube is let down 33 feet) only nine tenths of half way, then ſay as 9 : 10 : : 33 : 36$\frac{2}{3}$ feet, the depth of each column, which being once known, the number of columns of water is to be multiplied by this number of feet, whereby the depth of the ſea in feet will be known.

But ſince when the inſtrument has deſcended to the depth of 99 columns, or 99 times 33 feet, the air will be compreſſed into the one hundredth part of fifty inches, that is half an inch, the diviſions both ſome time before and after that will be ſo very ſmall, that the difference in depth of ſeveral columns of water, will not be ſenſible. So that an inſtrument of no greater length than this would ſcarcely give an eſtimate of half a mile's depth, that is, 2640 feet or 80 columns depth of water. The lengthening of this inſtrument to four, five or ten times this length, would obviate this defect, and make the difference of the degrees of deſcent much more ſenſible. But ſince it is impracticable to make a metalline tube of ſo great a length, and if it were made, it would be ſo unweildy as to be eaſily broken, I propoſe to obviate theſe difficulties by the following method *viz.*

Let there be a globoſe metalline body made of iron or copper, nearly of this form (*fig.* II.) K, L, M, N, Q, whoſe capacity within ſide may be equal to nine times the capacity of the metalline tube Z K L. Let this globoſe body be firmly ſcrewed to the metalline tube, at K L, with a leathern collar well ſoaked in

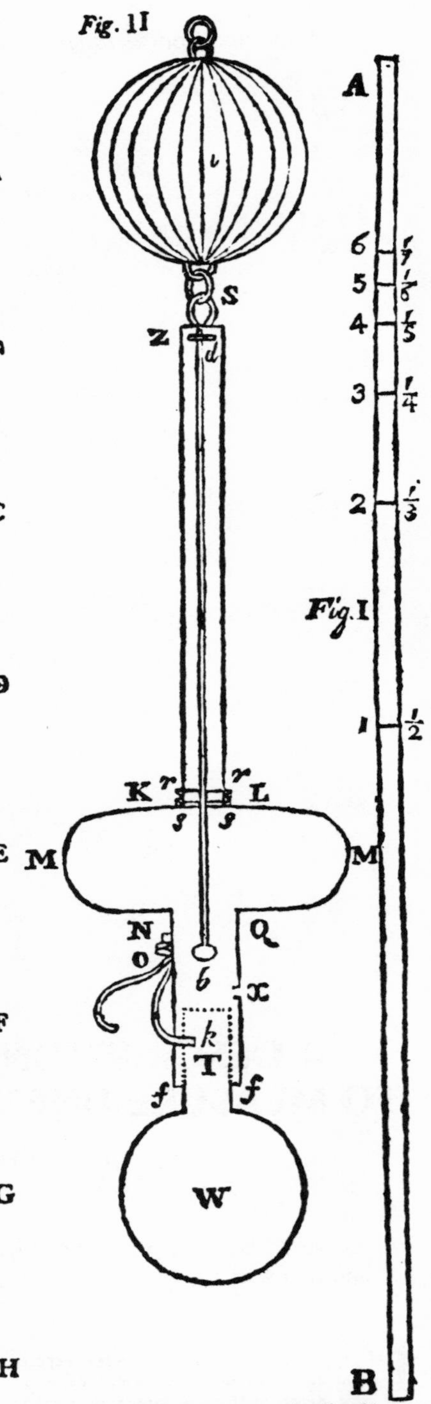

ſome unctuous matter, at the ſhoulder or joining to, thereby to ſecure that joint in the

most effectual manner. Let there be a small hole at *x* for the sea water freely to enter in at, and let some coloured oil, red, blue, or any other colour, be poured into the globose body, to fill it up to the hole *x*. Let there also be provided a slender iron, brass, or wooden rod *d b*, screwed, or somehow fasten'd into the metalline diameter *s s*, which diameter must also be made to screw in and out, thereby to take out the rod at pleasure. The rod must also have fastened to its upper end a small button *d*, which will prevent its being daubed, by falling against the sides of the tube.

The capacity of the tube must be estimated by pouring water in, when the rod and metalline diameter are fixed in their places.

Now, since the lower vessel is supposed to contain nine times as much air as the tube Z L, which is the same thing as if the tube is nine times as long; therefore the air in the globose vessel will not all be forced within the capacity of the tube till the vessel has descended to the depth of nine columns, or nine times 33 feet, for then the air will be compressed within one tenth of the space it at first took up.

Supposing therefore the instrument to have descended to the depth of 99 columns of water, or 99 times 33 feet, *viz.* 3267 feet, then the air will be compressed within one hundredth part of 500 inches (for the capacity of the whole vessel was supposed equal to a tube of that length) that is, within five inches of the top of the tube, and consequently the rod *d b* will be found tinged with the coloured oil within five inches of its top.

Suppose again the instrument to have descended to the depth of 199 columns of 33 feet each, then the air will be compressed within one 200th part of the whole, that is nearly within two and a half inches of the top of the tube. In this case the instrument will have descended 6567 feet; that is, a mile and a quarter, and 132 feet.

Suppose again the instrument to have descended to the depth of 399 columns, then the air will be compressed within one 400th part of the whole, that is, nearly within one inch and a quarter of the top of the tube. In this case the instrument will have descended two miles and a half, wanting 53 feet, which may probably be the greatest depth of the sea.

Hence we see, that if there were occasion to explore greater depths of the sea than this, it might be done with tolerable accuracy, by enlarging the capacity of the globose vessel K, L, M, N, Q, which might easily be done, without rendering it very cumbersome; for suppose the capacity of the tube Z K L were about three fourths of an inch, *viz.* common musket barrel bore, and that it were fifty inches long, if the globose vessel were nineteen times as big, it would not in that case exceed the bulk of three gallons; but the bigger the globose vessel, the greater care must be taken to secure well the screw joint *r r*, that no air pass that way, or water press in. And indeed I am not without some apprehensions, that when the air shall be compressed with a column of water of two or three miles, it may be forced thro' the pores of the iron; and if this should be the case, the most likely way to prevent this inconvenience would be to line the iron tube with one of glass.

The bigger the globose body is, the more weighty it ought to be, thereby the more effectually to keep it in a low depending posture, else the buoyancy of the air in it might raise it as high or higher than the upper part, whereby water rushing in to the top Z of the tube, no observations could be made, the rod being thereby wetted from end to end. When one experiment has been made, the rod and tube must be wiped very clean before another be repeated.

This sea-gage being thus prepared, a large buoy *i*, must be fixed to it, which ought to be of a large solid piece of fir, or any other light solid wood, well covered with tar, to prevent any water's being pressed into its sap vessels: For if the buoy be made of a bladder, or hollow globe, with its orifice inverted downwards, the air in them will be compressed to such a degree, at great depths, as thereby to make the buoyant body become specifically heavier than the sea water, which will prevent its re-ascending to the surface of the sea; for which reason also the buoy ought to be able to buoy up the instrument when full of water. Besides, if the buoy, when it rises again, do not appear some considerable height above water, it will not easily be discovered; for 'tis probable, that from great depths it may rise at a considerable distance from the ship, tho' in a calm. It will therefore be adviseable to fix on the top of the buoy broad fans of tin, painted white or black.

For greater accuracy it will be needful first to try this sea-gage, at several different depths, down to the greatest depth that a line can reach, thereby to discover, whether, or how much, the spring of the air is disturbed or condensed, not only by the great pressure of the incumbent water, but also by its degrees of warmth or coldness at great depths, and in what proportion at different known depths, and in different lengths of time; that allowance may accordingly be made for it at unfathomable depths.

And, because 'tis probable, that the temper of the air, when the experiment is made, will be either warmer or colder than that of the sea at a considerable depth, it will therefore be adviseable to let down the instrument with a line to a good depth, there to continue for some time, till the air in it may be supposed to come to the same temper with the sea-water; then the instrument is to be pulled up so far above water as to let the air freely pass thro' the hole *x*, either in or out, according as the former included air shall have either dilated or contracted; then let the instrument loose to drop down to the bottom of the sea, which it will do by means of a sinking weight of ballast, which must be fixed, in the following manner, to the bottom of the instrument, *viz.* W is a weight of ballaſt hanging by its ſhank T. in the ſocket *ff*, which ſocket is ſcrewed faſt to Q. The ſhank is retained in its place by the catch *k* of the ſpring O, while the machine is deſcending; but as ſoon as W touches the ground at the bottom of the ſea, the catch O *k*, ſinking by its deſcending force, a little below the upper part of the hole *k*, is thereby at liberty to fly back, and ſo lets go the weight; then the buoy riſes up to the ſurface of the water with the machine. — Springs might alſo be fixed on the inſide of the ſocket *ff*, ſo as to fly back in the ſame manner, and let go their hold as ſoon as the weight touches the ground.

The weight of the ſinking ballaſt W ought to be ſo proportioned as to be juſt ſufficient to ſink the machine at firſt, for, as the machine deſcends, it grows continually ſpecifically heavier, by reaſon of the condenſation of the air in its cavity, on which account its motion will be accelerated, as well as on account of the inceſſant action of the power of gravity upon it; ſo that if this gravitating power far exceeded the contrary renitency of the buoy, it would ſtrike the bottom of the ſea with ſo great a force, as might endanger the breaking of the machine.

It would therefore be adviſeable firſt to let down the buoy with ſomething of equal weight with the machine, and an iron rod intervening between the machine and the buoy, thereby to gueſs by the bending, or not bending of the rod, with what degree of force it might ſtrike the bottom of the ſea. And if the force ſhould be found to be great, I believe it might be adviſeable to fix a pole between the machine and the ballaſt of ſuch a degree of ſtrength that it might break before it could give reſiſtance enough to hurt the machine; this would greatly break its force againſt the ground.

It would be adviſeable alſo to keep an exact account of the machines ſtay under water, which may be done by a watch, which beats ſeconds, or by a pendulum vibrating ſeconds, which muſt be three feet, three inches, and one fifth of an inch long, between the middle of the bob and the upper end of the line.

Dr *Hook*, in the *Philoſ. Tranſ. Lowthorp*'s *Abridgment*, vol. I. p. 258, found upon trial that a leaden ball which weighed two pounds being fixed to a wooden ball of the ſame weight, and both let down in 14 fathom wa- they reached the bottom in 17 ſeconds, and the detached wooden ball aſcended to the ſurface of the water in 17 ſeconds more; ſo that if the machine deſcended and aſcended to greater depths with the ſame velocity, it would reach to the depth of a mile in 17 minutes, and reaſcend in the like time. But ſince the buoyant body may return faſter or ſlower to the ſurface of the water than it deſcended, therefore eſtimates from the time of the body's keeping under water will be very uncertain; yet, when frequently compared with the eſtimate that is made from the height of the water in the gage-tube, a rule may perhaps be formed from thence, eſpecially if the whole machine be always the ſame, and the ſinking ballaſt be always of the ſame weight and ſize; as ſuppoſe the ballaſt were put into globular earthen veſſels, made all of the ſame diameter.

That the ſea is not many miles deep is probable from hence, that all the great oceans are here and there interſperſed with iſlands; an argument, that though as far as the ſounding line has reached, the ſea is found to be deeper and deeper the farther from the ſhore, though with ſome unevenneſs, yet it would come to a great depth indeed, if it continued ſo from one boundary of the vaſt ocean to the other; but which the interſperſed iſlands prove to be otherwiſe.

But if we ſuppoſe the ſea to deepen from its ſhores in nearly the ſame proportion that land riſes from the ſhores, then, from the following eſtimate, the greateſt depth of the ſea will not exceed five or ſix miles. For ſince ſlow rivers are found to have a fall of about a foot in a mile; if we ſuppoſe the river *Niger* in *Africa*, (which is one of the longeſt rivers in the world, and runs about 3000 miles in length) to fall at the rate of four feet each mile; then its whole fall, from its riſe to its diſcharging itſelf into the ſea, will be 2.27 miles: If it fall at the rate of 6 feet each mile, then its whole fall will be 3.41 miles. If eight feet each mile, its fall will be 4.54 miles. But if the fall be ſet at ten feet each mile, then the whole fall of the river will be 5.73 miles, which is a large allowance, and may therefore well include the height of the mountains from whoſe ſides thoſe ſprings break forth.

If we ſuppoſe that the whole quantity of earth, which is above the level of the ſeas were equal to the whole bulk of water contained in the baſin of the ſeas; then, ſince the ſum of the expanſe of all the ſeas is conſiderably more than the ſum of the ſurface of all the earth on this globe, the general depth of the ſea muſt therefore be conſiderably leſs than the general height of the earth above the ſurface of the ſea.

Captain *Ellis*, in the latitude 25° 13′ north, longitude 25° 12′ weſt, let down with a line a ſea-gage, to take the different degrees of warmth and ſaltneſs of the ſea at different depths, of 5346 feet, that is a mile and 66 feet, which is the greateſt depth that I have known a line has been let down. He found the water ſalter and ſalter, and cooler and cooler, in proportion to the depths, to 3900 feet, from whence the mercury in the thermometer came up at 53 degrees, that is 21 degrees above the freezing point, about the degree of coolneſs of a good cellar, but no cooler, tho' he let it down 2446 feet lower. The warmth of the ſurface water of the ſea was then 84, or 52 degrees above the freezing point. *Philoſ. Tranſ.* for the years 1751 and 1752, *p.* 213.

The ingenious Mr *Eraſmus King*, when he was with the late right hon. the earl of *Baltimore*, in the gulph of *Finland*, took the depth of the ſea, with ſuch an inſtrument, which was found to be exactly true, when meaſured afterwards with a plumb line. And if any one ſhall think fit to try the experiment, it will be moſt adviſeable to begin the tryal in calm weather, firſt in ſhallow ſeas, and thence to proceed gradually deeper and deeper, till they come to unfathomable depths.

[Note. *The foregoing paper was drawn up about the year 1732, or 33, for the late* Colin Campbell, *Esq; who employed the ingenious Mr* Francis Hawkſbee *to make the machine, which was tried in various depths of the* Thames, *anſwered very well, and always returned, leaving the balaſt behind. It was ſoon after put aboard the ſhip in which Mr* Campbell *ſailed for* Jamaica; *and in a clear calm day was by him let down into the ſea, not many leagues from* Bermudas, *ſeveral other ſhips being in company, and a good look out ordered from them all; yet it was not ſeen to return, though they waited for it between three and four hours. This account, and the paper, the Communicator had from Mr* Campbell *himſelf, before his laſt departure, with his deſire, that it ſhould be publiſhed ſome time or other.*

What relates to Capt. Ellis's *and Mr* King's *experiments is added for the affinity they have to the foregoing.*]

30

A DESCRIPTION OF A SEA GAGE, TO MEASURE UNFATHOMABLE DEPTHS

Rev. Stephen Hales, D.D. F.R.S.

In my *Vegetable Statics,* under experiment 89, I proposed a method for finding out the depth of the sea where it is unfathomable; which method the ingenious Dr Desaguliers put in practice before the Royal Society, with a machine which he contrived, the description of which he has given in the *Philosophical Transactions*, numb. 405. I shall here give a more particular description how to prepare and graduate this sea gage.

Suppose A B (*Fig.* I.) to be an iron tube or musket-barrel of any length, as 50 inches, having its upper end A well closed up. If this tube be let down in this position about 33 feet into the sea, a column of water of that height is nearly equal to the middle weight of our atmosphere, and consequently from a known property of the air's elasticity, it will be compressed into half the space it took up before; so that the water will ascend half way up the tube; and if the tube be let down 33 feet deeper, the air will be compressed into one third of its first dimensions, and so on, one fourth, one fifth, one sixth etc. the air being constantly compressible, in proportion to the incumbent weight; whence, by knowing to what height the water has ascended in the tube, we may readily know to what depth the tube has descended into the sea.

Now to measure the depth of one of these columns of sea-water; first, by a line, let the iron tube, with a weight at its bottom, sink about 33 feet, which depth, in salt water, will nearly answer to the weight of the air at a mean height of the barometer; then draw up the tube, and observe how far the water rose. If 33 feet of water is equal to one atmosphere, then will the water rise so high as to fill exactly one half of the tube. But if the water rises higher or lower than half way, then, by the rule of three, say, as the number to which the water rises, is to one, so is thirty-three, to the number of feet measuring the depth of the column required. For example, suppose the water rises (when the tube is let down 33 feet) only ninetenths of half way, then say as 9:10::33:36½ feet, the depth of each column, which being once known, the number of columns of water is to be multiplied by this number of feet, whereby the depth of the sea in feet will be known.

But since when the instrument has descended to the depth of 99 columns, or 99 times 33 feet, the air will be compressed into the one hundredth part of fifty inches, that is half an inch, the divisions both some time before and after that will be so very small, that the difference in depth of several columns of water, will not be sensible. So that an instrument of no greater length than this would scarcely give an estimate of half a mile's depth, that is, 2640 feet or 80 columns depth of water. The lengthening of this instrument to four, five or ten times this length, would obviate this defect, and make the difference of the degrees of descent much more sensible. But since it is impracticable to make a metalline tube of so great a length, and if it were made, it would be so unweildy as to be easily broken, I propose to obviate these difficulties by the following method *viz.*

Let there be a globose metalline body made of iron or copper, nearly of this form (*fig.* II.) K, L, M, N, Q, whose capacity within side may be equal to nine times the capacity of the metalline tube Z K L. Let this globose body be firmly screwed to the metalline tube, at K L, with a leathern collar well soaked in some unctuous matter, at the shoulder or joining to, thereby to secure that joint in the most effectual manner. Let there be a small hole at *x* for the sea water freely to enter in at, and let some coloured oil, red, blue, or any other colour, be poured into the globose body, to fill it up to the hole *x*. Let there also be provided a slender iron, brass, or wooden rod *d b*, screwed, or somehow fasten'd into the metalline diameter *s s*, which diameter must also be made to screw in and out, thereby to take out the rod at pleasure. The rod must also have fastened to its upper end a small button *d*, which will prevent its being daubed, by falling against the sides of the tube.

The capacity of the tube must be estimated by pouring water in, when the rod and metalline diameter are fixed in their place.

Now, since the lower vessel is supposed to contain nine times as much air as the tube Z L, which is the same thing as if the tube is nine times as long; therefore the air in the globose vessel will not all be forced within the capacity of the tube till the vessel has descended to the depth of nine columns, or nine times 33 feet, for then the air will be compressed within one tenth of the space it at first took up.

Supposing therefore the instrument to have descended to the depth of 99 columns of water, or 99 times 33 feet, *viz.* 3267 feet, then the air will be compressed within one hundredth part of 500 inches (for the capacity of the whole vessel was supposed equal to a tube of that length) that is, within five inches of the top of the tube, and consequently the rod *d b* will be found tinged with the coloured oil within five inches of its top.

Suppose again the instrument to have descended to the depth of 199 columns of 33 feet each, then the air will be compressed within one 200th part of the whole, that is nearly within two and a half inches of the top of the tube. In this case the instrument will have descended 6567 feet; that is, a mile and a quarter, and 132 feet.

Suppose again the instrument to have descended to the depth of 399 columns, then the air will be compressed within one 400th part of the whole, that is, nearly within one inch and a quarter of the top of the tube. In this case the instrument will have descended two miles and a half, wanting 53 feet, which may probably be the greatest depth of the sea.

Hence we see, that if there were occasion to explore greater depths of the sea than this, it might be done with tolerable accuracy, by enlarging the capacity of the globose vessel K, L, M, N, Q, which might easily be done, without rendering it very cumbersome; for suppose the capacity of the tube Z K L were about three fourths of an inch, *viz.* common musket barrel bore, and that it were fifty inches long, if the globose vessel were nineteen times as big, it would not in that case exceed the bulk of three gallons; but the bigger the globose vessel, the greater care must be taken to secure well the screw joint *r r*, that no air pass that way, or water press in. And indeed

I am not without some apprehensions, that when the air shall be compressed with a column of water of two or three miles, it may be forced thro' the pores of the iron; and if this should be the case, the most likely way to prevent this inconvenience would be to line the iron tube with one of glass.

The bigger the globose body is, the more weighty it ought to be, thereby the more effectually to keep it in a low depending posture, else the buoyancy of the air in it might raise it as high or higher than the upper part, whereby water rushing in to the top Z of the tube, no observations could be made, the rod being thereby wetted from end to end. When one experiment has been made, the rod and tube must be wiped very clean before another be repeated.

This sea gage being thus prepared, a large buoy *i*, must be fixed to it, which ought to be of a large solid piece of fir, or any other light solid wood, well covered with tar, to prevent any water's being pressed into its sap vessels: For if the buoy be made of a bladder, or hollow globe, with its orifice inverted downwards, the air in them will be compressed to such a degree, at great depths, as thereby to make the buoyant body become specifically heavier than the sea water, which will prevent its re-ascending to the surface of the sea; for which reason also the buoy ought to be able to buoy up the instrument when full of water. Besides, if the buoy, when it rises again, do not appear some considerable height above water, it will not easily be discovered; for 'tis probable, that from great depths it may rise at a considerable distance from the ship, tho' in a calm. It will therefore be adviseable to fix on the top of the buoy broad fans of tin, painted white or black.

For greater accuracy it will be needful first to try this sea-gage, at several different depths, down to the greatest depth that a line can reach, thereby to discover, whether, or how much, the spring of the air is disturbed or condensed, not only by the great pressure of the incumbent water, but also by its degrees of warmth or coldness at great depths, and in what proportion at different known depths, and in different lengths of time; that allowance may accordingly be made for it at unfathomable depths.

And, because 'tis probable, that the temper of the air, when the experiment is made, will be either warmer or colder than that of the sea at a considerable depth, it will therefore be adviseable to let down the instrument with a line to a good depth, there to continue for some time, till the air in it may be supposed to come to the same temper with the sea-water; then the instrument is to be pulled up so far above water as to let the air freely pass thro' the hole *x*, either in or out, according as the former included air shall have either dilated or contracted; then let the instrument loose to drop down to the bottom of the sea, which it will do by means of a sinking weight of ballast, which must be fixed, in the following manner, to the bottom of the instrument, *viz.* W is a weight of ballast hanging by its shank T, in the socket *f f*, which socket is screwed fast to N Q. The shank is retained in its place by the catch *k* of the spring O, while the machine is descending; but as soon as W touches the ground at the bottom of the sea, the catch O *k*, sinking by its descending force, a little below the upper part of the hole *k*, is thereby at liberty to fly back, and so lets go the weight; then the buoy rises up to the surface of the water with the machine. Springs might also be fixed on the inside of the socket *f f*, so as to fly back in the same manner, and let go their hold as soon as the weight touches the ground.

The weight of the sinking ballast W ought to be so proportioned as to be just sufficient to sink the machine at first, for, as the machine descends, it grows continually specifically heavier, by reason of the condensation of the air in its cavity, on which account its motion will be accelerated, as well as on account of the incessant action of the power of gravity upon it; so that if this gravitating power far exceeded the contrary renitency of the buoy, it would strike the bottom of the sea with so great a force, as might endanger the breaking of the machine.

It would therefore be adviseable first to let down the buoy with something of equal weight with the machine, and an iron rod intervening between the machine and the buoy, thereby to guess by the bending, or not bending of the rod, with what degree of force it might strike the bottom of the sea. And if the force would be found to be great, I believe it might be adviseable to fix a pole between the machine and the ballast of such a degree of strength that it might break before it could give resistance enough to hurt the machine; this would greatly break its force against the ground.

It would be adviseable also to keep an exact account of the machines stay under water, which may be done by a watch, which beats seconds, or by a pendulum vibrating seconds, which must be three feet, three inches, and one fifth of an inch long, between the middle of the bob and the upper end of the line.

Dr *Hook*, in the *Philos. Trans. Lowthorp's Abridgment*, vol. I p. 258, found upon trial that a leaden ball which weighed two pounds being fixed to a wooden ball of the same weight, and both let down in 14 fathom wa[ter] they reached the bottom in 17 seconds, and the detached wooden ball ascended to the surface of the water in 17 seconds more; so that if the machine descended and ascended to greater depths with the same velocity, it would reach to the depth of a mile in 17 minutes, and reascend in the like time. But since the buoyant body may return faster or slower to the surface of the water than it descended, therefore estimates from the time of the body's keeping under water will be very uncertain; yet, when frequently compared with the estimate W is made from the height of the water in the gage-tube, a rule may perhaps be formed from thence, especially if the whole machine be always the same, and the sinking ballast be always of the same weight and size; as suppose the ballast were put into globular earthen vessels, made all of the same diameter.

That the sea is not many miles deep is probable from hence, that all the great oceans are here and there interspersed with islands; an argument, that though as far as the sounding line has reached, the sea is found to be deeper and deeper the farther from the shore, though with some unevenness, yet it would come to a great depth indeed, if it continued so from one boundary of the vast ocean to the other; but which the interspersed islands prove to be otherwise.

But if we suppose the sea to deepen from its shores in nearly the same proportion that land rises from the shores, then, from the following estimate, the greatest depth of the sea will not exceed five or six miles, for since slow rivers are found to have a fall of about a foot in a mile; if we suppose the river *Niger* in *Africa*, (which is one of the longest rivers in the world, and runs about 3000 miles in length) to fall at the rate of four feet each mile; then its whole fall, from its rise to its discharging itself into the sea, will be 2.27 miles: If it fall at the rate of 6 feet each mile, then its whole fall will be 3.41 miles. If eight feet each mile, its fall will be 4.54 miles. But if the fall be set at ten feet each mile, then the whole fall of the river will be 5.73 miles, which is a large allowance, and may therefore well include the height of the mountains from whose sides those springs break forth.

If we suppose that the whole quantity of earth, which is above the level of the seas were equal to the whole bulk of water contained in the basin of the seas; then, since the sum of the expanse of all the seas is considerably more than the sum of the surface of all the earth on this globe, the general depth of the sea must therefore be considerably less than the general height of the earth above the surface of the sea.

Captain *Ellis*, in the latitude 25° 13' north, longitude 25° 12' west, let down with a line a sea-gage, to take the different degrees of warmth and saltness of the sea at different depths, of 5346 feet, that is a mile and 66 feet, which is the greatest depth that I have known a line has been let down. He found the water salter and salter, and cooler and cooler, in proportion to the depths, to 3900 feet, from whence the mercury in the thermometer came up at 53 degrees, that is 21 degrees above the freezing point, about the degree of coolness of a good cellar, but no cooler, tho' he let it down 2446 feet lower. The warmth of the surface water of the sea was then 84, or 52 degrees above the freezing point. *Philos. Trans.* for the years 1751 and 1752, p. 213.

The ingenious Mr *Erasmus King*, when he was with the late right hon. the earl of *Baltimore*, in the gulph of *Finland*, took the depth of the sea, with such an instrument, which was found to be exactly true, when measured afterwards with a plumb line. And if any one shall think fit to try the experiment, it will be most adviseable to begin the tryal in calm weather, first in shallow seas, and thence to proceed gradually deeper and deeper, till they come to unfathomable depths.

[Note, *The foregoing paper was drawn up about the year* 1732, *or* 33, *for the late* Colin Campbell, *Esq; who employed the ingenious Mr* Francis Hawksbee *to make the machine, which was tried in various depths of the* Thames, *answered very well, and always returned, leaving the balast behind. It was soon after put aboard the ship in which Mr* Campbell *sailed for* Jamaica; *and in a clear calm day was by him let down into the sea, not many leagues from* Bermudas, *several other ships being in company, and a good look out ordered from them all; yet it was not seen to return, though they waited for it between three and four hours. This account, and the paper, the Communicator had from Mr* Campbell *himself, before his last departure, with his desire, that it should be published some time or other.*

What relates to Capt. Ellis's *and Mr* King's *experiments is added for the affinity they have to the foregoing.*]

31

Reprinted from *Edinburgh New Philos. J.* 35:340-341 (July-Oct. 1843)

On the Temperature limiting the Distribution of Corals. **By JAMES D. DANA, Geologist of the United States Exploring Expedition. Read before the Association of American Geologists and Naturalists, at Albany, April 29, 1843.**

I have before stated to the Association, that the temperature limiting the distribution of corals in the ocean is not far from 66° F. On ascertaining the influence of temperature on the growth of corals, I was at once enabled to explain the singular fact, that no coral occurs at the Gallapagos, although under the Equator, while growing reefs have formed the Bermudas in latitude 33°, four or five degrees beyond the usual coral limits. In justice to myself, I may state here, that this explanation, which was published some two years since by another, was originally derived from my manuscripts, which were laid open most confidingly for his perusal, while at the Sandwich Islands in 1840.* The anomalies which the Gallapagos and Bermudas seemed to present, were dwelt upon at some length in the manuscript, and attributed in the *latter* case to the influence of the warm waters of the Gulf Stream; in the *former* to the southern current up the South American coast, whose cold waters reduce the ocean temperature about the Gallapagos to 60° F. during some seasons, although twenty degrees to the west, the waters stand at 84° F. *Extratropical* currents, like that which flows by the Gallapagos, are found on the western coasts of both continents, both north and south of the Equator, and *intratropical* currents are as distinctly traceable on the eastern coasts.† In consequence of these currents, the coral zone is contracted on the western coasts and expanded on the eastern; it is reduced to a width of sixteen degrees on the western coast of America, and of but twelve degrees on the

---

* The publication here alluded to, we understand, refers to an Article, by Mr J. P. Couthouy, which appeared last year in the Boston Journal of Natural History.

† The existence of these great oceanic currents was first pointed out to me by our distinguished meteorologist Mr William C. Redfield, who kindly furnished me with charts of the same before the sailing of the Expedition.

east coast of America; while in mid-ocean it is at least fifty-six degrees wide, and about sixty-four degrees on the east coast of Asia and New Holland. The peculiar bend of the east coast of South America carries off to the northward much of the usual south intratropical current, and it is therefore less distinct in its effects than the *northern* intratropical or Gulf Stream.

We have hence the remarkable fact, that the coral zone is fifty degrees wider on the eastern than on the western coasts of our continents. Such is the effect of the ocean currents in limiting the distribution of marine animals. These facts will be brought out more fully in the reports of the Exploring Expedition. The important bearing of these facts upon the distribution of fossil species is too apparent to require more than a passing remark. The many anomalies which have called out speculations as to our globe's passing through areas in space of unequal temperatures, are explained without such an hypothesis. Instead of looking to space for a cause, we need not extend our vision beyond the coasts of our continents.—*American Journal of Science and Arts*, vol. xlv., No. I. p. 130., July 1843.

32

Reprinted from *Edinburgh New Philos. J.* 35:341-345 (July-Oct. 1843)

On the Areas of Subsidence in the Pacific, as indicated by the Distribution of Coral Islands. By James D. Dana, Geologist of the United States Exploring Expedition. Read before the Association of American Geologists and Naturalists, at Albany, April 29, 1843.

The theory of Mr Darwin, with regard to the formation of atolls, or annular coral islands, has been fully confirmed by the investigations of the Exploring Expedition; but his regions of subsidence and elevation, and the conclusion that these changes are now in progress, appear to have been deduced without sufficient examination. Observations at a single point of time cannot determine whether such changes are in progress; they can only assure us with regard to the past. A series of examinations, for years in succession, is necessary to enable us to arrive at the grand deduction, that the land in any part of our globe is now undergoing a gradual change of elevation. The views of Mr Darwin, respecting the rise of the South American coast, as well as that of the Pacific and East Indies, may well be received with some hesitation. According to my own observations, regions, in which his theory would require a subsidence, have actually experienced an elevation at some recent period. I might instance several examples of this elevation in various parts of the Pacific. Suffice it to say here, that I found nothing to support the principle laid down by him, that islands with a barrier reef are subsiding, while those with only a fringing reef are rising; indeed, facts most stubbornly deny it. Without entering upon the discussion of these facts, which, as they will appear in the government publications, I am not at liberty to dwell upon here, I propose to point out what are the regions of subsidence which the coral islands in the Pacific indicate as having been in progress during their formation.

Before proceeding, I may be excused for adding here a few words in explanation of Mr Darwin's theory, with regard to the formation of coral islands. He rejects the unfounded hypothesis that coral islands are built upon the craters of extinct volcanoes, and proposes the following theory in its stead, which is supported by a minute as well as general survey of the facts. The coral belt or atoll, he supposes to have been originally a barrier reef around a high island, like the reef round many islands in the Pacific. When the reef commenced, it could not have been extended to a lower depth than 100 or 120 feet, for this is the limit of the reef forming corals. But if the island gradually subsided--so gradually that the corals could by their growth keep themselves at the surface, the reef might finally attain any thickness, according to the extent of the subsidence. In this manner, subsidence might finally submerge the whole island, and leave nothing but the reef at the surface. Mr Darwin points to instances in which only the mountain tops now remain above the ocean. Carry the process a little farther, and we have the coral belt surrounding its little sea—the usual condition of the coral island.

This theory, as is seen, supposes extensive subsidence. And so, we remark, must every theory; for, without it, we could only have reefs 120 feet in depth, instead of the great thickness they are believed to possess. It is my present object to fix the area of this subsidence, and suggest something with regard to the extent of it in different parts of the ocean. On examining a map of the Pacific, between the Sandwich Islands and the Society group, we find a large area just north of the Equator with scarcely an island. To the south, the islands increase in number; and off Tahiti, to the northward and eastward, they become so numerous, and are so crowded together, as to form a true archipelago. They are all, too, coral islands throughout this interval. This, then, is a rather remarkable fact in the distribution of these islands. But let us look farther.

If we draw a line running nearly E.S.E., from New Ireland, near New Guinea, just by Rotumah, Wallis's Island, Samoa or the Navigators, the Society Islands, and thence bending southward a little, to the Gambier group, we shall have all the islands to the north of it, with two or three exceptions, purely coral, while those to the south are very generally high basaltic islands. These basaltic islands are bordered by reefs, and these reefs are most extensive about the islands nearest this line. In the Feejees, the north-eastern part of the group contains some coral rings, while the north-western consists of large basaltic islands with barrier reefs.

Again, to the north of this boundary line, the islands *farthest* from it are usually small, in many instances mere points of reef, a fraction of a mile in diameter, while some of the coral islands *near* the same line are thirty or forty miles in length.

Now, a growing coral island or atoll will gradually become smaller in diameter as subsidence goes on, and by the same process must finally be reduced to a mere spot of reef, or, if the subsidence is too rapid, that is, more rapid than the growth of the coral, the island will become wholly submerged and leave nothing at the surface.

On these principles I base my conclusions. Along the equator, as explained, there is a large area containing few islands, and these small, while farther north, the coral islands are numerous and large: Is this not evidence, that the subsidence was either more rapid or carried on for a longer period in the former region than in the latter, where they are numerous and large?

Near the boundary line pointed out, stand some of these coral rings, enclosing mountain tops as islets,—as at the Gambier group. Does not this indicate, that the subsidence was less here than among the islands *purely coral* to the north? And greater, than south of the line, where the reefs are more contracted, and the high islands larger and more elevated? Washington Island (coral), in lat. 5° N., is the last spot of land as we recede from our boundary line to the north-east. Beyond is a bare sea, to the Sandwich Islands. Is not this an area where the subsidence was too rapid for the corals to keep the islands at the surface?

It appears that, during this era, the Pacific from 30° N. to 30° S., and perhaps beyond, was one vast region of subsidence: that subsidence took place most rapidly over the bare area between the Sandwich Islands and the Equator, and less and less so, as we go from this to the south-south-west. At the boundary line pointed out, it was not sufficient to submerge many of the mountain summits; and south of this, the effect was still less.

This area covers at least 5000 miles in longitude, and 3000 in latitude. The seas about the north-west coast of New Holland, shew by their reefs a contemporaneous subsidence, and they should probably be included, as well as some parts of the East Indies. Fifteen millions of square miles is not, then, an over-estimate of the extent of the region that participated in this subsidence.

The region of greatest subsidence lies nearly in a west-north-west line, for we may trace it along by Washington Island, far towards the Arctic Coast. The whole broad area of subsidence has nearly the same direction; for this is the course of the boundary line we have laid down as separating the high basaltic and the low coral islands. It is highly interesting to observe, that the trend of the principal groups of islands in the Pacific corresponds nearly with this course. The low or coral Archipelago, the Society Islands, the Navigators, and the Sandwich Islands, lie in the same general direction, nearly west-north-west, and east-south-east. It should be remarked, that the Sandwich group does not contain merely the seven or eight islands usually so-called; eight or ten others stretch off the line to the north; some, small rocky islets, and others, coral, and the whole belong evidently to one series. I will not say that there is a connection between the trend of these groups and the area of subsidence, yet it looks much like it.

A further point may be worthy of consideration. The Sandwich group consists of basaltic islands of various ages. The island at the north-west extremity, Tauai, is evidently more ancient than the others, as its rocks, its gorges, and broken mountains, indicate. By the same kind of evidence, it is placed beyond doubt, that igneous eruptions on these islands continued to be more and more recent, as we go from the north-west to the south-east: at the present time the great active volcano is at the south-east extremity of Hawaii, the south-east island. The fires have gradually become extinct from the north-westward, and now only burn on the south-west point of the group. At the Navigators, and I believe also at the Society group, the reverse was true; the north-west island was last extinct. Is there any connection between this and the fact, that low islands are numerous north-north-west of the Sandwich Islands, and south-south-east of the Society? Does it indicate any thing with regard to the character or the subsidence in these regions?

The time of these changes we cannot definitely ascertain; neither when the subsidence ceased, for it appears to be no longer in progress. The latter part of the tertiary and the succeeding ages may have witnessed it. Although I am by no means confident of any connection, yet for those who would find a balance-motion in the changes, I would suggest that the tertiary rocks of the Andes and North America indicate great elevation since their deposition; and possibly during this great Pacific subsidence, America, the other scale of the balance, was in part undergoing as great or greater elevation.

But why if the Western American coast was rising, do we find no corals on its tropical shores to indicate it? The cold extratropical currents of the ocean furnish us with a satisfactory reply.—*American Journal of Science and Arts,* vol. xlv. No. I. p. 131, July 1843.

33

Reprinted from *Naut. Mag.* 22(7):393-396 (Aug. 1853)

Ocean Soundings: *The Deepest of the Deep Sea Soundings discussed.—By M. F. Maury, LL.D., Lieut. U.S.N.*

Since the great sounding of 5,700 fathoms was made by Lieut. Walsh, commanding U.S. schooner *Taney*, Nov. 15th, 1849, lat. 31° 59′ N., long. 58° 43′ W., three other casts have been taken, each with a greater length of line out, but all, I think, more or less doubtful as to the real "up and down" depth of the ocean. One of these casts was of 8,300 fathoms by Lieut. I. P. Parker, of the U.S. frigate *Congress*, 4th April 1852, lat. 35° 35′ S., long. 45° 10′ W.; another of 7,706 fathoms by Captain Denham, of H.M.S. *Herald*, 30th Oct., 1852, lat. 36° 49′ S., long. 37° 06′ W.; and the other of 6,600 fathoms, by Lieut. O. H. Berryman, commanding U.S. brig *Dolphin*, 12th February, 1853, lat. 32° 55′ N., long. 47° 58′ W.

The first two casts, it will be observed, were made within 400 miles of each other and with the same twine, for Commodore McKeever supplied, from the stock on board the *Congress*, 15,000 fathoms to the *Herald*. The plummet used by Captain Denham was a 9lb. lead. It is much to be regretted that he did not use a 32lb shot, for then, his line being the same, his sounding might have been compared with our own with far greater satisfaction.

Captain Denham's last 706 fathoms (from 7,000 to 7,706) went out at the rate of four-fifths of a mile per hour. He had a 9lb sinker. Now, let us ask any sailor who is familiar with the resistance made by lines when towed through the water, whether, in his opinion, a force of 9lbs could tow eight miles length of line, three-tenths of an inch in circumference, at the rate of four-fifths of a mile the hour. Moreover, his 8th thousand fathoms went out faster than his 5th; surely a 9lb. lead would not drag 7,000 fathoms and upwards through the water faster than it would drag 4,000.

It is probable that there is in all parts of the deep sea one or more under-currents of greater or less velocity.

Suppose where Captain Denham sounded there had been but one, and that that had a rate of only one-tenth of a mile per hour: the line then that his 9lb sinker had to tow through the water instead of being straight was probably a curve, and, possibly, a curve of several convolutions.

Parker, of the *Congress*, gives the time of every 500 fathoms after the first 300 had gone out; Denham, of the *Herald*, gives the time of every 100 fathoms from the beginning; Berryman, of the *Dolphin*, gives the time for every 500 for the first 1,500 fathoms, then for every 200 till he reached 2,500 fathoms, then for 400, then for 1,000, then for 100, and so on at irregular intervals, which impairs the value of his results. Denham's is the best in this respect. To have them all for like intervals, I compute Berryman's to make them correspond with Parker's times and intervals, arranging Denham's accordingly.

Let us now compare the times of the three casts together, that we may see the difference of rate at which the same line ran out, as Parker's and Denham's, to sinkers of different weights, as well as the depths at which a uniformity of rate begins to appear.

Time of Running Out.

Length of Line in Fathoms.	8,300 fths. 32lb shot Congress		7,706 fths. 9lb Herald		6,600 fths. 46lb Dolphin		
	m	s	m	s	m	s	
From 300 to 800 fathoms	8	45	14	20	12	.06	
" 800 " 1,300 "	11	00	18	25	12	51	
" 1,300 " 1,800 "	13	00	19	30	15	07	
" 1,800 " 2,300 "	15	00	22	00	20	07	
" 2,300 " 2,800 "	19	00	23	50	24	11	
" 2,800 " 3,300 "	37	00	28	20	25	53	
" 3,300 " 3,800 "	51	00	39	20	28	00	} 1,000 faths.
" 3,800 " 4,300 "	28	00	43	40	34	00	
" 4,300 " 4,800 "	33	15	42	25	47	22	
" 4,800 " 5,300 "	34	45	47	50	52	16	
" 5,300 " 5,800 "	34	00	53	50	64	50	
" 5,800 " 6,300 "	34	30	55	05	70	32	
" 6,300 " 6,800 "	21	30	53	55	72	34	
" 6,800 " 7,300 "	27	00	52	25			
" 7,300 " 7,800 "	38	30	44	14			
" 7,800 " 8,300 "	21	00					

I do not recollect the size of the *Dolphin's* twine; it is evident, however, that this as well as other sounding twine requires force to pull it from the real and to drag it down through the depths of the ocean; that the deeper the plummet and the greater the length of line to be dragged after it, the greater the resistance and, therefore, the slower the rate at which the line goes out.

Hence we may deduce a rule which, as a general rule, may be taken as correct, viz.: that when the line ceases to go out at something like a regularly decreasing rate there is no reliance to be put upon the sounding after the change, and that when the rate of going out becomes uniform the plummet has probably ceased to drag the line down, and the force which continues to take the sounding line out is due to the wind, currents, heave of the sea, or drift,—one, some, or all.

Let us apply this rule to these casts. That of the *Congress* fulfilled these conditions as to a tolerably regular decreasing rate to the 2,800 fathoms' mark. The rates after that indicate pretty clearly that, whatever might have been the agent which continued to take the line out, it was not the sinking of the 32lb shot. There is an appearance of too much uniformity in the rate after that. Therefore, I infer that when the 2,800 fathoms mark went out the shot was probably on or near the bottom, and that where this sounding was made the ocean, instead of being some 8,300 fathoms deep, is not more than 3,000.

The *Herald's* plummet fulfilled the conditions generally of a decreasing rate until the 4,300 fathoms mark went out; and after this the rate becomes so uniform as to justify the conclusion that the 9lb sinker used had then ceased or nearly ceased to descend, if it were not already on the bottom.

The care with which Captain Denham observed every 100 fathoms' mark, and timed it as it went out, enables us to detect probably more closely in his sounding than in either of the others, the time when his plummet ceased to sink. From 100 to 700 fathoms, each 100 fathoms' mark required between two and three minutes to go out; from 700 to 1,600, each mark required between three and four minutes; from 1,600 to 2,700, each mark required between four and five minutes; from 2,700 to 3,000, each required between five and six minutes. Here the times begin to become irregular, the 3,200 and 3,300 marks each took between six and seven minutes to go out. After this there is no more regularity as to the *increasing* times. Every 100 fathoms' mark thereafter appears to have a rate of its own, varying from seven to twelve minutes, but now fast, now slow, and in such a manner as to justify the inference that the ocean, where the *Herald* reports 7,706 fathoms, is probably not more than 4,000 fathoms deep.

The *Dolphin* had the heaviest plummet and the largest line. The time required with her for each of the first 500 fathoms' marks to run out was longer than the *Congress* but shorter than the *Herald*. But after the 4,300 fathoms' mark of the *Herald* went out, then the *Herald's* line was the swifter. How shall we account for this? If it be supposed that the *Herald* had not reached bottom we are forced into the absurdity of maintaining that a 9lb lead can drag 4,300 fathoms of line, and upwards, down through the water faster than a 46lb lead can. To avoid such an absurdity, I suppose the *Herald's* plummet to have touched bottom, and that an under current then continued to act upon the line, when it assumed the condition of equal lengths in equal times; whereas the *Dolphin's* continued to decrease its rate and to go down slower and slower till the 6,300 fathoms' mark went out. She sent down 6,600 fathoms; the interval, therefore, from 6,300 to 6,800 is computed. But the inference is, that the weight ceased to go down about the time the 6,300 fathoms' mark went out;—that the ocean here is about 6,000 fathoms deep, *and that this is the greatest depth ever yet reached by the plummet.*

National Observatory, Washington, June 7th, 1853.

34

Reprinted from *Philos. Soc. Glasgow Proc.* 9(1):111–117 (1873–1875)

On Deep-Sea Sounding by Piano-forte Wire. By Professor Sir William Thomson, LL.D., F.R.S.

[Read before the Society, March 18, 1874.]

[*Editor's Note:* The plates accompanying this article have not been reproduced because of limitations of space.]

Sir William Thomson said—I have now to bring before the Society a new process for deep-sea sounding, which has been practised for more than a year with much success. At a Meeting (March 18, 1873) of the Institution of Engineers in Scotland last session, an apparatus of this kind was shewn, and the whole process of using the wire for deep-sea sounding was elaborately explained, so that it is not now necessary for me to enter on details of this subject. I will briefly remark that the great advantage of steel wire over hemp rope, is the comparatively small resistance which the wire experiences from the water. The great amount of resistance which rope meets from the water is well known to sailors. The resistance of even so short a length as 50 fathoms of ordinary deep-sea sounding line going through the water, is very considerable. Common observation in the use of the Massey log, shews that the resistance of a few fathoms of the rope is greater than the resistance of the log itself. The deep-sea lead goes down with great rapidity for 10, 20, or 30 fathoms, but after that depth the resistance of the rope begins to tell seriously, and at 50 or 150 fathoms the resistance to the rope moving through the water is very great. In sounding at these depths the great force required to haul in the rope, at any considerable speed, is well known. The greater part of that force is due to the resistance that the water opposes to the rope hauled through it, and a very small portion is due to the lead. The small area of the wire surface and its smoothness, as compared with the ordinary hemp sounding line, give it a great advantage in these respects. Then, as regards strength, this wire, which is Messrs. Webster and Horsfall's piano-forte wire, No. 22 gauge, bears 230 pounds, and weighs 14½ pounds per 1000 fathoms. The wire I shewed to the Institution of Engineers last year, was not the ordinary quality of piano-forte wire; it was a wire made specially for sounding by

Messrs. Richard Johnson & Nephew, of Manchester. They succeeded in making a wire three miles in length, from crucible steel, which has many good qualities, particularly that of strength; but its temper was not equal to that of the best piano-forte wire, which, however, can only be produced in much shorter lengths. The form of splice which I had then designed (but only imperfectly tested) has since proved so successful that I do not now care so much for length; Messrs. Webster and Horsfall give me lengths of about 100 fathoms.

In upwards of one hundred soundings on the East and North coasts of Brazil, in the Pacific, and in the Bay of Biscay, in depths of from 500 to 2,700 fathoms, partly with Johnson's special wire, and partly with Webster and Horsfall's, there has in no one instance been a failure of the splice. The splice is made very easily, and in a few minutes.

There is a considerable difference in the mechanism now shewn to the Philosophical Society from that I shewed to the Institution of Engineers last session, but a far more important difference is in the wire. Wire of an inferior quality is brittle at places, and breaks when it kinks. I believe not a single case of this has happened with the Webster and Horsfall piano-forte wire now used.

[Sir William Thomson, at this stage, referring to his new sounding machine, shewn to the meeting, proceeded to explain the nature of its various parts, and its action as a whole. The machine which he shewed is represented in the present report by the two accompanying Plates, one of which shews it by elevations, sections, &c.; and the other of which is a perspective drawing made from a photograph of the apparatus itself. By study of these drawings with the aid of the brief explanatory notes written upon them, together with the explanations here following, as noted from the lecture, the reader may arrive at a good conception of the nature of the apparatus, and of the sounding processes for which it is adapted. He continued as as follows:—]

The wire is coiled on a large wheel, which is made as light as possible, so that, when the weight reaches the bottom, the inertia of the wheel may not shoot the wire out so far as to let it coil on the bottom. The avoidance of such coiling of the wire on the bottom is the chief condition requisite to provide against the possibility of kinks; and for this reason a short piece of hemp line, about five fathoms in length, is interposed between the wire and the sounding-weight; so that, although a little of the hemp line may coil on the bottom, the wire line may be quite prevented from reaching the bottom. There is a clamp, or a ring, attached directly at the bottom of the wire, so as to form the coupling or junction between the wire and the hemp line. The clamp is about 4 or 5 pounds in weight, and either it, or the ring, which may be considerably lighter, and may be used instead of the clamp, suffices to keep the wire tight when the lead is on the bottom, and the hemp line is slackened. The art of deep-sea sounding is to put such a resistance on the wheel as shall secure that the moment the weight reaches the bottom the wheel will stop. By "the moment" I mean within two or three seconds of time. Lightness of the wheel is necessary for this. Whatever length of wire is estimated as necessary to reach the bottom is coiled on the wheel. For a series of deep-sea soundings in depths exceeding 1000 fathoms, it is convenient to keep a length of 3000 fathoms coiled on the wheel. When we do not get bottom with 3000 fathoms, the process of splicing on a new length of wire ready coiled on a second wheel, is done in a very short time—two minutes at most. The friction brake which you see, is simpler in construction than that shewn to the Engineers last session. This is simply a return to the form of brake which I used in June, 1872, when I first made a deep-sea sounding with piano-forte wire in the Bay of Biscay, in 2,700 fathoms. The process of sounding is this:—Such a weight is applied to the brake as shall apply to the wire leaving the wheel a resistance exceeding by 10 pounds the weight of the wire out. Thus, commencing with a resistance of 10 pounds, when 100 fathoms are out add 1¼ pounds to the brake-resistance. For every 1000 fathoms of wire out, add 12 pounds to the brake-resistance, because the weight of 1000 fathoms of the wire in water is about 12 pounds. The only failures in deep-sea soundings with piano-forte wire hitherto made have been owing to neglect of this essential condition. The circumference of the wheel is a fathom (with a slight correction for the increased diameter from the quantity of wire on). Hence for every 250 turns of the wheel, add three pounds weight to the brake-resistance. The action down to 2,700 fathoms is perfect. I cannot speak beyond that from personal observation, but nevertheless I can confidently say, that soundings in the greatest depths of the sea which have been hitherto sounded, can be made with this apparatus with perfect ease. Getting back the weight is the most difficult part of the process. In ordinary deep-sea soundings there is a trigger apparatus, by which the weight (300 pounds or 400 pounds of iron) may be detached, and then the rope, with only a tube bringing up a specimen of the bottom, is hauled up. In sounding with the wire apparatus up to 2,700 fathoms, I use a 30 pounds lead sinker, and I have not found it necessary to use the detaching apparatus, but in depths exceeding 4,000 fathoms, I might be induced to use it. The process of recovering the wire and sinker involves a little difficulty, from the fact that coiling so great a length of wire on the drum under so great a tension, applies a prodigious force to the wheel. Suppose there is a pull of 50 pounds on the wire as it is coiled round the wheel, then we have the halves of the wheel on two sides of any diameter, pressed together with a force of 100 pounds by one coil, or one fathom, of the wire. One thousand times that gives 100,000 pounds pressing the wheel together for about 1000 fathoms, so that if the wheel did not yield a little, we should have (after hauling in lead and wire from a depth of 3,000 fathoms) a pressure of more than 100 tons pressing it together. In point of fact, the wheel yields to some degree. Practically I found that in hauling in the line from 2,700 fathoms the wheel was squeezed out of shape, and came to be something like the shape of the old-fashioned, three-cornered cocked hat. This part of the process has undergone very considerable alterations since 1872. The United States navy have taken up with great ardour the system of deep-sea soundings by piano-forte wire, and have been most successful in practising it. They have worked it out in a way somewhat different from that which I have followed. They get over the difficulty of hauling in by strengthening the wheel, and detaching the weight. I prefer not to lose the lead. If the sounding is less than 1,000 fathoms, the wheel will not suffer. I sounded off Funchal, Madeira, last summer, in 1,200 fathoms, and coiled the wire in safety direct on the wheel. Although the wheel shewed signs of distress, it was not seriously injured. But if we had to haul up from 2,000 fathoms, it would probably be seriously damaged. Therefore, if the depth exceeds 1,000 fathoms we must either strengthen the wheel or detach the lead, or both: or we must relieve the wheel of a great part of the strain. I prefer the latter remedy. The weight of the wire is 14½ pounds per nautical mile in air, and in water about 12 pounds per nautical mile. In a depth of three miles, therefore, we have 36 pounds of wire, 30 pounds of lead, and about 4 pounds of ring or clamp:—in all, 70 pounds. In preparation for hauling in, a spun yarn stopper, attached to the lower framing of the sounding machine projecting over the taffrail, or to the taffrail itself, is applied to the wire hanging down below, to hold the wire up and relieve the wheel from the necessity of performing that duty: or otherwise, two men, with thick leather gloves, can easily hold the wire up.* A little of the wire is then paid out from the wheel, and the wheel with its framing is run inboard about five feet on slides which carry its framing [see the accompanying drawings†], and the slack wire is carried once or twice round a grooved drum below, which overhangs the bearings of its own axle, so as to allow the loop or the two loops of the wire to be got on. Two handles attached to the shaft of this drum, worked by one man on each or two men on each, take from two-thirds to nine-tenths of the strain off the wire before it reaches its own wheel, on which it is coiled by one man or two men working on handles attached to its shaft.

If the ship is hove to when the wire is being hauled in, the wire will generally stream to one side (if out by the stern, which is the position I now prefer). By having the bearing of the aftermost wheel, an oblique fork turning round a horizontal axis (like the *castor* of a piece of furniture laid on its side), the wire is hauled in with ease though streaming to either side, at any angle. [See the drawings.‡] This castor wheel is a very important addition to the hauling-in gear. (In paying out, the wire runs direct from its own wheel into the sea.)

But it is not necessary to keep the ship hove to during the whole time of hauling in the wire. When the depth exceeds 3,000 fathoms, it will, no doubt, be generally found convenient to keep the ship hove to until a few hundred fathoms of the wire have been brought on board. When the length out does not exceed 2,500 fathoms, the ship may be driven ahead slowly, with gradually increasing speed. When the length of wire out does not exceed 1,500 fathoms, the ship may be safely driven ahead at five or six knots. The last 500 fathoms may be got on board with ease and safety, though the ship is going ahead at ten or twelve knots. Thus, by the use of wire, a great

* The spun yarn stopper is to be seen in the accompanying perspective drawing, shewn as hanging ready for use.

† The Side Elevation shews the sounding wheel projecting over the taffrail in the position for paying out the wire: and the perspective drawing shews it as run inboard in the position for hauling up.

‡ In respect to the arrangement of framing for bearing the castor axle of the forked piece in which the castor wheel or pulley runs, the side elevation is purposely made to suggest an improvement from the arrangement actually existing in the machine shewn at the meeting, and represented in the perspective drawing. The improvement consists merely in lengthening the castor axle, and providing for it two bearings, instead of its having only one, as was the case in the machine shewn at the meeting, and as is exhibited in the perspective drawing from a photograph of the machine itself.

saving of time is effected; for in the ordinary process the hemp rope must be kept as nearly as possible up and down, until the whole length out does not exceed a few hundred fathoms.

Approximate soundings, of great use in ordinary navigation, may be obtained in depths of 200 fathoms, or less, with remarkable ease, without reducing the speed of the ship below five or six knots, even when the wire is being paid out. For this purpose let the weight fall direct from the wire wheel over the taffrail, with a brake-resistance of from five to ten pounds. The moment of its reaching the bottom is indicated by a sudden decrease in the speed of rotation of the wheel. The moment this is observed, a man standing at the wheel grasps it with his two hands, and stops it. Not more than three or four hundred fathoms of wire having run out, the hauling-in is easy. In following this process I have generally found it convenient to arm the lead with a proper mixture of tallow and wax, in the usual manner, to bring up specimens from the bottom. The actual depth is, of course, less than the length of wire run out. The difference, to be subtracted from the length of wire out to find the true depth, may be generally estimated with considerable accuracy after some experience. The estimation of it is assisted by considering that the true depth is always, as we see from the annexed diagram, greater than $l-a$ and less than $\sqrt{l^2-a^2}$, where l denotes the length of wire out,

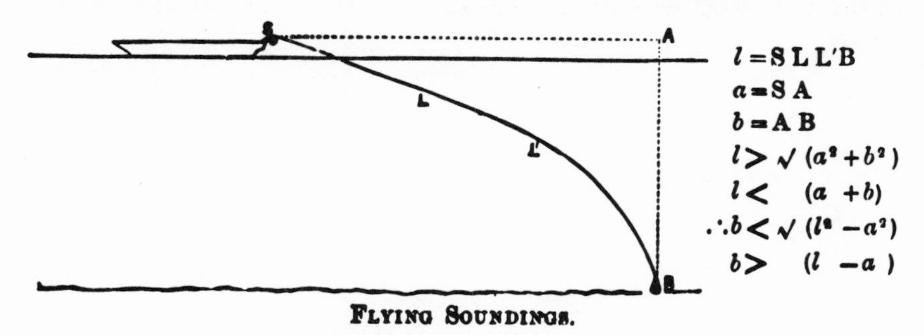

Flying Soundings.

and a the space travelled by the ship, diminished by the space travelled horizontally by the sinker during the time of its going to the bottom.

The contrast between the ease with which the wire and sinker are got on board from a depth of 200 fathoms by a single man, or by two men, in this process, and the labour of hauling in the ordinary deep-sea lead and line, by four or five men, when soundings are taken in the ordinary way from a ship going through the water at four or five knots in depths of from 30 to 60 fathoms, is remarkable. Professor Jenkin and I found this process of great value on board the

"Hooper," during the laying of the Western and Brazilian Telegraph Company's cables between Para, Pernambuco, Bahia, and Rio Janeiro. I am now having constructed, for the purposes of navigation, a small wire wheel of 12 inches diameter, to have 400 fathoms of piano-forte wire coiled on it, for flying soundings in depths of from 5 to 200 fathoms, without any reduction of the speed of the ship, or, at all events, without reducing it below five or six knots.

During the whole process of sounding, we are continually reminded of the original purpose of the wire by the sounds it gives out. A person of a musical ear can tell within a few pounds what pull is on the wire by the note it sounds in the length between the castor pulley at the stern and the haul-in drum, which is about five feet inboard of it.

Mr. JAMES R. NAPIER said that there were some very ingenious and simple contrivances for overcoming difficulties about Sir William Thomson's sounding machine, as indeed there were about all his inventions. The two which struck him most were—1st, the application of a definite weight to the friction wheel, whereby almost the exact instant of the weight's reaching the bottom could be ascertained by the sudden stopping of the revolutions of the coil; and the other was, his method of preventing the compressing force of the wire from accumulating at each revolution, so as to crush the wheel.

35

Reprinted from pp. 516–529 of the *Royal Soc. Edinburgh Proc.*
12:495–529 (1882–1884)

ON THE NOMENCLATURE, ORIGIN, AND DISTRIBUTION OF DEEP-SEA DEPOSITS

J. Murray and A. Renard

[*Editor's Note:* In the original, material precedes this excerpt.]

Geographical and Bathymetrical Distribution.—In the preceding pages we have confined our remarks essentially to the lithological nature of the deep-sea deposits, including in this term the dead shells and skeletons of organisms. From this point of view it has been possible to define the sediments and to give them distinctive names. We now proceed to consider their geographical and bathymetrical distribution, and the relations which exist between the mineralogical and organic composition, and the different areas of the ocean in which they are formed.

A cursory glance at the geographical distribution shows that the deposits which we have designated MUDS and SANDS are situated at various depths at no great distance from the land, while the ORGANIC OOZES and RED CLAYS occupy the abysmal regions of the ocean basins far from land. Leaving out of view the coral and volcanic muds and sands which are found principally around oceanic islands, we notice that our blue muds, green muds and sands, red muds, together with all the coast and shore formations, are situated along the margins of the continents and in enclosed and partially enclosed seas. The chief characteristic of these deposits is the presence in them of continental débris. The blue muds are found in all the deeper parts of the regions just indicated, and especially near the embouchures of rivers. Red muds do not differ much from blue muds except in colour, due to the presence of ferruginous matter in great abundance, and we find them under the same conditions as the blue muds. The green muds and sands occupy, as a rule, portions of the coast where detrital matter from rivers is not apparently accumulating at a rapid rate, viz., on such places as the Agulhas Bank, off the east coast of Australia, off the coast of Spain, and at various points along the coast of America.

Let us cast a glance at the region occupied by terrigenous deposits, in which we include all truly littoral formations. This region extends from high-water mark down, it may be, to a depth of over four miles, and in a horizontal direction from 60 to perhaps 300 miles seawards, and includes, in the view we take, all inland seas, such as the North Sea, Norwegian Sea, Mediterranean Sea, Red Sea, China Sea, Japan Sea, Carribean Sea, and many others. It is the region of change and of variety with respect to light, temperature, motion, and biological conditions. In the surface waters the temperature ranges from 80° Fahr. in the tropics, to 28° Fahr. in the polar regions. Below the surface, down to the nearly ice-cold water found at the lower limits of the region in the deep sea, there is in the tropics an equally great range of temperature. Plants and animals are abundant near the shore, and animals extend in relatively great abundance down to the lower limits of this region which is now covered by these terrigenous deposits. The specific gravity of the water varies much, owing to mixture with river water or great local evaporation, and this variation in its turn affects the fauna and flora. In the terrigenous region tides and currents produce their maximum effect, and these influences can in some instances be traced to a depth of 300 fathoms, or nearly 2000 feet. The upper or continental margin of the region is clearly defined by the high-water mark of the coast-line, which is constantly changing through breaker action, elevation, and subsidence. The lower or abysmal margin is less clearly marked out. It passes in most cases insensibly into the abysmal region, but may be regarded as ending when the mineral particles from the neighbouring continents begin to disappear from the deposits, which then pass into an organic ooze or a red clay.

Contrast with these those conditions which prevail in the abysmal region in which occur the organic oozes and red clay, the distribution of which will presently be considered. This area comprises vast undulating plains from two to five miles beneath the surface of the sea, the average being about three miles, here and there interrupted by huge volcanic cones (the oceanic islands). No sunlight ever reaches these deep cold tracts. The range of temperature over them is not more than 7°, viz., from 31° to 38° Fahr., and is apparently constant throughout the whole year in each locality. Plant life is absent, and although animals belonging to all the great types are present, there is no great variety of form or abundance of individuals. Change of any kind is exceedingly slow.

What is the distribution of deposits in this abysmal region of the earth's surface? In the tropical and temperate zones of the great oceans, which occupy about 110° of latitude between the two polar zones, at depths where the action of the waves is not felt, and at points to which the terrigenous materials do not extend, there are now forming vast accumulations of *Globigerina* and other pelagic Foraminifera, coccoliths, rhabdoliths, shells of pelagic Molluscs, and remains of other organisms. These deposits may perhaps be called the sediments of median depths and of warmer zones, because they diminish in great depths and tend to disappear towards the poles. This fact is evidently in relation with the surface temperature of the ocean, and shows that pelagic Foraminifera and Molluscs live in the superficial waters of the sea, whence their dead shells fall to the bottom. Globigerina ooze is not found in enclosed seas nor in polar latitudes. In the Southern Hemisphere it has not been met with beyond the 50th parallel. In the Atlantic it is deposited upon the bottom at a very high latitude below the warm waters of the Gulf Stream, and is not observed under the cold descending polar current which runs south in the same latitude. These facts are readily explained, if we admit that this ooze is formed chiefly by the shells of surface organisms, which require an elevated temperature and a wide expanse of sea. But as long as the conditions of the surface are the same we would expect the deposits at the bottom also to remain the same. In showing that such is not the case, we are led to take into account an agent which is in direct correlation with the depth. We may regard it as established that the majority of the calcareous organisms, which make up the Globigerina and Pteropod oozes, live in the surface waters, and we may also take for granted that there is always a specific identity between the calcareous organisms which live at the surface, and the shells of these pelagic creatures found at the bottom. This observation will permit us to place in relation the organic deposits and those which are directly or indirectly the result of the chemical activity of the ocean. Globigerina ooze is found in the tropical zone at depths which do not exceed 2400 fathoms, but when depths of 3000 fathoms are explored in this zone of the Atlantic and Pacific, there is found an argillaceous deposit without, in many instances, any trace of calcareous organisms. When we descend from the "submarine plateaux" to depths which exceed 2250 fathoms the Globigerina ooze gradually disappears, passing into a greyish marl, and finally is wholly replaced by an argillaceous material which covers the bottom at all depths greater than 2900 fathoms.

The transition between the calcareous formations and the argillaceous ones takes place by almost insensible degrees. The thinner and more delicate shells disappear first. The thicker and larger shells lose little by little the sharpness of their contour, and appear to undergo a profound alteration. They assume a brownish colour, and break up in proportion as the calcareous constituent disappears. The red clay predominates more and more as the calcareous element diminishes in the deposit.

If we now recollect that the most important elements of the organic deposits have descended from the superficial waters, and that the variations in contour of the bottom of the sea cannot of themselves prevent the débris of animals and plants from accumulating upon the bottom, their absence in the red clay areas can only be explained by a decomposition, under the action of a cause which we must seek to discover.

Pteropod ooze, it will be remembered, is a calcareous organic deposit, in which the remains of Pteropods and other pelagic Mollusca are present, though they do not always form a preponderating constituent, and it has been found that their presence is in correlation with the bathymetrical distribution.

In studying the nature of the calcareous elements which are deposited in the pelagic areas, it has been noticed that, like the shells of the Foraminifera, those of the Thecosomatous Pteropoda, which live everywhere in the superficial waters, especially in the tropics, become fewer in number as the depth from which the sediments are derived increases. We have just observed that the shells of Foraminifera disappear gradually as we descend along a series of soundings from a point where the Globigerina ooze has abundance of carbonate of lime, towards deeper regions; but we notice also that when the sounding-rod brings up a graduated series of sediments from a declivity descending into deep water, among the calcareous shells those of the Pteropods and Heteropods disappear first in proportion as the depth increases. At depths less than 1400 fathoms in the tropics a Pteropod ooze is found with abundant remains of Heteropods and Pteropods; deeper soundings then give a Globigerina ooze without these molluscan remains; and in still greater depths, as before mentioned, there is a red clay in which calcareous organisms are nearly, if not quite, absent.

In this manner, then, it is shown that the remains of calcareous organisms are completely eliminated in the greatest depths of the ocean. For if such be not the case, why do we find all these shells at the bottom in the shallower depths, and not at all in the greater depths, although they are equally abundant on the surface at both places? There is reason to think that this solution of calcareous shells is due to the presence of carbonic acid throughout all depths of ocean water. It is well known that this substance, dissolved in water, is an energetic solvent of calcareous matter. The investigations of Buchanan and Dittmar have shown that carbonic acid exists in a free state in sea water, and in the second place, Dittmar's analyses show that deep-sea water contains more lime than surface water. This is a confirmation of the theory which regards carbonic acid as the agent concerned in the total or partial solution of the surface shells before or immediately after they reach the bottom of the ocean, and is likewise in relation with the fact that in high latitudes where fewer calcareous organisms are found at the surface, their remains are removed at lesser depths than where these organisms are in greater abundance. It is not improbable that sea water itself may have some effect in the solution of carbonate of lime, and further, that the immense pressure to which water is subjected in great depths, may have an influence on its chemical activity. We await the result of further researches on this point, which have been undertaken in connection with the "Challenger" Reports. We are aware that objections have been raised to the explanation here advanced, on account of the alkalinity of sea water, but we may remark that alkalinity presents no difficulty which need be here considered.*

This interpretation permits us to explain how the remains of Diatoms and Radiolarians (surface organisms like the Foraminifera) are found in greater abundance in the red clay than in a Globigerina ooze. The action which suffices to dissolve the calcareous matter has little or no effect upon the silica, and so the siliceous shells accumulate. Nor is this view of the case opposed to the distribution of the Pteropod ooze. At first we should expect that the Foraminifera shells, being smaller, would disappear from a deposit before the Pteropod shells; but if we remember that the latter are very thin

* Dittmar, *Phys. Chem. Chall. Exp.*, Part i., 1884.

and delicate, and, for the quantity of carbonate of lime present, offer a larger surface to the action of the solvent than the thicker, though smaller, Globigerina shells, we shall see the explanation of this apparent anomaly.

It remains now to point out the area occupied by the red clay. We have seen how it passes at its margins into organic calcareous oozes, found in the lesser depths of the abysmal regions, or into the siliceous organic oozes or terrigenous deposits. In its typical form the red clay occupies a larger area than any of the other true deep-sea deposits, covering the bottom in vast regions of the North and South Pacific, Atlantic, and Indian Oceans. As above remarked, this clay may be said to be universally distributed over the floor of the oceanic basins; but it only appears as a true deposit at points where the siliceous and calcareous organisms do not conceal its proper characters.

Having now indicated its distribution, we must consider the mode of its formation, and give, in addition, a concise description of the minerals and of the organic remains which are commonly associated with it. The origin of these vast deposits of clay is a problem of the highest interest. It was at first supposed that these sediments were composed of microscopic particles arising from the disintegration of the rocks by rivers and by the waves on the coasts. It was believed that the matters held in suspension were carried far and wide by currents, and gradually fell to the bottom of the sea. But the uniformity of composition presented by these deposits was a great objection to this view. It could be shown, as we have mentioned above, that mineral particles, even of the smallest dimensions, continually set adrift upon disturbed waters must, owing to a property of sea water, eventually be precipitated at no great distance from land. It has also been supposed that these argillaceous deposits owe their origin to the inorganic residue of the calcareous shells which are dissolved away in deep water, but this view has no foundation in fact. Everything seems to show that the formation of the clay is due to the decomposition of fragmentary volcanic products, whose presence can be detected over the whole floor of the ocean.

These volcanic materials are derived from floating pumice and volcanic ashes ejected to great distances by terrestrial volcanoes, and carried far by the winds. It is also known that beds of lava and of tufa are laid down upon the bottom of the sea. This assemblage of pyrogenic rocks, rich in aluminous silicates, decomposes under the chemical action of the water, and gives rise, in the same way as do terrestrial volcanic rocks, to argillaceous matters, according to reactions, which we can always observe on the surface of the globe, and which are too well known to need special mention here.

The detailed microscopic examination of hundreds of soundings has shown that we can always demonstrate in the argillaceous matter the presence of pumice, of lapilli, of silicates, and other volcanic minerals in various stages of decomposition.

As we have shown in another paper,* the deposit most widely distributed over the bed of modern seas is due to the decomposition of the products of the internal activity of the globe, and the final result of the chemical action of sea water is seen in the formation of this argillaceous matter, which is found everywhere in deep-sea deposits, sometimes concealed by the abundance of siliceous or calcareous organisms, sometimes appearing with its own proper characteristics associated with mineral substances, some of which allow us to appreciate the extreme slowness of its formation, or whose presence corroborates the theory advanced to explain its origin.

In the places where this red clay attains its most typical development, we may follow, step by step, the transformation of the volcanic fragments into argillaceous matter. It may be said to be the direct product of the decomposition of the basic rocks, represented by volcanic glasses, such as hyalomelan and tachylite. This decomposition, in spite of the temperature approximating to zero (32° F.), gives rise, as an ultimate product, to clearly crystallised minerals, which may be considered the most remarkable products of the chemical action of the sea upon the volcanic matters undergoing decomposition. These microscopic crystals are zeolites lying free in the deposit, and are met with in greatest abundance in the typical red clay areas of the central Pacific. They are simple, twinned, or spheroidal groups which scarcely exceed half a millimetre in diameter. The crystallographic and chemical study of them shows that they must be referred to Christianite. It is known how easily the zeolites crystallise in the pores of eruptive rocks in process of decomposition; and the crystals of Christianite, which we

* "On Cosmic and Volcanic Dust," *Proc. Roy. Soc. Edin.*, 1883-84.

observe in considerable quantities in the clay of the centre of the Pacific, have been formed at the expense of the decomposing volcanic matters spread out upon the bed of that ocean.

In connection with this formation of zeolites, reference may be made to a chemical process whose principal seat is the red clay areas, and which gives rise to nodules of manganiferous iron. This substance is almost universally distributed in oceanic sediments, yet it is not so much of the areas of its abundance that we intend to speak as to the fact of its occurrence in the red clay, because this association tends to show a common relation of origin. It is exactly in those regions where there is an accumulation of pyroxenic lavas in decomposition, containing silicates with a base of manganese and iron, such for example as augite, hornblende, olivine, magnetite, and basic glasses, that manganese nodules occur in greatest numbers. In the regions where the sedimentary action, mechanical and organic, is, as it were, suspended, and where, as will appear in the sequel, everything shows an extreme slowness of deposition,—in these calm waters favourable to chemical reactions, ferro-manganiferous substances form concretions around organic and inorganic centres.

These concentrations of ferric and manganic oxides, mixed with argillaceous materials, whose form and dimensions are extremely variable, belong generally to the earthy variety or wad, but pass sometimes, though rarely, into varieties of hydrated oxide of manganese with distinct indications of radially fibrous crystallisation. The interpretation to which we are led, in order to explain this formation of manganese nodules, is the same as that which is admitted in explanation of the formation of coatings of this material on the surface of terrestrial rocks. These salts of manganese and iron, dissolved in water by carbonic acid, then precipitated in the form of carbonate of protoxide of iron and manganese, become oxidised, and give rise in the calm and deep oceanic regions to more or less pure ferro-manganiferous concretions. At the same time it must be admitted that rivers may bring to the ocean a contribution of these same substances.

Among the bodies which, in certain regions where red clay predominates, serve as centres for these manganiferous nodules, are the remains of vertebrates. These remains are the hardest parts of the skeleton—tympanic bones of whales, beaks of Ziphius, teeth of sharks; and just as the calcareous shells are eliminated in the depths, so all the remains of the larger vertebrates are absent except the most resistant portions. These bones often serve as a centre for the manganese-iron concretions, being frequently surrounded by layers several centimetres in thickness. In the same dredgings on the red clay areas, some sharks' teeth and cetacean ear-bones, some of which belong to extinct species, are surrounded with thick layers of the manganese, and others with merely a slight coating. We will make use of these facts to establish the conclusions which terminate this paper.

In these red clays there occur, in addition, the greatest number of cosmic metallic spherules, or chondres, the nature and characters of which we have pointed out elsewhere.* We merely indicate their presence here, as we will support our conclusions by a reference to their distribution.

Reviewing, then, the distribution of oceanic deposits, we may summarise thus:—

(1) The terrigenous deposits, the blue muds, green muds and sands, red muds, volcanic muds and sands, coral muds and sands, are met with in those regions of the ocean nearest to land. With the exception of the volcanic muds and sands, and coral muds and sands, around oceanic islands, these deposits are found only along the borders of continents and continental islands, and in enclosed and partially enclosed seas.

(2) The organic oozes and red clay are confined to the abysmal regions of the ocean basins; a Pteropod ooze is met with in tropical and subtropical regions in depths less than 1500 fathoms, a Globigerina ooze in the same regions between the depths of 500 and 2800 fathoms, a Radiolarian ooze in the central portions of the Pacific at depths greater than 2500 fathoms, a Diatom ooze in the Southern Ocean south of the latitude of 45° South, a red clay anywhere within the latitudes of 45° north and south at depths greater than 2200 fathoms.

Conclusions.—All the facts and details enumerated in the foregoing pages point to certain conclusions which are of considerable geological interest, and which appear to be warranted by the present state of our investigations.

* "On Cosmic and Volcanic Dust," *Proc. Roy. Soc. Edin.*, 1883-4.

We have said that the débris carried away from the land accumulates at the bottom of the sea before reaching the abysmal regions of the ocean. It is only in exceptional cases that the finest terrigenous materials are transported several hundred miles from the shores. In place of layers formed of pebbles and clastic elements with grains of considerable dimensions, which play so large a part in the composition of emerged lands, the great areas of the ocean basins are covered by the microscopic remains of pelagic organisms, or by the deposits coming from the alteration of volcanic products. The distinctive elements that appear in the river and coast sediments are, properly speaking, wanting in the great depths far distant from the coasts. To such a degree is this the case that in a great number of soundings, from the centre of the Pacific for example, we have not been able to distinguish mineral particles on which the mechanical action of water had left its imprint, and quartz is so rare that it may be said to be absent. It is sufficient to indicate these facts in order to make apparent the profound differences which separate the deposits of the abysmal areas of the ocean basins from the series of rocks in the geological formations. As regards the vast deposits of red clay, with its manganese concretions, its zeolites, cosmic dust, and remains of vertebrates, and the organic oozes which are spread out over the bed of the central Pacific, Atlantic, and Indian oceans, have they their analogues in the geological series of rocks? If it be proved that in the sedimentary strata the pelagic sediments are not represented, it follows that deep and extended oceans like those of the present day cannot formerly have occupied the areas of the present continents, and as a corollary the great lines of the ocean basins and continents must have been marked out from the earliest geological ages. We thus get a new confirmation of the opinion of the permanence of the continental areas.

But without asserting in a positive manner that the terrestrial areas and the areas covered by the waters of the great ocean basins have had their main lines marked out since the commencement of geological history, it is, nevertheless, a fact, proved by the evidence derived from a study of the pelagic sediments, that these areas have a great antiquity. The accumulation of sharks' teeth, of the ear-bones of cetaceans, of manganese concretions, of zeolites, of volcanic material in an advanced state of decomposition, and of cosmic dust, at points far removed from the continents, prove this. There is no reason for supposing that the parts of the ocean where these vertebrate remains are found are more frequented by sharks or cetaceans than other regions where they are never or only rarely dredged from the deposits at the bottom. When we remember also that these ear-bones, teeth of sharks, and volcanic fragments, are sometimes incrusted with two centimetres of manganese oxide, while others have a mere coating, and that some of the bones and teeth belong to extinct species, we may conclude with great certainty that the clays of these oceanic basins have accumulated with extreme slowness. It is indeed almost beyond question that the red clay regions of the central Pacific contain accumulations belonging to geological ages different from our own. The great antiquity of these formations is likewise confirmed in a striking manner by the presence of cosmic fragments, the nature of which we have described.* In order to account for the accumulation of all these substances in such relatively great abundance in the areas where they were dredged, it is necessary to suppose the oceanic basins to have remained the same for a vast period of time.

The sharks' teeth, ear-bones, manganese nodules, altered volcanic fragments, zeolites, and cosmic dust, are met with in greatest abundance in the red clays of the central Pacific, at that point on the earth's surface farthest removed from continental land. They are less abundant in the Radiolarian ooze, are rare in the Globigerina, Diatom, and Pteropod oozes, and they have been dredged only in a few instances in the terrigenous deposits close to the shore. These substances are present in all the deposits, but owing to the abundance of other matters in the more rapidly forming deposits their presence is masked, and the chance of dredging them is reduced. We may then regard the greater or less abundance of these materials, which are so characteristic of a true red clay, as being a measure of the relative rate of accumulation of the marine sediments in which they lie. The terrigenous deposits accumulate most rapidly, then follow in order Pteropod ooze, Globigerina ooze, Diatom ooze, Radiolarian ooze, and, slowest of all, red clay.

* "On Cosmic and Volcanic Dust," *Proc. Roy. Soc. Edin.*

From the data now advanced it appears possible to deduce other conclusions important from a geological point of view. In the deposits due essentially to the action of the ocean, we are at once struck by the great variety of sediments which may accumulate in regions where the external conditions are almost identical. Again marine faunas and floras, at least those of the surface, differ greatly, both with respect to species and to relative abundance of individuals, in different regions of the ocean; and as their remains determine the character of the deposit in many instances, it is legitimate to conclude that the occurrence of organisms of a different nature in several beds is not an argument against the synchronism of the layers which contain them.

The small extent occupied by littoral formations, especially those of an arenaceous nature, shown by our investigations, and the relatively slow rate at which such deposits are formed along a stable coast, are matters of importance.

In the present state of things there does not appear to be anything to account for the enormous thickness of the clastic sediments making up certain geological formations, unless we consider the exceptional cases of erosion which are brought into play when a coast is undergoing constant elevation or subsidence. Great movements of the land are doubtless necessary for the formation of thick beds of transported matter like sandstones and conglomerates.

In this connection may be noted the fact that in certain regions of the deep sea no appreciable formation is now taking place. Hence the absence, in the sedimentary series, of a layer representing a definite horizon must not always be interpreted as proof either of the emergence of the bottom of the sea during the corresponding period, or of an ulterior erosion. Arenaceous formations of great thickness require seas of no great extent and coasts subject to frequent oscillations, which permit the shores to advance and retire. Along these, through all periods of the earth's history, the great marine sedimentary phenomena have taken place.

The continental geological formations, when compared with marine deposits of modern seas and oceans, present no analogues to the red clays, Radiolarian, Globigerina, Pteropod, and Diatom oozes. On the other hand, the terrigenous deposits of our lakes, shallow seas, enclosed seas, and the shores of the continents, reveal the equivalents of our chalks, greensands, sandstones, conglomerates, shales, marls, and other sedimentary formations. Such formations as certain tertiary deposits of Italy, Radiolarian earth from Barbadoes, and portions of the Chalk where pelagic conditions are indicated, must be regarded as having been laid down rather along the border of a continent than in a true oceanic area. On the other hand, the argillaceous and calcareous rocks, recently discovered by Dr Guppy, in the upraised coral islands in the Solomon group, are nearly identical with the volcanic muds, and probably also with the Pteropod and Globigerina oozes of the Pacific.

Regions situated similarly to enclosed and shallow seas and the borders of the present continents appear to have been, throughout all geological ages, the theatre of the greatest and most remarkable changes; in short, all, or nearly all, the sedimentary rocks of the continents would seem to have been built up in areas like those now occupied by the terrigenous deposits, which we may designate "*the transitional or critical area of the earth's surface.*" This area occupies, we estimate, about two-eighths of the earth's surface, while the continental and abysmal areas occupy each about three-eighths.

During each era of the earth's history, the borders of some lands have sunk beneath the sea and been covered by marine sediments; while in other parts the terrigenous deposits have been elevated into dry land, and have carried with them a record of the organisms which flourished in the sea of the time. In this transitional area there has been throughout a continuity of geological and biological phenomena.

From these considerations it will be evident that the character of a deposit is determined much more by distance from the shore of a continent than by actual depth; and the same would appear to be the case with respect to the fauna spread over the floor of the present oceans. Dredgings near the shores of continents, in depths of 1000, 2000, or 3000 fathoms, are more productive both in species and individuals than dredgings at similar depths several hundred miles seawards. Again, among the few species dredged in the abysmal areas furthest removed from land, the majority show archaic characters, or belong to groups which have a wide distribution *in time* as well as over the floor of the present oceans. Such are the

Hexactinellida, Brachiopoda, Stalked Crinoids and other Echinoderms, &c.

As already mentioned, the transitional area is that which now shows the greatest variety in respect to biological and physical conditions, and in past time it has been subject to the most frequent and the greatest amount of change. The animals now living in this area may be regarded as the greatly modified descendants of those which have lived in similar regions in past geological ages, and some of whose ancestors have been preserved in the sedimentary rocks as fossils. On the other hand, many of the animals dredged in the abysmal regions are most probably also the descendants of animals which lived in the shallower waters of former geological periods, but descended into deep water to escape the severe struggle for existence which must always have obtained in those depths affected by light, heat, motion, and other conditions. Having found existence possible in the less favourable and deeper water, they may be regarded as having slowly spread themselves over the floor of the ocean, but without undergoing great modifications, owing to the extreme uniformity of the conditions and the absence of competition. Or we may suppose that in the depressions which have taken place near coasts, some species have been gradually carried down to deep water, have accommodated themselves to the new conditions, and have gradually migrated to the regions far from land. A few species may thus have migrated to the deep sea during each geological period. In this way the origin and distribution of the deep-sea fauna in the present oceans may in some measure be explained. In like manner, the pelagic fauna and flora of the ocean is most probably derived originally from the shore and shallow water. During each period of the earth's history a few animals and plants have been carried to sea, and have ultimately adopted a pelagic mode of life.

Without insisting strongly on the correctness of some of these deductions and conclusions, we present them for the consideration of naturalists and geologists, as the result of a long, careful, but as yet incomplete, investigation.

36

Reprinted from *Nature* **190**(4779):854–857 (June 3, 1961)

CONTINENT AND OCEAN BASIN EVOLUTION BY SPREADING OF THE SEA FLOOR

By ROBERT S. DIETZ,

U.S. Navy Electronics Laboratory, San Diego 52, California

ANY concept of crustal evolution must be based on an Earth model involving assumptions not fully established regarding the nature of the Earth's outer shells and mantle processes. The concept proposed here, which can be termed the 'spreading sea-floor theory', is largely intuitive, having been derived through an attempt to interpret sea-floor bathymetry. Although no entirely new proposals need be postulated regarding crustal structure, the concept requires the acceptance of a specific crustal model, in some ways at variance with the present consensus of opinion. Since the model follows from

the concept, no attempt is made to defend it. The assumed model is as follows:

(1) Large-scale thermal convection cells, fuelled by the decay of radioactive minerals, operate in the mantle. They must provide the primary diastrophic forces affecting the lithosphere.

(2) The sequence of crustal layers beneath the oceans is markedly different from that beneath the continents and is quite simple (Fig. 1). On an average 4·5 km. of water overlies 0·3 km. of unconsolidated sediments (layer 1). Underlying this is layer 2, consisting of about 2·0 km. of mixed volcanics and lithified sediments. Beneath this is the layer 3 (5 km. thick), commonly called the basalt layer and supposedly forming a world-encircling cap of effusive basic volcanics over the Earth's mantle from which it is separated by the Mohorovičić seismic discontinuity. Instead we must accept the growing opinion that the 'Moho' marks a change of phase rather than a chemical boundary, that is, layer 3 is chemically the same as the mantle rock but petrographically different with low-pressure phase minerals above the Moho and high-pressure minerals below. This change of phase may be either from eclogite to gabbro[1], or from peridotite to serpentine[2]; its exact nature is not vital to our concept, but we can tentatively accept the eclogite–gabbro transition as it has more adherents. Common usage requires that we reserve the term 'mantle' for the substance beneath the Moho, but in point of fact, the gabbro layer (as a change of phase) is also a part of the mantle—a sort of 'exo-mantle'. Except for a very thin veneer, then, the sea floor is the exposed mantle of the Earth in this larger sense.

(3) It is relevant to speak of the strength and rigidity of the Earth's outer shell. The term 'crust' has been effectively pre-empted from its classical meaning by seismological usage applying it to the layer above the Moho, that is, the sial in continental regions and the 'basaltic' layer under the oceans so that the continents have a thick crust and the ocean basins a thin crust. Used in this now accepted sense, any implications equating the crust with rigidity must be dismissed. For considerations of convective creep and tectonic yielding, we must refer to a lithosphere and an asthenosphere. Deviations from isostasy prove that approximately the outer 70 km. of the Earth (under the continents and ocean basins alike) is moderately strong and rigid even over time-spans of 100,000 years or more; this outer rind is the lithosphere. Beneath lies the asthenosphere separated from the lithosphere by the level of no strain or isopiestic level; it is a domain of rock plasticity and flowage where any stresses are quickly removed. No seismic discontinuity marks the isopiestic level and very likely it is actually a zone of uniform composition showing a gradual transition in strength as pressure and temperature rise; and in spite of the lithosphere's rigidity, to speak of it as a crust or shell greatly exaggerates its strength. Because of its grand dimensions, for model similitude we must think of it as weak[3]. If convection currents are operating 'subcrustally', as is commonly written, they would be expected to shear below the lithosphere and not beneath the 'crust' as this term is now used.

(4) As gravity data have shown, the continents are low-density tabular masses of sial—a 'basement complex' of granitic rocks about 35 km. thick with a thin sedimentary veneer. Since they are buoyant and float high hydrostatically in the sima, they are analogous to icebergs in the ocean. This analogy has additional merit in that convection of the sima cannot enter the sial. But the analogy gives the wrong impression of relative strength of sial and sima; the continental lithosphere is no stronger than the oceanic lithosphere, so it is mechanically impossible for the sial to 'sail through the sima' as Wegnerian continental drift proposes. The temperature and pressure are too high at the base of the sial to permit a gabbroic layer above the Moho; instead, there may be an abrupt transition from granite to eclogite.

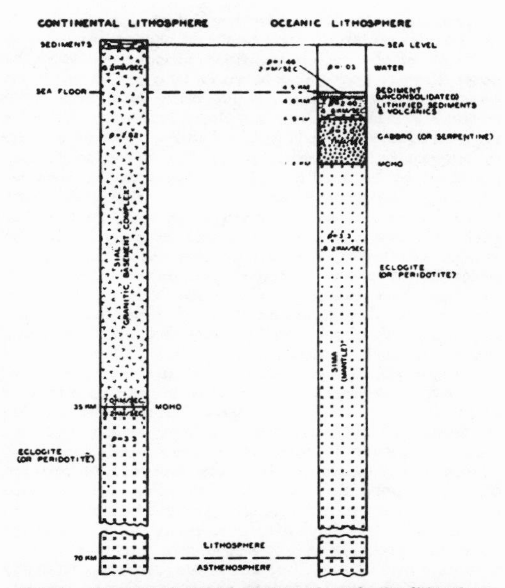

Fig. 1. Generalized crustal sections through the lithosphere beneath the continents and the ocean basins as presumed in this paper. Seismic velocities and densities are shown for the various layers

Spreading Sea Floor Theory

Owing to the small strength of the lithosphere and the gradual transition in rigidity between it and the asthenosphere, the lithosphere is not a boundary to convection circulation, and neither is the Moho beneath the oceans because this is not a density boundary but simply a change of phase. Thus the oceanic 'crust' (the gabbroic layer) is almost wholly coupled with the convective overturn of the mantle creeping at a rate of a few cm./yr. Since the sea floor is covered by only a thin veneer of sediments with some mixed-in effusives, it is essentially the outcropping mantle. So the sea floor marks the tops of the convection cells and slowly spreads from zones of divergence to those of convergence. These cells have dimensions of several thousands of kilometres; some cells are quite active now while others are dead or dormant. They have changed position with geological time causing new tectonic patterns.

The gross structures of the sea floor are direct expressions of this convection. The median rises[4,5] mark the up-welling sites or divergences; the trenches are associated with the convergences or down-welling sites; and the fracture zones[6] mark shears between regions of slow and fast creep. The high heat-flow under the rises[7] is indicative of the ascending convection currents as also are the groups of volcanic seamounts which dot the backs of these rises.

Much of the minor sea-floor topography may be even directly ascribable to spreading of the sea floor. Great expanses of rough topography skirt both sides of the Mid-Atlantic Rift; similarly there are extensive regions of abyssal hills in the Pacific. The roughness is suggestive of youth, so it has commonly been assumed to be simply volcanic topography because the larger seamounts are volcanic. But this interpretation is not at all convincing, and no one has given this view formality by publishing a definitive study. Actually, the topography resembles neither volcanic flows nor incipient volcanoes. Can it not be that these expanses of abyssal hills are a 'chaos topography' developed as strips of juvenile sea-floor (by a process which can be visualized only as mixed intrusion and extrusion) and then placed under rupturing stresses as the sea floor moves outward?

The median position of the rises cannot be a matter of chance, so it might be supposed that the continents in some manner control the convection pattern. But the reverse is considered true: conditions deep within the mantle control the convective pattern without regard for continent positions. By viscous drag, the continents initially are moved along with the sima until they attain a position of dynamic balance overlying a convergence. There the continents come to rest, but the sima continues to shear under and descend beneath them; so the continents generally cover the down-welling sites. If new upwells do happen to rise under a continental mass, it tends to be rifted. Thus, the entire North and South Atlantic Ocean marks an ancient rift which separated North and South America from Europe and Africa. Another such rift has opened up the Mediterranean. The axis of the East Pacific Rise now seems to be invading the North American continent, underlying the Gulf of California and California[9]. Similarly, the Indian Ocean Rise may extend into the African Rift Valleys, tending to fragment that continent.

The sialic continents, floating on the sima, provide a density barrier to convection circulation—unlike the Moho, which involves merely a change of phase. The convection circulation thus shears beneath the continents so that the sial is only partially coupled through drag forces. Since the continents are normally resting over convergences, so that convective spreading is moving toward them from opposite sides, the continents are placed consequently under compression. They tend to buckle, which accounts for alpine folding, thrust faulting, and similar compressional effects so characteristic of the continents. In contrast, the ocean basins are simultaneously domains of tension. If the continental block is drifted along with the sima, the margin is tectonically stable (Atlantic type). But if the sima is slipping under the sialic block, marginal mountains tend to form (Pacific type) owing to drag forces.

Implications of the Concept

Ad hoc hypotheses are likely to be wrong. On the other hand, one which is consonant with our broader understanding of the history of the Earth may have merit. While the thought of a highly mobile sea floor may seem alarming at first, it does little violence to geological history.

Volumetric changes of the Earth. Geologists have traditionally recognized that compression of the continents (and they assumed of the ocean floors as well) was the principal tectonic problem. It was supposed that the Earth was cooling and shrinking. But recently, geologists have been impressed by tensional structures, especially on the ocean floor. To account for sea floor rifting, Heezen[10], for example, has advocated an expanding Earth, a doubling of the diameter. Carey's[11] tectonic analysis has resulted in the need for a twenty-fold increase in volume of the Earth. Spreading of the sea floor offers the less-radical answer that the Earth's volume has remained constant. By creep from median upwellings, the ocean basins are mostly under tension, while the continents, normally balanced against sima creepage from opposite sides, are under compression.

The geological record is replete with transgressions and regressions of the sea, but these have been shallow and not catastrophic; fluctuations in sea-level as severe as those of the Pleistocene are abnormal. The spreading concept does no violence to this order of things, unlike dilation or contraction of the Earth. The volumetric capacity of the oceans is fully conserved.

Continental Drift. The spreading concept envisages limited continental drifting, with the sial blocks initially being rafted to down-welling sites and then being stablized in a balanced field of opposing drag forces. The sea floor is held to be more mobile and to migrate freely even after the continents come to rest. The sial moves largely *en bloc*, but the sea floor spreads more differentially.

Former scepticism about continental drift is rapidly vanishing, especially due to the palæomagnetic findings and new tectonic analyses. A principal objection to Wegener's continental drift hypothesis was that it was physically impossible for a continent to 'sail like a ship' through the sima; and nowhere is there any sea floor deformation ascribable to an on-coming continent. Sea floor spreading obviates this difficulty: continents never move through the sima—they either move along with it or stand still while the sima shears beneath them. The buoyancy of the continents, rather than their being stronger than the sima, accounts for this. Drag associated with the shearing could account for alpine folding and related compressional tectonic structures on the continents.

Persistent freeboard of the continents. A satisfactory theory of crustal evolution must explain why the continents have stood high throughout geological time in spite of constant erosional de-levelling. Many geologists believe that new buoyancy is added to continents through the gravitative differentiation from the mantle. Spreading of the sea floor provides a mechanism whereby the continents are placed over the down-wells where new sial would tend to collect, even though the convection is entirely a mantle process and the role of the continents is passive. It also follows that the clastic detritus swept into the deep sea from the continents is not permanently lost. Rather, it is carried slowly towards, and then beneath, the continents, where it is granitized and added anew to the sialic blocks.

Youth of the ocean floor. It follows paradoxically from the spreading concept that, although the ocean basins are old, the sea floor is young—much younger than the rocks of the continents. Marine sediments, seamounts, and other structures slowly impinge against the sialic blocks and are destroyed by underriding them. Pre-Cambrian and perhaps even most Palæozoic rocks should prove absent from the ocean floors; and Mohole drilling should not reveal the great missing sequence of the Lipalian interval

(Pre-Cambrian to Cambrian) as hoped for by some. All this may seem surprising, but marine geological evidence supports the concept.

On his discovery of the guyots of the Pacific, Hess[12] supposed these were Pre-Cambrian features protected from erosion by the cover of the sea. But Hamilton[13] proved the guyots of the Mid-Pacific Mountains were Cretaceous, and these seem to be among the oldest of the seamount groups. In an analysis of the various seamount groups of the western Pacific, I was forced to conclude that none of them was older than mid-Mesozoic. The young age of the seamounts has been puzzling; certainly they can neither erode away nor subside completely. Also, there seem to be too few volcanic seamounts, if the present population represents the entire number built over the past hundred million years or more. The puzzle dissolves if sea floor spreading has operated. Modern examples of impinging groups of seamounts may be the western end of the Caroline Islands, the Wake–Marcus Seamounts, and the Magellan Seamounts[14]. All may be moving into the western Pacific trenches. Seamount *GA*-1 south of Alaska may be moving into the Aleutian Trench[15].

The sedimentary layers under the sea also appear to be young. No fossiliferous rocks older than Cretaceous have yet been dredged from any ocean basin. Radioactive dating of a basalt from the Mid-Atlantic ridge gave a Tertiary age[16]. Kuenen[17] estimated that the ocean basins should contain on an average about 3·0 km. of sedimentary rocks assuming the basins are 200 million years old. But seismic reflexions indicate an average of only 0·3 km. of the unconsolidated sediments. Hamilton[18], however, believes that much of layer 2 may be lithified sediments. If *all* layer 2 is lithified sediments, Hamilton finds that the ocean basins may be Palæozoic or late Pre-Cambrian in age—but not Archæan. But very likely layer 2 includes much effusive material and sedimentary products of sea floor weathering. In summing up, the evidence from the sediments, although still fragmentary, suggests that the sea floors may be not older than Palæozoic or even Mesozoic.

Spreading and magnetic anomalies. Vacquier, V., *et al.* (in the press) recently have completed excellent sea-floor magnetic surveys off the west coast of North America. A striking north–south lineation shows up which seems to reveal a stress pattern (Mason, R. G., and Raff, A. D., in the press). Such interpretation would fit into spreading concept with the lineations being developed normal to the direction of convection creep. The lineation is interrupted by Menard's[6] three fracture-zones, and anomalies indicate shearing offsets of as much as 640 nautical miles in the case of the remarkable Mendocinco Escarpment[19]. Great mobility of the sea floor is thus suggested. The offsets have no significant expression after they strike the continental block; so apparently they may slip under the continent without any strong coupling. Another aspect is that the anomalies smooth out and virtually disappear under the continental shelf; so the sea floor may dive under the sial and lose magnetism by being heated above the Curie point.

By considering an Earth crustal model only slightly at variance with that commonly accepted, a novel concept of the evolution of continents and ocean basins has been suggested which seems to fit the 'facts' of marine geology. If this concept were correct, it would be most useful to apply the term 'crust', which now has a confusion of meanings, only to any layer which overlies and caps the convective circulation of the mantle. The sialic continental blocks do this, so they form the true crust. The ocean floor seemingly does not, so the ocean basin is 'crustless'.

I wish to express my appreciation to E. L. Hamilton, F. P. Shepard, H. W. Menard, V. Vacquier, R. Von Herzen and A. D. Raff for critical discussions.

[1] Kennedy, G. C., *Amer. Sci.*, **47**, 4, 491 (1959).
[2] Hess, H. H., *Abst. Bull. Geol. Soc. Amer.*, **71**, Pt. 2, 12, 2097 (1960).
[3] Griggs, D. A., *Amer. J. Sci.*, **237**, 611 (1939).
[4] Ewing, M., and Heezen, B. C., *Amer. Geophys. Union Geophys. Mon. No.* 1, 75 (1956).
[5] Menard, H. W., *Bull. Geol. Soc. Amer.*, **69**, 9, 1179 (1958).
[6] Menard, H. W., *Bull. Geol. Soc. Amer.*, **66**, 1149 (1955).
[7] Von Herzen, R. P., *Nature*, **183**, 882 (1959).
[8] Menard, H. W., *Science*, **132**, 1737 (1960).
[9] Heezen, B. C., *Sci. Amer.*, Oct. 2, 14 (1960).
[10] Heezen, B. C., Preprints, First Intern. Ocean. Cong., 26 (1959).
[11] Carey, W. S., *The Tectonic Approach to Continental Drift: in Continental Drift—A Symposium*, 177 (Univ. Tasmania, 1958.)
[12] Hess, H. H., *Amer. J. Sci.*, **244**, 772 (1946).
[13] Hamilton, E. L., *Geol. Soc. Amer. Mem.*, **64**, 97 (1956).
[14] Dietz, R. S., *Bull. Geol. Soc. Amer.*, **65**, 1199 (1954).
[15] Menard, H. W., and Dietz, R. S., *Bull. Geol. Soc. Amer.*, **62**, 1263 (1951).
[16] Carr, D., and Kulp, J., *Bull. Geol. Soc. Amer.*, **64**, 2, 263 (1953).
[17] Kuenen, Ph., *Marine Geology* (John Wiley and Sons, New York, 1950).
[18] Hamilton, E. L., *Bull. Geol. Soc. Amer.*, **70**, 1399 (1959); *J. Sed. Petrol.*, **30**, 3, 370 (1960).
[19] Menard, H. W., and Dietz, R. S., *J. Geol.*, **60**, 3 (1952).

Part VII

MARINE BIOLOGY

Editor's Comments on Papers 37, 38, and 39

37 FORBES, E.
On the Light thrown on Geology by Submarine Researches

38 HUXLEY
On Some Organisms Living at Great Depths in the North Atlantic Ocean

39 FORBES, D.
The Depths of the Sea

Aristotle studied the marine life of the Aegean Sea (Aristotle 1908–1952) but there was a long gap between him and the founders of modern marine biology. It was not until the eighteenth century that scientists turned to marine life as a whole, though earlier works had been published on fish and whales, for example, Rondelet's *Histoire Entière des Poissons* (1558) and Sir Robert Sibbald's observations (1692) on whales stranded on Scottish shores. One of the pioneers was Marsigli whose *Histoire Physique de la Mer* (1725) is in fact mainly about the marine life of the Gulf of Lions (Pérès 1968). Later in the same century the dredge was used by another Italian scientist, Vitaliano Donati (1758), working in the Adriatic, and by the Danish scientist O. F. Muller (Wolff 1967). However, it was in the early nineteenth century that marine biology began its great period of expansion. Perhaps it was the work of systematization and classification initiated by Linnaeus, Buffon, and Cuvier, together with the growing interest in biology that caused biologists to turn with increasing enthusiasm to the sea.

The number of professional biologists in the early nineteenth century was very small. There was as well a small number of those rich enough to work full time on what interested them. Therefore, particularly in the earlier part of the century, marine biologists were a heteorogeneous assembly. They included characters as diverse as the aristocratic Scottish lawyer and antiquarian Sir John Dalyell (Yonge 1972), the poor minister John Fleming (note the reference to marine invertebrates in the paper on tides, Paper 15) and the wealthy London solicitor and leading conchologist John Gwyn Jeffreys. Some, like Fleming, eventually achieved academic position but many continued working on their own stretch of coastline and particular groups of creatures and are remembered today only by the specialist.

Early workers in France included A. Risso and J. B. Verany in the south (Trégouboff 1968), and J. V. Audouin and Henri Milne Edwards who, in the 1820s, carried out researches on the coasts of Normandy and Brittany (Théodoridès 1968). In Norway pioneer work was carried out by Hans Strom and Bishop Gunnerus but the modern study of marine biology began with Michael Sars (Sivertsen 1968). Sars' career was not unlike Fleming's. He began as a clergyman in a remote seaside parish. His zoological work won him international recognition and he ended his days as a professor at the University of Christiania (Oslo). In the United States, the Swiss scientist J. L. Agassiz, Professor of Natural History at Harvard, and Count L. F. de Pourtalès, who joined the Coast Survey in 1848, were among pioneers in marine zoology (Scheltema 1972).

The newly formed British Association for the Advancement of Science began giving financial support for dredging expeditions in the 1840s. The official voyages of discovery also provided an opportunity for observing marine life and large collections were made on the great expeditions of the mid-nineteenth century by people like J. D. Dana, naturalist on the U.S. Exploring Expedition of 1838–1842. The Russian explorer Admiral Bellingshausen, one of the early visitors to Antarctic waters, made use of a tow net and observed the diurnal migration of plankton (Debenham 1945). A naturalist working in Ireland, J. Vaughan Thompson, was also using a plankton net in the 1820s (Hedgpeth 1957). Survey vessels provided a unique opportunity for some British scientists, like Darwin in the *Beagle,* 1831–1836. Much scientific work was also performed by ships' surgeons, perhaps the most famous being T. H. Huxley who sailed to Australasia in HMS *Rattlesnake* in the late 1840s. Perhaps the most significant occasion for marine biology was when Captain Thomas Graves of HMS *Beacon* invited the young Manx biologist Edward Forbes, 1815–1854, to join his ship (Merriman 1965; Wilson and Geikie 1865) as naturalist in the Mediterranean.

During his researches into the marine fauna of British coasts, Forbes had observed what other biologists had already noted for the intertidal zone (Hedgpeth 1957)—that the different species always occurred in the same faunal assemblage and at the same range of depth—and extended the principle to deeper water. In Paper 37 he relates how in the Mediterranean he found eight zones of life between the surface and a depth of 230 fathoms, corresponding to those he had found off Britain. He observed that, among the many other interesting points in the paper, the number of species was less in the lower than in the upper zones, getting successively fewer with depth. Plant life had dwindled to a single species by 100 fathoms. Forbes concluded: "We may fairly infer, then, that as there is a zero of vegetable life so there is one of animal life." Forbes first thought the zero would be at 300 fathoms but T. A. B. Spratt (1848), one of the *Beacon's* officers, who continued Forbes' work after he had returned home, soon obtained specimens at slightly over that depth so the limit was shifted to 400 fathoms.

The idea of the "azoic zone," as the supposedly lifeless depths of the ocean were called, seemed so logical, considering the darkness, cold, and intense pressure at great depths, that many scientists readily accepted the idea, not withstanding the fact that contradictory evidence was already in existence. In 1818, on Sir John Ross's expedition in search of the North-West Passage, a starfish had been brought up clinging to the sounding line below the 800 fathom mark (J. Ross 1819). Sir John's nephew, Sir James Clark Ross (1847), never doubted the existence of life in the deep ocean and himself successfully dredged at depths of over 300 fathoms in the Antarctic on his voyage in the *Erebus* and *Terror* during 1839–1843. About the same time that Forbes was in the Mediterranean, the French scientist Aimé (1845) was making soundings off the coast of Algeria and brought up creatures from depths as great as 1800 m. As time went on the list grew. In spite of this, it is interesting to observe and hard to explain how firmly the idea of the azoic zone persisted.

During the 1850s, soundings were made in the North Atlantic to locate a route for the trans-Atlantic telegraph cable (see Part VI). The first specimens of the deep-sea floor were obtained with the apparatus designed by Midshipman Brooke, assistant to Maury (1855). Maury sent some of the samples to J. W. Bailey and the German biologist C. G. Ehrenberg. They found that the samples consisted largely of the shelly remains of a single species of foraminifera, of the genus *Globigerina.* Bailey believed that the foraminifera lived at the surface of the sea, and that their remains fell to the bottom after death. On the other hand Ehrenberg, already (1844) skeptical of the azoic zone, and T. H. Huxley (1858) and William King (Deacon 1971, ch. 13), who analyzed samples collected by British ships, believed that these creatures lived on the sea bed. The controversy was finally settled during the *Challenger* expedition. John Murray showed (Paper 35) that the populations of globigerina and of diatoms that were living in the surface layers, were reflected in the character of the deposits on the sea bed below.

It was while he was reexamining some of the original sea-bed samples a few years later that Thomas Henry Huxley, 1825–1895, thought he had observed a new form of life. He discovered jelly-like masses he had not observed before and tests seemed to show that they were made of organic material. Huxley concluded with a certain amount of caution that he had found a form of protoplasm and named it

Bathybius Haeckelii after the eminent German zoologist who had worked on single-celled animals (Paper 38). Here, too, it was the *Challenger* expedition that provided the real explanation. J. Y. Buchanan, the chemist, reported that the substance Huxley had observed was formed by a chemical reaction between the sea water in the sediment and the alcohol used to preserve it (Huxley 1875; Buchanan 1913, p. ix).

Although willing to admit that simple forms of life might inhabit the sea bed, biologists generally still believed that no higher forms would be found there. It took a long time for this view to be shaken, even though evidence continued to pile up against it. In 1860 George Wallich (1862), naturalist on the cable survey expedition of HMS *Bulldog,* reported starfish brought up from over 1200 fathoms. In 1868 Michael Sars (1869) published a list giving many species found at depths of over 300 fathoms. His son G. O. Sars (1872) said that he had rejected the idea of the azoic zone by 1850. G. O. Sars himself dredged successfully at up to 450 fathoms while engaged in fishery research. This work aroused the interest of two British biologists, W. B. Carpenter and C. Wyville Thomson. They approached the Admiralty and were allowed to make cruises in HMS *Lightning* in 1868 and in HMS *Porcupine* in 1869 and 1870. On the first expedition in the *Porcupine,* Thomson successfully dredged organisms from nearly 2500 fathoms. This meant that the idea of the azoic zone could no longer be sustained, as David Forbes, 1828–1876, brother of Edward, pointed out (see Paper 39). The *Challenger* and other expeditions soon showed that advanced animal life was to be found even in the greatest depths of the oceans (Thomson 1880).

The major portion of the biological work of these expeditions was the collection and identification of new species. It was the exception to have a special theme such as the German Plankton Expedition of 1889. Alexander Agassiz (1888, vol. 1, pp. 28–31) made a great improvement by introducing wire rope for dredging. This was both stronger and less bulky than the rope used hitherto. Much effort went into the design of better nets and trawls, in particular to secure opening and closing at prearranged depths. The results of the first expeditions gave a very imperfect idea of the distribution with depth of free-swimming creatures and Agassiz went so far as to revive the idea of an azoic zone and claim that life would be found only in the top few hundred fathoms and on or just above the sea bed.

Marine biologists continued their work in shallow water, working increasingly at marine biological stations. The movement to create such institutions began in the 1840s. One of the first was established by the Belgian scientist P. J. Van Beneden near Ostend in 1842 (Charlier and Leloup 1968). It was during the 1870s and 1880s that the great expansion took place with the foundation, among others, of the Stazione Zoologica at Naples by Anton Dohrn in 1873 (Herdman 1923) and the marine biological laboratories at Woods Hole in 1881 (Lillie 1944) and Plymouth in 1888 (Kofoid 1910), still leaders in the field today.

In common with other well-established sciences, biology was beginning at this time to offer a better chance for a professional career. Marine biology was well represented and specialists in the field, such as Wyville Thomson and Nansen, occupied university chairs of zoology or natural history. Much of the work at the marine stations was seasonal and during spring and summer vacations university staff would descend with their students for field work on the neighboring shores. Many stations were linked to universities though others were maintained by societies or by wealthy or energetic private individuals. The work of the marine stations was largely biological though chemical or geological work was sometimes included. John Murray's station at Granton, on the Firth of Forth, was an example of an institution where a wider variety of work was attempted.

The growth of the marine stations and of the discipline of marine biology was greatly assisted by the onset of governmental concern about the decline in fisheries, which became acute during the last quarter of the nineteenth century. The problem was so serious that state intervention was thought necessary and government bodies were set up in Germany, Norway (Solhaug and Saetersdal 1972), and other countries. In the United States, the U.S. Fish Commission began work in 1870 (Galtsoff 1962) and its ship the *Albatross* worked on both the Atlantic and Pacific coasts (Hedgpeth 1945). In the United Kingdom, the Scottish Fisheries Board began scientific research in 1882 and money also went to biologists working in the universities and for the marine biological associations. Much of the work of I.C.E.S. was on fishery problems (Went 1972a, b). There was considerable difference of opinion as to the cause of the decline in catches and some scientists, such as W. C. McIntosh of St. Andrew's University (Merriman 1972), resolutely believed that man's activities could not effect the vast resources of the ocean. Conclusive evidence that they had done so was supplied by the increase in supplies after the interruption of fishing in the North Sea during both world wars (Hardy 1959, pp. 247–253). Since World War II, commissions of enquiry and conventions restricting the size of catch, type of net and mesh etc., have multiplied (Cushing 1972).

By the 1920s the decline in whale stocks was also causing concern. The slower swimming whales of the Arctic had been decimated by the early nineteenth century and of the Southern and Pacific oceans by the 1870s. The introduction of powered and later of factory ships meant that the faster species soon shared the same fate. In 1925 ships of the

Discovery Committee began visiting the Southern Ocean. Their work up to and just after World War II was to study the biology of whales and their life history, subjects almost totally unexplored by science up to that time (Mackintosh 1946).

During the 1930s, it was a matter for concern that so much of the effort of marine biologists generally was going into the description of species and so little into the study of their life cycles and ecology (Vaughan 1937). Today, some people feel that although there is still a vast amount to do in them, taxonomic work is the basis of progress in these fields and that the pendulum has perhaps swung too far in the opposite direction.

37

Reprinted from *Edinburgh New Philos. J.* 36:318-327 (1844)

***On the Light thrown on Geology by Submarine Researches; being the Substance of a Communication made to the Royal Institution of Great Britain, Friday Evening, the* 23*d February* 1844.** By **Edward Forbes, F.L.S., M.W.S., &c. Prof. Bot. King's College, London. Communicated by the Author.**

About the middle of the last century, certain Italian naturalists* sought to explain the arrangement and disposition of organic remains in the strata of their country, by an examination of the distribution of living beings on the bed of the Adriatic Sea. They sought in the bed of the present sea for an explanation of the phenomena presented by the upheaved beds of former seas. The instrument, by means of which they conducted their researches, was the common oyster-dredge. The results they obtained bore importantly on Geology; but since their time, little has been done in the same line of research,—the geologist has been fully occupied above water, and the naturalist has pursued his studies with far too little reference to their bearing on geological questions, and on the history of animals and plants *in time*. The dredge, when used, has been almost entirely restricted to the search after rare animals, by the more adventurous among zoologists.

Convinced that inquiries of the kind referred to, if conducted with equal reference to all the natural history sciences,

* Marsili and Donati, and after them Soldani.

and to their mutual connection, must lead to results still more important than those which have been obtained, I have, for several years, conducted submarine researches by means of the dredge. In the present communication, I shall give a brief account of some of the more remarkable facts and conclusions to which they have led, and as briefly point out their bearings on the science of geology.

I. *Living beings are not distributed indifferently on the bed of the sea, but certain species live in certain parts, according to the depth, so that the sea-bed presents a series of zones or regions, each peopled by its peculiar inhabitants.*—Every person who has walked between high and low water-marks on the British coasts, when the tide was out, must have observed, that the animals and plants which inhabit that space, do not live on all parts of it alike, but that particular kinds reach only to certain distances from its extremities. Thus the species of *Auricula* are met with only at the very margin of high water mark, along with *Littorina cœrulescens,* and *saxatilis*, *Velutina otis*, *Kellia rubra*, *Balani*, &c.; and among the plants, the yellow *Chondrus crispus* (*Carrigeen*, or Iceland moss of the shops), and *Corallina officinalis*. These are succeeded by other forms of animals and plants, such as *Littorina littorea*, *Purpura lapillus*, *Trochi*, *Actineæ*, *Porphyra laciniata* (Laver, Sloke), and *Ulvæ*. Towards the margin of low water, *Lottia testudinaria*, *Solen siliqua*, and the Dulse, *Rhodomenia palmata*, with numerous Zoophytes and Ascidian molluscs, indicate a third belt of life, connected, however, with the two others, by certain species common to all three, such as *Patella vulgata*, and *Mytilus edulis*. These sub-divisions of the sea-bed, exposed at ebb-tide, have long attracted attention on the coasts of our own country, and on those of France, where they have been observed by Audouin and Milne Edwards, and of Norway, where that admirable observer Sars has defined them with great accuracy.

Now this subdivision of the tract between tide-marks into zones of animal life, is a representation in miniature of the entire bed of the sea. The result of my observations, first in the British seas,* and more lately in the Ægean, has been to define a series of zones or regions in depth, and to ascertain *specifically* the animal and vegetable inhabitants of each. Regarding the tract between tide-marks as one region, which I have termed the *Littoral Zone*, we find a series of equivalent regions, succeeding it in depth. In the British seas, the littoral zone is succeeded by the region of Laminariæ, filled by forests of broad-leaved Fuci, among which live some of the most brilliantly coloured and elegant inhabitants of the ocean. This is the chosen habitat of *Lacunœ*, of *Rissoœ*, and of *Nudibranchous mollusca*. A belt generally of mud or gravel, in which numerous bivalve mollusca live, intervenes between the laminarian zone (in which the Flora of the sea appears to have its maximum), and the region of Corallines, which, ranging from a depth of from 20 to 40 fathoms, abounds in beautiful flexible zoophytes and in numerous species of Mollusca and Crustacea, to be procured only by means of the dredge. The great banks of Monomyarious Mollusca, which occur in many districts of the Northern Seas, are for the most part included in this region, and afford the zoologist his richest treasures. Deeper still is a region as yet but little explored, from which we draw up the more massy corals found on our shores, accompanied by shellfish of the class *Brachiopoda*. In the Eastern Mediterranean (where, through the invaluable assistance afforded by Captain Graves, and the Mediterranean Survey, I have been enabled to define the regions in depth, to an extent, and with a precision which, without similar aid, cannot be hoped for in the British seas), between the surface and the depth of 230 fathoms, the lowest point I had an opportunity of examining, there are eight well-defined zones, corresponding in part, and presenting similar characters with those which I have enumerated as presented by the sea-bed in the North. The details of these will be given in the forthcoming volume of the Transactions of the

* The first notice of these was published in the Edinburgh Academic Annual for 1840.

British Association, to which body I had the honour of presenting a report on the subject, at the last meeting.

When we examine the distribution and association of organic remains, in the upheaved beds of tertiary seas, we find the zones of depth as evident as they are in the present ocean. I have proved this to my own satisfaction, by a minute comparison of the newer Pliocene strata of Rhodes, where that formation attains a great thickness, with the present state of the neighbouring sea, and carrying on the comparison through the more recent tertiaries with the more ancient, have found indubitable evidences of the same phenomena. The strata of the cretaceous system yield similar evidences, and doubtless, in all time, the element of depth exercised a most important influence in regulating the distribution of animal life in the sea. If so, as our researches extend, we may hope eventually to ascertain the probable depth, or, at any rate, the region of depth, in which a given stratum containing organic remains was deposited. Every geologist will at once admit, that such a result would contribute materially to the history of sedimentary formations, and to the progress of geological science.

II. *The number of species is much less in the lower zones than in the upper. Vegetables disappear below a certain depth, and the diminution in the number of animal species indicates a zero not far distant.*—This conclusion is founded on my Ægean researches. Vegetables become fewer and fewer in the lower zones; and dwindle to a single species,—a *nullipora,* at the depth of 100 fathoms. Although the lower zones have a much greater vertical range than the higher, the number of animal species is infinitely greater in the latter. The lowest region (the 8th) in the Mediterranean, exceeds in extent all the other regions together; yet its fauna is comparatively small, and at the lowest portion explored, the number of species of testacea found was only eight. In the littoral zone, there were above 150 species. We may fairly infer, then, that as there is a zero of vegetable life, so is there one of animal life. In the sea, the vertical range of animals is greater than that of vegetables;—on the land, the reverse is the case. The geological application of this fact, of a zero of life in the ocean, is evident. All deposits formed below that zero, will be void, or almost void, of organic contents. The greater part of the sea is far deeper than the point zero; consequently, the greater part of deposits forming, will be void of organic remains. Hence we have no right to infer that any sedimentary formation, in which we find few or no traces of animal life, was formed either before animals were created, or at a time when the sea was less prolific in life than it now is. *It might have been formed in a very deep sea.* And that such was the case in regard to some of our older rocks, such as the great slates, is rendered the more probable, seeing that the few fossils we find in them, belong to tribes which, at present, have their maximum in the lowest regions of animal life, such as the Brachiopoda, and Pteropoda, of which, though free swimmers in the ocean, the remains accumulate only in very deep deposits. The uppermost deposits, those in which organic remains would be most abundant, are those most liable to disappear, in consequence of the destroying action of denudation. The great and almost nonfossiliferous strata of Scaglia, which form so large a part of the south of Europe and of Western Asia, were probably, for the most part, formed below the zero of life. The few fossils they contain, chiefly nummulites, correspond to the foraminifera which now abound mostly in the lowest region of animals. There is no occasion to attribute to metamorphic action the absence of traces of living beings in such rocks.

III. *The number of northern forms of animals and plants is not the same in all the zones of depth, but increases either positively, or by representation, as we descend.*—The association of species in the littoral zone is that most characteristic of the geographical region we are exploring; but the lower zones have their faunas and floras modified by the presence of species which, in more northern seas, are characteristic of the littoral zones. Of course, this remark applies only to the northern hemisphere; though, from analogy, we may expect to find such *inversely* the case also in the southern. The law, put in the abstract, appears to be, that *parallels in* *depth are equivalent to parallels in latitude*, corresponding to a well-known law in the distribution of terrestrial organic beings, viz. that *parallels in elevation are equivalent to parallels in latitude :* for example, as we ascend mountains in tropical countries, we find the successive belts of vegetation more and more northern or southern (according to the hemisphere) in character, either by identity of species, or by representation of forms by similar forms ; so in the sea, as we descend, we find a similar representation of climates in parallels of latitude in depth. The possibility of such a representation has been hypothetically anticipated in regard to marine animals by Sir Henry De La Beche,* and to marine plants by Lamouroux. To me it has been a great pleasure to confirm the felicitous speculations of those distinguished observers. The fact of such a representation has an important geological application. It warns us that all climatal inferences drawn from the number of northern forms in strata containing assemblages of organic remains, are fallacious, unless the element of depth be taken into consideration. But the influence of that element once ascertained (and I have already shewn the possibility of doing so), our inferences assume a value to which they could not otherwise pretend. In this way, I have no doubt, the per-centage test of Mr Lyell will become one of the most important aids in geology and natural history generally ; and, in fact, the most valuable conclusions to which I arrived by the reduction of my observations in the Ægean, were attained through the employment of Mr Lyell's method.

IV. *All varieties of sea-bottom are not equally capable of sustaining animal and vegetable life.*—In all the zones of depth there are occasionally more or less desert tracts, usually of sand or mud. The few animals which frequent such tracts are mostly soft and unpreservable. In some muddy and sandy districts, however, worms are very numerous, and to such places many fishes resort for food. The scarcity of remains of testacea in sandstones, the tracks of worms on ripple-marked sandstones, which had evidently been deposited in a

* Ten years ago, in his " Researches in Theoretical Geology."

shallow sea, and the fish remains often found in such rocks, are explained, in a great measure, by these facts.

V. *Beds of marine animals do not increase to an indefinite extent. Each species is adapted to live on certain sorts of sea-bottom only. It may die out in consequence of its own increase changing the ground.*—Thus, a bed of scallops, *Pecten opercularis*, for example, or of oysters having increased to such an extent that the ground is completely changed, in consequence of the accumulation of the remains of dead scallops or oysters, becomes unfitted for the further sustenance of the tribe. The young cease to be developed there, and the race dies out, and becomes silted up or imbedded in sediment, when the ground being renewed, it may be succeeded either by a fresh colony of scallops, or by some other species or assemblage of species. This "rotation of crops," as it were, is continually going on in the bed of the sea, and affords a very simple explanation of the alternation of fossiliferous and nonfossiliferous strata; organic remains in rocks being very rarely scattered through their substance, but arranged in layers of various thickness, interstratified with layers containing few or no fossils. Such interstratification may, in certain cases, be caused in another way, to-wit, by the elevation or subsidence of the sea-bottom, and the consequent destruction of the inhabitants of one region of depth, and the substitution of those of another. It is by such effects of oscillation of level, we may account for the repetition, at intervals, in certain formations of strata indicating the same region of depth.

VI. *Animals having the greatest ranges in depth have usually a great geographical, or else a great geological range, or both.*—I found that such of the Mediterranean testacea as occur both in the existing sea, and in the neighbouring tertiaries, were such as had the power of living in several of the zones in depth, or else had a wide geographical distribution, frequently both. The same holds true of the testacea in the tertiary strata of Great Britain. The cause is obvious: such species as had the widest horizontal and vertical ranges in space, are exactly such as would live longest in time, since they would be much more likely to be independent of catastrophes and destroying influences, than such as had a more limited distribution. In the cretaceous system, also, we find that such species as lived through several epochs of that era, are the few which are common to the cretaceous rocks of Europe, Asia, and America. Count D'Archiac and M. De Verneuil, in their excellent remarks on the fauna of the Palæozoic rocks, appended to Mr Murchison and Professor Sedgwick's valuable memoir on the Rhenish Provinces, have come to the conclusion that the fossils common to the most distant localities, are such as have the greatest vertical range. My observations on the existing testacea and their fossil analogues, lead to the same inference. It is very interesting thus to find a general truth coming out, as it were, in the same shape, from independent inquiries at the two ends of time.

VII. *Mollusca migrate in their larva state, but cease to exist at a certain period of their metamorphosis, if they do not meet with favourable conditions for their development;* i. e. *if they do not reach the particular zone of depth in which they are adapted to live as perfect animals.*

This proposition, which, as far as I am aware, is now put forward for the first time, includes two or three assertions which require explanation and proof, before I can expect the whole to be received. First, *that mollusca migrate.* In the fourth volume of the Annals of Natural History (1840), I gave a zoo-geological account of a shell-bank in the Irish Sea, being a brief summary of the results of seven years' observations at a particular season of the year. In that paper, I made known the appearance, after a time, of certain mollusca on the coasts of the Isle of Man, which had not previously inhabited those shores. They were species of limpet, about which there could be no mistake, and one was a littoral species. At that time, I could not account for their appearance. Many similar facts have since come to my knowledge, and fishermen are familiar with what they call "shifting" of shell-beds, which they erroneously attribute to the moving away and swimming off of a whole body of shell-fish, such as mussels and oysters. Even the *Pectens,* much less the testacea just named, have very little power of progressing to any distance, when fully developed. The "shifting" or migration is accomplished by the young animals when in a larva state. This brings me to a second point, which needs explanation. *All mollusca undergo a metamorphosis* either in the egg, or out of the egg, but, for the most part, among the marine species out of the egg. The relations of the metamorphoses of the several tribes are not yet fully made out; but sufficient is now known to warrant the generalization. In one great class of mollusca, the *Gasteropoda*, all appear to commence life under the same form, both of shell and animal, viz. a very simple, spiral, helicoid shell, and an animal furnished with two ciliated wings or lobes, by which it can swim freely through the fluid in which it is contained. *At this stage of the animal's existence, it is in a state corresponding to the permanent state of a Pteropod,** and the form is alike whether it be afterwards a shelled or shell-less species. (This the observations of Dalyell, Sars, Alder and Hancock, Allman, and others prove, and I have seen it myself.) It is in this form that most species migrate, swimming with ease through the sea. Part of the journey may be performed sometimes by the strings of eggs which fill the sea at certain seasons, and are wafted by currents. My friend, Lieut. Spratt, R.N., has lately forwarded me a drawing of a chain of eggs of mollusca, taken eighty miles from shore, and which, on being hatched, produced shelled larvæ of the forms which I have described. If they reach the region and ground, of which the perfect animal is a member, then they develope and flourish; but if the period of their development arrives before they have reached their destination, they perish, and their fragile shells sink into the depths of the sea. Millions and millions must thus perish, and every handful of the fine mud brought up from the eighth zone of depth in the Mediterranean, is literally filled with hundreds of these curious exuviæ of the larvæ of mollusca.†

---

* It is not improbable that the form of the larva of the Pteropod, when it shall be known, will be found to be that of an Ascidian polype, even as the larva of the Tunicata presents us with the representation of a hydroid polype.

† The nucleus of the shells of the Cephalopoda is a spiral-univalve

Were it not for the law which permits of the development of these larvæ only in the region of which the adult is a true native, the zones of depth would long ago have been confounded with each other, and the very existence of the zones of depth is the strongest proof of the existence of the law. Our confidence in their fixity, which the knowledge of the fact *that mollusca migrate* might at first shake, is thus restored, and with it our confidence in the inferences applicable to geology which we draw from submarine researches.

Some of the facts advanced in this communication are new, some of them have been stated before: but all, for which no authority is given, whether new or old, are put forth as the results of personal observation.

---

similar in form to the undeveloped shells above alluded to, and it is yet to be seen whether all Cephalopoda do not commence their existence under a *spiral-shelled Pteropodous form.*

38

Reprinted from *Quart. J. Microsc. Sci.* 8:203-212 (1868)

[*Editor's Note:* The plates accompanying this article have been omitted because of limitations of space.]

On some Organisms living *at* Great Depths *in the* North Atlantic Ocean. By Professor Huxley, F.R.S.

In the year 1857, H.M.S. "Cyclops," under the command of Captain Dayman, was despatched by the Admiralty to ascertain the depth of the sea and the nature of the bottom in that part of the North Atlantic in which it was proposed to lay the telegraph cable, and which is now commonly known as the "Telegraph plateau."

The specimens of mud brought up were sent to me for examination, and a brief account of the results of my observations is given in 'Appendix A' of Captain Dayman's Report, which was published in 1858 under the title of "Deep-Sea Soundings in the North Atlantic Ocean." In this Appendix (p. 64) the following passage occurs:

"But I find in almost all these deposits a multitude of very curious rounded bodies, to all appearance consisting of several concentric layers surrounding a minute clear centre, and looking, at first sight, somewhat like single cells of the plant *Protococcus;* as these bodies, however, are rapidly and completely dissolved by dilute acids, they cannot be organic, and I will, for convenience sake, simply call them coccoliths."

In 1860, Dr. Wallich accompanied Sir Leopold McClintock in H.M.S. "Bulldog," which was employed in taking a line of soundings between the Faröe Islands, Greenland, and Labrador; and, on his return, printed, for private circulation, some "Notes on the presence of Animal Life at vast depths in the Sea." In addition to the coccoliths noted by me, Dr. Wallich discovered peculiar spheroidal bodies, which he terms "coccospheres," in the ooze of the deep-sea mud, and he throws out the suggestion that the coccoliths proceed from the coccospheres. In 1861, the same writer published a paper in the 'Annals of Natural History,' entitled "Researches on some novel Phases of Organic Life, and on the Boring Powers of minute Annelids at great depths in the Sea." In this paper Dr. Wallich figures the coccoliths and the coccospheres, and suggests that the coccoliths are identical with certain bodies which had been observed by Mr. Sorby, F.R.S., in chalk.

The 'Annals' for September of the same year (1861) contains a very important paper by the last-named writer, "On the Organic Origin of the so-called 'Crystalloids' of the Chalk," from which I must quote several passages. Mr. Sorby thus commences his remarks:

"The appearance of Dr. Wallich's interesting paper published in this magazine (vol. viii, p. 52), in which he alludes to my having found in chalk objects similar to coccoliths, induces me to give an account of my researches on the subject. I do not claim the discovery of such bodies in the chalk, but to have been the first to point out (1) that they are not the result of crystalline action; (2) that they are identical with the objects described as coccoliths by Professor Huxley; and (3) that these are not single separate individuals, but portions of larger cells."

In respect of the statement which I have numbered (1), Mr. Sorby observes:

"By examining the fine granular matter of loose, unconsolidated chalk in water, and causing the ovoid bodies to turn round, I found that they are not flat discs, as described and figured by Ehrenberg, but, as shown in the oblique side view (fig. 5), *concave* on one side, and *convex* on the other, and indeed of precisely such a form as would result from cutting out oval watch-glasses from a moderately thick, hollow glass sphere, whose diameter was a few times greater than their own. This is a shape so entirely unlike anything due to crystalline, or any other force, acting independently of organization—so different to that of such round bodies, formed of minute radiating crystals, as can be made artificially, and do really occur in some natural deposits—and pointed so clearly to their having been derived from small hollow spheres, that I felt persuaded that such was their origin."

Mr. Sorby then states that, having received some specimens of Atlantic mud from me, he at once perceived the identity of the ovoid bodies of the chalk with the structures which I had called coccoliths, and found that, as he had predicted several years before, "the ovoid bodies were really derived from small hollow spheres, on which they occur, separated from each other at definite intervals."

The coccospheres themselves, Mr. Sorby thinks, may be

"an independent kind of organism, related to, but not the mere rudimentary form of, Foraminifera."

"With respect to the coccoliths, their optical character proves that they have an extremely fine, radiating, crystalline structure, as if they had grown by the deposition of carbonate of lime on an elongated central nucleus, in accordance with the oval-ringed structure shown in fig. 1 (magnified 800 linear)."

I am not aware that anything has been added to our knowledge of the "coccoliths" and "coccospheres" since the publication of Mr. Sorby's and Dr. Wallich's researches. Quite recently I have had occasion to re-examine specimens of Atlantic mud, which were placed in spirits in 1857, and have since remained in my possession. I have employed higher magnifying powers than I formerly worked with, or than subsequent observers seem to have used, my great help having been an excellent $\frac{1}{12}$th by Ross, which easily gives a magnifying power of 1200 diameters, and renders obvious many details hardly decipherable with the $\frac{1}{6}$th inch objective which I used in 1857.

The sticky or viscid character of the fresh mud from the bottom of the Atlantic is noted by Captain Dayman.* "Between the 15th and 45th degrees of west longitude lies the deepest part of the ocean, the bottom of which is almost wholly composed of the same kind of soft, mealy substance, which, for want of a better name, I have called ooze. This substance is remarkably sticky, having been found to adhere to the sounding rod and line (as has been stated above) through its passage from the bottom to the surface—in some instances from a depth of more than 2000 fathoms."

This stickiness of the deep-sea mud arises, I suppose, from the circumstance that, in addition to the *Globigerinæ* of all sizes which are its chief constituents, it contains innumerable lumps of a transparent, gelatinous substance. These lumps are of all sizes, from patches visible with the naked eye to excessively minute particles. When one of these is submitted to microscopical analysis it exhibits—imbedded in a transparent, colourless, and structureless matrix—granules, coccoliths, and foreign bodies.

The *granules* vary in size from $\frac{1}{40000}$th of an inch to $\frac{1}{5000}$th, and are aggregated together into heaps of various sizes and shapes (Pl. IV, fig. 1), some having the form of mere irregular streaks, but others possessing a more definitely limited

* Loc. cit., p. 9.

oval or rounded figure (fig. 1 *c*). Some of the heaps attain $\frac{1}{1000}$th of an inch or more in diameter, while others have not more than a third or a fourth of that size. The smallest granules are rounded; of the larger, many are biconcave oval discs, others are rod-like,* the largest are irregular.

Solution of iodine stains the granules yellow, while it does not affect the matrix. Dilute acetic acid rapidly dissolves all but the finest and some of the coarsest granules, but apparently has no effect on the matrix. Moderately strong solution of caustic soda causes the matrix to swell up. The granules are little affected by weak alkalies, but are dissolved by strong solutions of caustic soda or potash.

I have been unable to discover any nucleus in the midst of the heaps of granules, and they exhibit no trace of a membranous envelope. It occasionally happens that a granule-heap contains nothing but granules (fig. 1 *a*), but, in the majority of cases, more or fewer coccoliths lie upon, or in the midst of, the granules. In the latter case the coccoliths are almost always small and incompletely developed (fig. 1 *b*, *c*).

The *coccoliths* are exceedingly singular bodies. My own account of them, quoted above, is extremely imperfect, and in some respects erroneous. And though Mr. Sorby's description is a great improvement on mine, it leaves much to be said.

I find that two distinct kinds of bodies have been described by myself and others under the name of coccoliths. I shall term one kind *Discolithus*, and the other *Cyatholithus*.

The *Discolithi* (fig. 2) are oval discoidal bodies, with a thick, strongly refracting rim, and a thinner central portion, the greater part of which is occupied by a slightly opaque, as it were, cloud-like patch. The contour of this patch corresponds with that of the inner edge of the rim, from which it is separated by a transparent zone. In general, the discoliths are slightly convex on one side, slightly concave on the other, and the rim is raised into a prominent ridge on the more convex side, so that an edge view exhibits the appearance shown in fig. 2 *d*.

The commonest size of these bodies is between $\frac{1}{4000}$th and $\frac{1}{5000}$th of an inch in long diameter; but they may be found, on the one hand, rising to $\frac{1}{2700}$th of an inch in length, (fig. 2 *f*), and, on the other, sinking to $\frac{1}{11000}$th (fig. 2 *a*). The last mentioned are hardly distinguishable from some of

* These apparent rods are not merely edge views of disks.

the granules of the granule-heaps. The largest discoliths are commonly free, but the smaller and smallest are very generally found imbedded among the granules.

The second kind of coccolith (fig. 4 *a—m*), when full grown, has an oval contour, convex upon one face, and flat or concave upon the other. Left to themselves, they lie upon one or other of these faces, and in that aspect appear to be composed of two concentric zones (fig. 4 *d*, 2, 3) surrounding a central corpuscle (fig. 4 *d*, 1). The central corpuscle is oval, and has thick walls; in its centre is a clear and transparent space. Immediately surrounding this corpuscle is a broad zone (2), which often appears more or less distinctly granulated, and sometimes has an almost moniliform margin. Beyond this appears a narrower zone (3), which is generally clear, transparent, and structureless, but sometimes exhibits well-marked striæ, which follow the direction of radii from the centre. Strong pressure occasionally causes this zone to break up into fragments bounded by radial lines.

Sometimes, as Dr. Wallich has already observed, the clear space is divided into two (fig. 1 *e*). This appears to occur only in the largest of these bodies, but I have never observed any further subdivision of the clear centre, nor any tendency to divide on the part of the body itself.

A lateral view of any of these bodies (fig. 4 *f—i*) shows that it is by no means the concentrically laminated concretion it at first appears to be, but that it has a very singular and, so far as I know, unique structure. Supposing it to rest upon its convex surface, it consists of a lower plate, shaped like a deep saucer or watch-glass; of an upper plate, which is sometimes flat, sometimes more or less watch-glass-shaped; of the oval, thick-walled, flattened corpuscle, which connects the centres of these two plates; and of an intermediate substance, which is closely connected with the under surface of the upper plate, or more or less fills up the interval between the two plates, and often has a coarsely granular margin. The upper plate always has a less diameter than the lower, and is not wider than the intermediate substance. It is this last which gives rise to the broad granular zone in the face view.

Suppose a couple of watch-glasses, one rather smaller and much flatter than the other; turn the convex side of the former to the concave side of the latter, interpose between the centre of the two a hollow spheroid of wax, and press them together —these will represent the upper and lower plates and the central corpuscle. Then pour some plaster of Paris into the interval left between the watch-glasses, and that will take the place of the intermediate substance. I do not wish to imply, however, that the intermediate substance is something totally distinct from the upper and lower plates. One would naturally expect to find protoplasm between the two plates; and the granular aspect which the intermediate substance frequently possesses is such as a layer of protoplasm might assume. But I have not been able to satisfy myself completely of the presence of a layer of this kind, or to make sure that the intermediate substance has other than an optical existence.

From their double-cup shape I propose to call the coccoliths of this form *Cyatholithi*. They are stained, but not very strongly, by iodine, which chiefly affects the intermediate substance. Strong acids dissolve them at once, and leave no trace behind; but by very weak acetic acid the calcareous matter which they contain is gradually dissolved, the central corpuscle rapidly loses its strongly refracting character, and nothing remains but an extremely delicate, finely granulated, membranous framework of the same size as the cyatholith.

Alkalies, even tolerably strong solution of caustic soda, affect these bodies but slowly. If very strong solutions of caustic soda or potash are employed, especially if aided by heat, the cyatholiths, like the discoliths, are completely destroyed, their carbonate of lime being dissolved out, and afterwards deposited usually in hexagonal plates, but sometimes in globules and dumb-bells.

The *Cyatholithi* are traceable from the full size just described, the largest of which are about $\frac{1}{1600}$th of an inch long, down to a diameter of $\frac{1}{8000}$th of an inch. Their structure remains substantially the same, but those of $\frac{1}{3000}$th of an inch in diameter and below it are always circular instead of oval; the central corpuscle, instead of being oval, is circular, and the granular zone becomes very delicate. In the smallest the upper plate is a flat disc, and the lower is but very slightly convex (fig. 1 *f*). I am not sure that in these very small cyatholiths any intermediate substance exists, apart from the under or inner surface of the upper disc. When their flat sides are turned to the eye, these young cyatholiths are extraordinarily like nucleated cells; and it is only by carefully studying side views, when the small cyatholiths remind one of minute shirt-studs, that one acquires an insight into their real nature. The central corpuscles in these smallest cyatholiths are often less than $\frac{1}{40000}$th of an inch in diameter, and are not distinguishable optically from some of the granules of the granule-heaps.

The *coccospheres* occur very sparingly in proportion to the coccoliths. At a rough guess, I should say that there is not one of the former to several thousand of the latter. And owing to their rarity, and to the impossibility of separating them from the other components of the Atlantic mud, it is very difficult to subject them to a thorough examination.

The coccospheres are of two types—the one compact, and the other loose in texture. The largest of the former type which I have met with measured about $\frac{1}{1300}$th of an inch in diameter (fig. 6 *e*). They are hollow, irregularly flattened spheroids, with a thick transparent wall, which sometimes appears laminated. In this wall a number of oval bodies (1), very much like the "corpuscles" of the cyatholiths, are set, and each of these answers to one of the flattened facets of the spheroidal wall. The corpuscles, which are about $\frac{1}{4500}$th of an inch long, are placed at tolerably equal distances, and each is surrounded by a contour line of corresponding form. The contour lines surrounding adjacent corpuscles meet and overlap more or less, sometimes appearing more or less polygonal. Between the contour line and the margin of the corpuscle the wall of the spheroid is clear and transparent. There is no trace of anything answering to the granular zone of the cyatholiths.

Coccospheres of the compact type of $\frac{1}{1700}$th to $\frac{1}{8000}$th of an inch in diameter occur under two forms, being sometimes mere reductions of that just described, while, in other cases (fig. 6, *c*), the corpuscles are round, and not more than half to a third as big ($\frac{1}{11000}$th of an inch), though their number does not seem to be greater. In still smaller coccospheres (fig. 6 *a*, *b*) the corpuscles and the contour lines become less and less distinct and more minute until, in the smallest which I have observed, and which is only $\frac{1}{4500}$th of an inch in diameter (fig. 6 *a*) they are hardly visible.

The coccospheres of the loose type of structure run from the same minuteness (fig. 7 *a*) up to nearly double the size of the largest of the compact type, viz. $\frac{1}{700}$th of an inch in diameter. The largest, of which I have only seen one specimen (fig. 7, *d*), is obviously made up of bodies resembling cyatholiths of the largest size in all particulars, except the absence of the granular zone, of which there is no trace. I could not clearly ascertain how they were held together, but a slight pressure sufficed to separate them.

The smaller ones (fig. 7 *b*, *c*, and *a*) are very similar to those of the compact type represented in figs. 6, *c* and *d*; but they are obviously, in the case of *b* and *c*, made up of bodies resembling cyatholiths (in all but the absence of the granular zone), aggregated by their flat faces round a common centre, and more or less closely coherent. In *a*, only the corpuscles can be distinctly made out.

Such, so far as I have been able to determine them, then, are the facts of structure to be observed in the gelatinous matter of the Atlantic mud, and in the coccoliths and coccospheres. I have hitherto said nothing about their meaning, as in an inquiry so difficult and fraught with interest as this, it seems to me to be in the highest degree important to keep the questions of fact and the questions of interpretation well apart.

I conceive that the granule-heaps and the transparent gelatinous matter in which they are imbedded represent masses of protoplasm. Take away the cysts which characterise the *Radiolaria*, and a dead *Sphærozoum* would very nearly resemble one of the masses of this deep-sea "Urschleim," which must, I think, be regarded as a new form of those simple animated beings which have recently been so well described by Haeckel in his 'Monographie der Moneren.'* I proposed to confer upon this new "Moner" the generic name of *Bathybius*, and to call it after the eminent Professor of Zoology in the University of Jena, *B. Haeckelii*.

From the manner in which the youngest *Discolithi* and *Cyatholithi* are found imbedded among the granules; from the resemblance of the youngest forms of the *Discolithi* and the smallest "corpuscles" of *Cyatholithus* to the granules; and from the absence of any evident means of maintaining an independent existence in either, I am led to believe that they are not independent organisms, but that they stand in the same relation to the protoplasm of *Bathybius* as the spicula of Sponges or of *Radiolaria* do to the soft part of those animals.

That the coccospheres are in some way or other closely connected with the cyatholiths seems very probable. Mr. Sorby's view is that the cyatholiths result from the breaking up of the coccospheres. If this were the case, however, I cannot but think that the coccospheres ought to be far more numerous than they really are.

The converse view, that the coccospheres are formed by the coalescence of the cyatholiths, seems to me to be quite as probable. If this be the case, the more compact variety of the coccospheres must be regarded as a more advanced stage of development of the loose form.

On either view it must not be forgotten that the components of the coccospheres are not identical with the free cyatholiths; but that, on the supposition of coalescence, the disappearance of the granular layer has to be accounted for;

* 'Jenaische Zeitschrift,' Bd. iv, Heft 1.

while, on the supposition that the coccospheres dehisce, it must be supposed that the granular layer appears after dehiscence; and, on both hypotheses, the fact that both coccospheres and cyatholiths are found of very various sizes proves that the assumed coalescence or dehiscence must take place at all periods of development, and is not to be regarded as the final developmental act of either coccosphere or cyatholith.

And, finally, there is a third possibility—that the differences between the components of the coccospheres and the cyatholiths are permanent, and that the coccospheres are from the first independent structures, comparable to the wheel-like spicula associated in the wall of the "seeds" of *Spongilla*, and perhaps enclosing a mass of protoplasm destined for reproductive purposes.

In addition to *Bathybius* and its associated discoliths, cyatholiths, and coccospheres, the Atlantic mud contains—

a. Masses of protoplasm surrounded by a thick but incomplete cyst, apparently of a membranous or but little calcified consistence, and resembling minute *Gromiæ*. It is possible that these are unfinished single chambers of *Globigerinæ*.

b. Globigerinæ of all sizes and ages, from a single chamber $\frac{1}{1500}$th of an inch in diameter, upwards. I may mention incidentally that very careful examination of the walls of the youngest forms of *Globigerina* with the $\frac{1}{12}$th leads me to withdraw the doubt I formerly expressed as to their perforation.

In the absence of any apparent reproductive process in *Globigerinæ*, is it possible that these may simply be, as it were, offsets, provided with a shell, of some such simple form of life as *Bathybius*, which multiplies only in its naked form?

c. Masses of protoplasm enclosed in a thin membrane.

d. A very few *Foraminifera* of other genera than *Globigerina*.

e. Radiolaria in considerable numbers.

f. Numerous *Coscinodisci* and a few other Diatoms.

g. Numerous very minute fragments of inorganic matter.

The *Radiolaria* and Diatoms are unquestionably derived from the surface of the sea; and in speculating upon the conditions of existence of *Bathybius* and *Globigerina*, these sources of supply must not be overlooked.

With the more complete view of the structure of the cyatholiths and discoliths which I had obtained, I turned to the chalk, and I am glad to have been enabled to verify Mr. Sorby's statements in every particular. The chalk contains cyatholiths and discoliths identical with those of the Atlantic soundings, except that they have a more dense look and coarser contours (figs. 3 and 5). In fact, I suspect that they are fossilized, and are more completely impregnated with carbonate of lime than the recent coccoliths.

I have once met with a coccosphere in the chalk; and, on the other hand, in one specimen of the Atlantic soundings I met with a disc with a central cross, just like the body from the chalk figured by Mr. Sorby (fig. 8).

39

Reprinted from *Nature* 1:100–101 (Nov. 25, 1869)

THE DEPTHS OF THE SEA

David Forbes

THE opening meeting of the Royal Society on Thursday last was attended by a numerous assemblage of men of science, especially attracted by the announcement that Dr. Carpenter, representing a committee consisting of Professor Wyville Thomson, Mr. Gwyn Jeffreys, and himself, would communicate the results of the deep-sea dredging explorations, carried out in the course of the past summer and autumn in the *Porcupine*, a vessel expressly fitted out and placed by the Government at the disposal of the committee for this purpose.

At the conclusion of Dr. Carpenter's lucid exposition, which was necessarily but a mere *résumé* of the report itself, it appeared quite evident that rumour had not at all exaggerated the scientific value of these explorations, for it is not too much to say that the results of this expedition must be classed with the most important which of late years have been brought before the notice of the scientific world.

More than a quarter of a century ago, the late Edward Forbes, one of the first naturalists who took the common oyster dredge from the hands of the fisherman to convert it into an instrument for extended scientific research, after employing it in the commencement along the shores of his native little Isle, and subsequently in the seas surrounding the British Islands, and in other parts of Europe, found, upon comparing his observations, that there appeared to be evidence in favour of the existence of a succession of natural zones of marine life according to depth, which zones, however, seemed to become more and more sterile in organisms in descending order; until at last it suggested itself that a zone might be arrived at, at a depth roughly estimated as exceeding 300 fathoms from the surface, containing but sparse traces of organic life, or even such an one as might be entitled to the appellation of Azoic.

This latter hypothesis was brought forward by him as a suggestion worthy of consideration, and not as a dogma or established principle, as he was fully aware that in the dredging explorations which he had been able to carry out up to that time, he had never reached so great a depth as even 300 fathoms, below which the sea-bottom was inferred to be comparatively or altogether sterile; on the contrary, whilst advancing the conclusions which seemed to be but natural deductions from the data then at his disposal, he continually kept pointing out that whether such an hypothesis was correct or not, it was of the highest importance to science to prosecute these researches further, so as to ascertain the true nature of the deep-sea bottom, for, to use his own words in his "History of the European Seas," "it is in its exploration that the finest field for marine discovery still remains."

Before the author of this suggestion had time or opportunity for carrying out such explorations as would have verified or disproved his hypothesis, he was unfortunately cut off by an early death; whilst the hypothesis, in the state in which he had left it, was without further investigation eagerly grasped at and accepted by men of science, both at home and abroad, for the special reason that it appeared to afford a simple explanation of various phenomena which had long remained enigmas to both palæontologists and geologists; as, for example, amongst others the occurrence, in various periods of the earth's history, of vast accumulations of sedimentary strata apparently altogether devoid of organic remains.

Although this hypothesis, when somewhat modified, may possibly be found to hold good in respect to certain forms and conditions of life, the results of some casts of the dredge made in depths of from 270 to 400 fathoms in Sir James Ross's Antartic Expedition, and subsequently, the deep-sea soundings described by Dr. Wallich as made in 1860, in the *Bulldog*, in vastly greater depths, demonstrated quite conclusively that it could no longer be retained as a generalisation.

It now appears strange to look back and observe what very little notice was taken of these new data; more especially of the important researches of Dr. Wallich on the North Atlantic sea-bed, which for years, if not all but overlooked, certainly do not appear to have received from zoölogists the full credit which they undoubtedly deserved: geologists and palæontologists were evidently loth to abandon an hypothesis which in many respects suited their requirements.

However long truth may remain dormant, it must eventually assert itself in science as in all other matters, and the advancing strides of Biology and Geology soon demanded that such problems should be definitely and conclusively solved, and that the depths of the sea also should be carefully searched for the missing links of evidence requisite to complete their respective chains of reasoning. This was not felt to be the case in England alone; already in Scandinavia we find the *savants* of Norway and Sweden working with their slender means in the right direction, and assisted by their Governments with a hearty good-will and determination which could not fail to ensure valuable results, such as have already been brought forward by Sars, Nordenschjold, Torrell, and others.

In England, men of science, equally impressed with the importance of this inquiry, wished, with an honourable pride, to see that the country which had so long claimed the empire of the sea, should, in a question of so purely marine investigation, do something worthy of herself; and, being fully alive to the impossibility of doing so without the aid of the Government, applied themselves first to the task of procuring such assistance. Since it is an acknowledged but melancholy fact, that science does not in England either obtain the high position in society, or the influence with the ruling powers of the country which is accorded to it on the Continent in general, it is a subject for congratulation that the urgent appeals made to the Government should have in this instance proved so successful; and after the Government had provided the ships and equipment necessary for the expeditions of last year and this, it is a further subject for congratulation that the direction of these scientific explorations should have been entrusted to such able men as Dr. Carpenter, Prof. Wyville Thomson, and Mr. Gwyn Jeffreys, who constitute the present committee.

The expedition of last year being the first of its kind, had, as might be anticipated, many difficulties to contend with; the ship itself, besides starting at a late season of the year, was ill suited to the undertaking, was provided with but extremely inefficient winding machinery, imperfect appliances and instruments, and moreover, the observers and their assistants had, as it were, to serve an apprenticeship in the management of such operations.

This year, besides being fortunate in securing unusually favourable weather during the major part of the operations, all the above-mentioned difficulties had been provided against; whilst, at the same time, the experience gained during the last year's cruise contributed very greatly to the complete success of the expedition as a whole.

As yet, it would be premature to attempt any description of the results of these explorations, for the Report which was commenced at the meeting of the Royal Society last Thursday is not yet concluded, but is to be continued at its next meeting; sufficient, however, has been already brought forward to prove satisfactorily the great importance of the data obtained to science in general. Besides corroborating, and in some respects correcting the conclusions deduced from the operations of the last year's expedition, many new facts and observations have been collected, whilst the supply of specimens and materials for examination which have been brought home will no doubt give full occupation to the members of the committee for some time to come, besides obliging them to bring to their assistance the services of the physicist, chemist, and mineralogist, each in their several departments.

The practicability of exploring even the deepest portions of the ocean bed may now be considered to be fully established; the conclusive proofs brought forward showing the existence of warm and cold areas of the deep-sea bottom, in close proximity to one another, each inhabited by its distinct and characteristic fauna, is as surprising as it is important in its scientific bearings, and particularly in its relations to geology and palæontology; whilst the investigations into the temperature of the different ocean zones, and the nature of the gases contained in the sea-water at various depths, are intensely interesting and suggestive.

The question as to the existence of an azoic ocean zone at *any* depth, must now be regarded as finally settled in the negative. The hypothesis which appeared to Edward Forbes to be warranted by all the data which the science of his day could supply, must now be abandoned; it is certain, however, that all who knew him will do his memory the justice of believing that, were he now alive, so far from regretting the necessity of withdrawing a suggestion which appeared to explain several important points in science now once more involved in obscurity, he would have been the first of the converts to the views now proved to be more correct, and the first to congratulate the members of the deep-sea dredging committee upon so successful and brilliant a termination of their labours.

BIBLIOGRAPHY

Agassiz, Alexander (1888) *A Contribution to American Thalassography.* Three Cruises of the United States Coast and Geodetic Survey Steamer *Blake* in the Gulf of Mexico, in the Caribbean Sea, and along the Atlantic Coast of the United States, from 1877 to 1880. Boston and New York, Houghton, Mifflin, 2 vols.

Aimé, Georges (1845)*Exploration scientifique de L'Algérie pendant les années 1840–1842: Physique générale.* Vol. 1. Recherches de physique générale sur la Méditerranée. Paris, Imprimerie Royale.

Aiton, E. J. (1954) Galileo's theory of the tides. *Annals Sci.,* **10**(1), 44–57.

Aleem, A. A. (1967) Concepts of currents, tides and winds among medieval Arab geographers in the Indian Ocean. *Deep-Sea Research,* **14**(4), 459–463.

Aleem, A. A. (1968) Ahmad Ibn Magid, Arab navigator of the 15th century, and his contributions to marine sciences. *Inst. Océanog. Monaco Bull.,* spec. no. 2, **2,** 565–580.

Aristotle (1908–1952) *The Works of Aristotle* Edited by J. A. Smith and W. D. Ross. Oxford, Clarendon Press, 12 vols.

Bache, A. D. (1856) Notice of earthquake waves on the western coast of the United States, on the 23rd and 25th of December, 1854. *Am. Jour. Sci. Arts,* ser 2, **21,** 37–43.

Barber, N. F. (1948) The magnetic field produced by earth currents flowing in an estuary or sea channel, *Royal Astron. Soc. Geophys. Monthly Notices,* Suppl. 5(7), 258–269.

Barber, N. F., and Ursell, F. (1948) The generation and propagation of ocean waves and swell, Pt. 1, Periods and velocities. *Royal Soc. London Philos. Trans.;* Ser. A, **240**(824), 527–560.

Bascom, Willard (1964) *Waves and Beaches. The Dynamics of the Ocean Surface.* New York, Anchor Books, Doubleday.

Beche, H. T. de la (1834) Researches in Theoretical Geology. London, Charles Knight.

Bergman, Torbern (1779–1790) *Opuscula Physica et Chemica.* Holmiae, Upsaliae and Aboae, 6 vols. Also in *Physical and Chemical Essays,* translated by Edmund Cullen, London, John Murray, 1784, 2 vols.

Bernoulli, Daniel, L. Euler, C. Maclaurin, and A. Cavalleri (1741) *Piéces qui ont remporté le prix de l'Académie Royale des Sciences en 1740.* Paris.

Birch, Thomas (1756–1757) *The History of the Royal Society of London for Improving of Natural Knowledge.* London, 4 vols.

Blagden, Sir Charles (1788) Experiments on the cooling of water below its freezing point. *Royal Soc. London Philos. Trans,* **78,** pt. 1, 125–146.

Boguslawski, Georg von (1884–1887) *Handbuch der Ozeanographie.* Stuttgart, J. Engelhorn.

Bolland, Richard (1704) A draught of the Streights of Gibraltar. With some observations upon the currents thereunto belonging. In: *A Collection of Voyages and Travels.* London, Awnsham and John Churchill, vol. 4, pp. 846–848.

Bourne, William (1578) *A Booke called the Treasure for Traueilers, deuided into fiue Books or partes, contaynyng very necessary matters, for all sortes of Trauailers, eyther by Sea or by Lande.* London, Thomas Woodcocke.

Boyle, Robert (1666) Other inquiries concerning the sea. *Royal Soc. London Philos. Trans,* **1**(18), 315–316. Reprinted in M. B. Deacon (1971), p. 410.

Boyle, Robert (1671) *Tracts written by the Honourable Robert Boyle.* About the Cosmicall Qualities of things. Cosmicall Suspitions. The Temperature of the Submarine Regions. The Temperature of the Subterraneall Regions. The Bottom of the Sea. Oxford, R. Davis.

Boyle, Robert (1673) *Tracts Consisting of Observations About the Saltness of the Sea:* An Account of a Statical Hygroscope And its Uses: Together with an Appendix about the Force of the Air's Moisture: A Fragment about the Natural and Preternatural State of Bodies. London, printed by E. Flesher for R. Davis, bookseller in Oxford.

Bridges, E. Lucas (1948) *Uttermost Part of the Earth.* London, Hodder & Stoughton.

Buchan, Alexander (1895) Report on oceanic circulation, based on the observations made on board HMS *Challenger,* and other observations. *Challenger* Report (q.v.), Vol. 2, *Summary of the Scientific Results,* appendix.

Buchanan, J. Y. (1886) On similarities in the physical geography of the great oceans. *Royal Geog. Soc. Proc.,* **8**(12), 753–768. Reprinted in Buchanan (1919), pp. 87–112.

Buchanan, J. Y. (1895) A retrospect of oceanography in the twenty years before 1895. *Internat. Geog. Cong., 6th, London 1895, Report* (1896), pp. 403–435. Reprinted in Buchanan (1919), pp. 28–86.

Buchanan, J. Y. (1913) *Scientific Papers.* London, Cambridge University Press.

Buchanan, J. Y. (1919) *Accounts Rendered of Work Done and Things Seen.* London, Cambridge University Press.

Burstyn, H. L. (1966) Early explanations of the role of the earth's rotation in the circulation of the atmosphere and the ocean. *Isis,* **57,** pt. 2, no. 188, 167–187.

Burstyn, H. L. (1968) Science and government in the nineteenth century: the *Challenger* expedition and its report. *Inst. Océanog. Monaco Bull.,* spec. no. 2, **2,** 603–611.

Burstyn, H. L. (1972) Pioneering in large-scale scientific organisation: the *Challenger* expedition and its report. I. Launching the expedition. *Royal Soc. Edinburgh Proc.,* ser. B, **72,** 47–61.

Carpenter, William B. (1868) Preliminary report of dredging operations in the seas to the north of the British Islands, carried on in HMS *Lightning,* by Dr. Carpenter and Dr. Wyville Thomson, Professor of Natural History in Queen's College, Belfast. *Royal Soc. [London] Proc.,* **17,** 168–200.

Carpenter, William B. (1872) Report on scientific researches carried on during the months of August, September, and October, 1871, in HM Surveying-ship *Shearwater. Royal Soc. [London] Proc.*, **20**(138), 535–644.

Carpenter, W. B., and J. Gwyn Jeffreys (1870) Report on deep-sea researches carried on during the months of July, August, and September 1870, in HM Surveying-ship *Porcupine. Royal Soc. [London] Proc.*, **19**(125), 146–221.

Carpenter, W. B., J. Gwyn Jeffreys, and C. Wyville Thomson (1869) Preliminary report of the scientific exploration of the deep sea in HM Surveying-vessel *Porcupine*, during the summer of 1869. *Royal Soc. [London] Proc.*, **18**(121), 397–492.

Cartwright, D. E. (1969) Deep sea tides, *Sci. Jour.*, **5**(1), 60–67. Reprinted in *Oceanography: Contemporary Readings in the Ocean Sciences*, edited by R. Gordon Pirie. New York, London, Toronto, Oxford University Press, 1973, pp. 148–158.

Cartwright, D. E. (1972) Some ocean tide measurements of the eighteenth century, and their relevance today. *Royal Soc. Edinburgh Proc.*, ser. B, **72**, 331–339.

Cavendish, Lord Charles (1757) A description of some thermometers for particular uses. *Royal Soc. London Philos. Trans.*, **50**, pt. 1, 300–310.

Challenger Report. *Report on the Scientific Results of the Voyage of HMS* Challenger *during the years 1872–76*, edited by C. Wyville Thomson and J. Murray: London, Her Majesty's Stationery Office, 1880–1895, 50 vols.

Chamisso, Adelbert von (1822) On the coral islands of the Pacific Ocean. *Edinburgh Philos. Jour.* **6**(11), 37–40.

Chappe d'Auteroche, J. B. (1772) *Voyage en Californie pour l'observation du passage de Vénus sur le disque du soleil, le 3 Juin 1769*, edited by Cassini fils. Paris, C. A. Jombert.

Charlier, R. H., and E. Leloup (1968) Brief summary of some oceanographic contributions in Belgium until 1922. *Inst. Océanog Monaco Bull.*, spec. no. 2, **1**, 293–310.

Clark-Kennedy, A. E. (1929) *Stephen Hales, D.D., F.R.S. An Eighteenth Century Biography*. Cambridge, Cambridge University Press.

Colladon, Daniel, and J. C. F. Sturm (1827) Mémoire sur la compression des liquides. *Annales Chimie*, ser. 2, **36**, 113–159, 225–257.

Cooper, L. H. N. (1967) Stratification in the deep ocean, *Sci. Progress*, **55**, 73–90 (1967).

Cornish, Vaughan (1934) *Ocean Waves and Kindred Geophysical Phenomena*. Cambridge, Cambridge University Press.

Couthouy, J. P. (1843–1844) Remarks upon coral formations in the Pacific; with suggestions as to the causes of their absence in the same parallels of latitude on the coast of South America. *Boston Jour. Nat. History*, **4**, 66–105, 137–162.

Cromwell, T., R. B. Montgomery, and E. D. Stroup (1954) Equatorial undercurrent in the Pacific Ocean revealed by new methods. *Science*, **119**(3097), 648–649.

Cuningham, William (1559) *The Cosmographical Glasse, conteinyng the pleasant Principles of Cosmographie, Geographie, Hydrographie, or Nauigation*. London.

Cushing, D. H. (1972) A history of some of the international fisheries commissions. *Royal Soc. Edinburgh Proc.*, ser. B, **73**, 361–390.

Dadić, Z. (1968) The history of the theories of the tide introduced by Yugoslav scientists until the 18th century. *Inst. Océanog Monaco Bull.*, spec. no. 2, **1**, 49–53.

Dana, J. D. (1845) On the composition of corals and the production of the phosphates, aluminates, silicates, and other minerals, by the metamorphic action of hot water. *Edinburgh New Philos. Jour.*, **39**, 293–295. This article had previously appeared in the *Am. Jour. Sci. Arts*.

Dana, J. D. (1853) *On Coral Reefs and Islands.* New York G. P. Putnam.

Darbyshire, J. (1962) Microseisms. *The Sea,* edited by M. N. Hill (q.v.), vol. 1, pp. 700–719.

Darwin, C. R. (1842) *The Structure and Distribution of Coral Reefs. The Geology of the Voyage of the* Beagle, Part 1. London, Smith, Elder & Co., London. Reprinted with a forward by H. W. Menard. Berkeley and Los Angeles, University of California Press, 1962.

Darwin, G. H. (1898) *The Tides and Kindred Phenomena in the Solar System.* London, John Murray.

Deacon, G. E. R. (1937) The hydrology of the Southern Ocean. *Discovery Rept.,* **15,** 1–124.

Deacon, G. E. R. (1968) Early scientific studies of the Antarctic Ocean *Inst. Océanog Monaco Bull.,* spec. no. 2, **1,** 269–278.

Deacon, G. E. R. (1972) Marine science. In *Growing Points in Science.* London, Her Majesty's Stationery Office, pp. 52–77.

Deacon, G. E. R., and J. W. S. Marr (1964) Les grandes expeditions océanographiques. *B. P. Rev.,* no. 16, 4–9.

Deacon, M. B. (1965) Founders of marine science in Britain: the work of the early fellows of the Royal Society. *Royal Soc. [London] Notes Rec.,* **20**(1), 28–50.

Deacon, M. B. (1968) Some early investigations of the currents in the Strait of Gibraltar. *Inst. Océanog. Monaco Bull.,* spec. no. 2, **1,** 63–74.

Deacon, M. B. (1971) *Scientists and the Sea, 1650–1900: A Study of Marine Science.* London and New York, Academic Press.

Deacon, M. B. (1972) The *Challenger* expedition and geology. *Royal Soc. Edinburgh Proc.,* ser. B, **72,** 145–153.

Debenham, F. (1945) *The Voyage of Captain Bellingshausen to the Antarctic Seas, 1819–1821.* Hakluyt Society, Cambridge University Press, 2nd series, vols. 91–92.

Defant, Albert (1961) *Physical Oceanography.* Oxford, London, New York, Paris, Pergamon Press, 2 vols.

Derham, William (1726) *The Philosophical Experiments and Observations of the late eminent Dr. Robert Hooke, SRS. and Geom. Prof. Gresh, and other eminent Virtuosos in his Time.* London, W. and J. Innys, 1726. (See also R. T. Gunther)

Derrotero de las Islas Antillas, de las costas de tierra firme, y de las del seno Mexicano, formado en la direccion de trabajos hidrográficos para intelligencia y uso de las cartas que ha publicado. Madrid, Imprenta Nacional, 2nd ed. (first published, 1810).

Descartes, René (1897–1909) *Oeuvres de Descartes.* edited by Charles Adam and Paul Tannery. Paris, Sous les auspices du Ministère de l'instruction publique, Léopold Cerf, 11 vols.

Destombes, M. (1968) 'Les plus anciens sondages portes sur les cartes nautiques aux XVIe et XVIIe siècles: contribution à l'histoire de l'océanographie. *Inst. Océanog. Monaco Bull.,* spec. no. 2, **1,** 199–222.

Dibner, B. (1964) *The Atlantic Cable.* New York, Toronto, London, Blaisdell Publishing, 2nd edition.

Directions for Observations and Experiments to be made by Masters of Ships, Pilots and other fit Persons in their Sea-Voyages (1667) *Royal Soc. Philos. Trans.,* **2,**(24), 433–448.

Donati, Vitaliano (1758) *Essai sur l'Histoire Naturelle de la Mer Adriatique.* La Haye. (First published, Venice 1750, in Italian.)

Drake, S. (1961) Galileo gleanings—X. Origin and fate of Galileo's theory of the tides. *Physis,* **3,** pt. 3, 185–193.

Drubba, H. and H. H. Rust, (1954) On the first echo-sounding experiment *Annals Sci.*, **10**(1), 28-32.

Duhem, P. (1913-1959) *Le Systéme du Monde: Histoire des doctrines cosmologiques de Platon á Copernic.* Paris, Hermann, 10 vols.

Dulong, P. L. and A. T. Petit, (1817) Recherches sur la mesure des températures et sur les lois de la communication de la chaleur. *Annales Chimie,* ser. 2, **7,** 113-154, 225-264, 337-367.

Dumont d'Urville, Jules (1842-1854) *Voyage au Pole Sud et dans l'Oceanie sur les corvettes l'*Astrolabe *et la* Zélée, *pendant les années 1837-1840.* Paris, 20 vols.

Ehrenberg, C. G. (1844) On microscopic life in the ocean at the South Pole, and at considerable depths. *Annals Mag. Nat. History,* **14**(90), 169-181. Translated from the article (1844), Vorläufige Nachricht über das kleinste Leben im Weltmeer, am Sudpol und in den Meerestiefen, *Proc. Royal Prussian Academy Sci.*, Berlin.

Ekman, V. W. (1905) On the influence of the earth's rotation on ocean-currents. *Arkiv Mathematik, Astronomi Fysik,* **2**(11). Reprinted (1963) by the Royal Swedish Academy of Sciences, Stockholm.

Ekman, V. W. (1906) On dead-water. *Scientific Results of the Norwegian North Polar Expedition, 1893-96:* Christiania, London, Leipzig, F. Nansen Fund for the Advancement of Science (1900-1906), 6 vols., vol. 5, no. 15.

Ellis, Henry (1751) A letter to the Rev. Dr. Hales, FRS from Captain Henry Ellis, FRS dated Jan. 7, 1750-51, at Cape Monte Africa, Ship *Earl of Hallifax. Royal Soc. Philos. Trans.* **47,** 211-214.

Ellis, John (1755) *An Essay towards a Natural History of the Corallines, and other marine productions of the like kind, commonly found on the coasts of Great Britain and Ireland.* To which is added the description of a large marine polype, taken near the North Pole, by the whale-fishers in the summer 1753: London, 1755.

Erman, G. A. (1828) Nouvelles recherches sur le maximum de densité de l'eau salée. *Annales Chimie,* ser. 2, **38,** 287-304.

Fairbridge, R. W., ed. (1966) *Encyclopedia of Earth Sciences,* vol. 1. *The Encyclopedia of Oceanography.* New York, Reinhold.

Fisher, Osmond (1889) *Physics of the Earth's Crust.* London and New York, Macmillan, 2nd edition.

Fitzroy, Robert (1836) Sketch of the surveying voyages of HMS *Adventure* and *Beagle. Geog. Journ.,* **6,** 311-343.

Flamsteed, John (1683) A correct tide-table, shewing the true times of the high-waters at London-Bridge, to every day in the year 1683. *Royal Soc. Philos. Trans.,* **13**(143), 10-11.

Forbes, Edward (1839) On a shell-bank in the Irish Sea, considered zoologically and geologically, *Annals Mag. Nat. Nistory,* **4,** 217-223.

Forbes, Edward (1840) On the associations of mollusca on the British coasts, considered with reference to Pleistocene geology. In *The Edinburgh Academic Annual for 1840.* Edinburgh, Adam and Charles Black, pp. 175-183.

Forbes, E. G. (1975) Greenwich Observatory: Origins and Early History (1675-1835). In E. G. Forbes, A. J. Meadows, and D. Howse, *Greenwich Observatory.* London, Taylor & Francis, 3 vols. Vol. 1.

Forchhammer, G. (1865) On the composition of sea-water in the different parts of the ocean. *Royal Soc. Philos. Trans.,* **155,** 203-262.

Franklin, Benjamin, W. Brownrigg, and J. Farish (1774) Of the stilling of waves by means of oil. Extracted from sundry letters between Benjamin Franklin, LL D.

FRS, William Brownrigg, MD FRS and the Reverend Mr. Farish. *Royal Soc. Philos. Trans.*, **64**, pt. 2, 445–460.

Galtsoff, P. S. (1962) *The Story of the Bureau of Commercial Fisheries Biological Laboratory, Woods Hole, Massachusetts.* U.S. Department of Interior, Fish and Wildlife Service and Bureau of Commercial Fisheries, Washington, D.C., circular 145.

Gibbon, Edward (1776) *The History of the Decline and Fall of the Roman Empire.* London, 1776–1788, 6 vols. Reprinted in (1966) Everyman's Library. London, J. M. Dent & Sons; and New York, E. P. Dutton, 6 vols. Vol. 1, p. 58.

Gilbert, William (1600) *De Magnete, magneticisque corporibus, et de magno magnete tellure;* Physiologia nova, plurimis & argumentis, & experimentis demonstrata. London, Peter Short. Reprinted (1958) in translation by S. P. Thompson, D. J. Price, ed. New York, Basic Books.

Gill, A. E. (1973) Circulation and bottom water production in the Weddell Sea. *Deep-Sea Research,* **20,** 111–140.

Gunther, R. T. (1921–1967) *Early Science in Oxford.* Vols 6–7 (1930) *The Life and Work of Robert Hooke.* Vol. 8 (1931) *The Cutler Lectures of Robert Hooke.* Oxford, printed for the subscribers, 15 vols.

Haeckel, E. (1868) Monographie der Moneren. *Jenaische Zeitsch. Medicin Naturw.*, **4,** 64–134.

Hales, Stephen (1751) A letter to the President, from Stephen Hales, DD & FRS. *Royal Soc. London Philos. Trans.*, **47,** 214–216.

Hallam, A. (1973) *A Revolution in the Earth Sciences.* From continental drift to plate tectonics. Oxford, Clarendon Press.

Halley, Edmond (1687) An estimate of the quantity of vapour raised out of the sea by the warmth of the sun; derived from an experiment shown before the Royal Society, at one of their late meetings. *Royal Soc. Philos. Trans.*, **16**(189), 366–370.

Halley, Edmond (1715) A short account of the cause of the saltness of the ocean, and of the several lakes that emit no rivers; with a proposal, by help thereof, to discover the age of the world. *Royal Soc. Philos. Trans.*, **29**(344), 296–300.

Hardy, A. C. (1959) *The Open Sea: Its Natural History.* Part 2. Fish and Fisheries. London, Collins.

Harris, R. A. (1898) *Manual of Tides. Part I.* Washington, Report of the Superintendent of the U.S. Coast and Geodetic Survey during the year 1897, appendix 8.

Haskell, D. C. (1942) "The United States Exploring Expedition, 1838–1842, and its publications, 1844–1874": The New York Public Library.

Hedgpeth, J. W. (1945) The United States Fish Commission Steamer *Albatross: Am. Neptune,* **5**(1), 5–26.

Hedgpeth, J. W., ed. (1957) Treatise on Marine Ecology and Paleoecology. I. Ecology. *Geol. Soc. America Mem.*, no. 67, Introduction.

Herdman, W. A. (1923) *Founders of Oceanography and Their Work: an Introduction to the Science of the Sea.* London, Edward Arnold.

Herschel, J. F. W. Unpublished correspondence at the Royal Society of London. W. B. Carpenter to Herschel, 26 January 1869.

Hill, M. N., ed. (1962–1963) *The Sea: ideas and observations on progress in the study of the seas.* New York and London, Interscience Publishers, 3 vols.

Humboldt, Alexander von (1814) *Voyage aux régions èquinoxiales du nouveau continent, fait en 1799–1804, par Alexandre de Humboldt et A. Bonpland.* Paris, F. Schoell, 3 vols. English translation by H. M. Williams (1814), London, 2 vols, Vol. 1, pp. 60–69.

Huxley, T. H. (1858) Report on samples of the sea bed. Appendix A. In J. Dayman, (1858) *Deep Sea Soundings in the North Atlantic Ocean between Ireland and Newfoundland made in HMS* Cyclops, *Lieut.-Commander Joseph Dayman, in June and July, 1857.* London, Her Majesty's Stationery Office, pp. 63-68.

Huxley, T. H. (1875) Notes from the *Challenger. Nature,* **12**(303), 315-316.

Jones, C. W. (1943) *Bedae Opera de Temporibus:* Cambridge, Massachusetts, Mediaeval Society of America, Publication no. 41.

Kofoid, C. A. (1910) The Biological Stations of Europe. *U.S. Bur. Ed. Bull.* no. 4, whole no. 440.

Krusenstern, A. J. (1813) *Voyage round the World, in the years 1803-1806, by order of his Imperial Majesty Alexander the First, on board the ships* Nadeshda *and* Neva. Translated by R. B. Hoppner, London, John Murray, 2 vols.

LaFond, E. C., and C. S. Cox (1962) Internal Waves. *The Sea.* Edited by M. N. Hill (q.v.), vol. 1, pp. 730-763.

Leighly, J. (1968) M. F. Maury in his time. *Inst. Océanog. Monaco Bull.*, spec. no. 2, **1**, 147-159.

Lenz, Emil (1831) Physikalische Beobachtungen angestellt auf einer Reise um die Welt unter dem Commando des Capitains Otto von Kotzebue in den Jahren 1823-26. *Mémoires de l'Académie Impériale des Sciences de St.-Pétersbourg:* ser. 6, Sciences mathématiques, physiques et naturelles, **1**, 221-341.

Lenz, Emil (1847) Bemerkungen über die Temperatur des Weltmeeres in verschiedenen Tiefen. *Bulletin de la Classe Physico-Mathématique de l'Académie Impériale des Sciences de St-Pétersbourg,* **5**, col. 65-74.

Lillie, F. R. (1944) *The Woods Hole Marine Biological Laboratory.* Chicago, Univeristy of Chicago Press.

Linklater, E. (1972) *The Voyage of the* Challenger. London, John Murray.

Linschoten, J. H. van (1598) *Discours of Voyages into ye Easte and West Indies.* Translated by W. Phillip. J. Wolfe, London.

Longuet-Higgins, M. S. (1950) A theory of the origin of microseisms. *Royal Soc. Philos. Trans.*, ser. A, **243**(857), 1-35.

Lubbock, J. W. (1835) Report on the Tides. London, *Report of the First & Second Meetings of the British Association for the Advancement of Science,* pp. 189-195.

Mackintosh, N. A. (1946) The natural history of whalebone whales, *Biol. Rev.* **21**, 60-74. Reprinted in *Smithson. Inst. Rept.*, 235-264 (1947).

Macquer, P. J. (1766) Mémoire sur la différente dissolubilité des sels neutres dans l'esprit de vin contenant des observations particulières sur plusieurs de ces sels. *Mélanges de Philosophie et de Mathématique de la Société Royale de Turin* (Miscellanea Taurinensia), Turin, 1762-1765, **3**, pt. 1, 1-30.

Malaguti, F. J., and J. Durocher (1859) Observations relatives á la présence de l'argent dans l'eau de la mer. *Acad. Sci. Comptes Rendus,* **49**, 536-537.

Mann, C. R. (1972) A review of the branching of the Gulf Stream system. *Royal Soc. Edinburgh, Proc.* ser. B, **72**, 341-349.

Marcet, A. J. G. (1807) An analysis of the waters of the Dead Sea and the River Jordan. *Royal Soc. Philos. Trans.,* **97**, pt. 2, 296-314.

Marcet, A. J. G. (1811) A chemical account of an aluminous chalybeate spring in the Isle of Wight. *Geol. Soc. London Trans.,* **1**, 213-248.

Marsigli, L. F. (1725) *Histoire Physique de la Mer.* Amsterdam.

Maskelyne, Nevil (1762a) Observations on the tides in the Island of St. Helena. *Royal Soc. Philos. Trans.,* **52**, pt. 2, 586-591.

Maskelyne, Nevil (1762b) Observations of the tides made in the harbour at James's Fort, St. Helena. *Royal Soc. Philos. Trans.,* **52**, pt. 2, 592-606.

Matthäus, W. (1968) The historical development of methods and instruments for the determination of depth-temperatures in the sea *in situ. Inst. Océanog. Monaco Bull.*, spec. no. 2, **1**, 35–47.

Matthäus, W. (1972) On the history of recording tide gauges. *Royal Soc. Edinburgh Proc.*, ser. B, **73**, 25–34.

Maury, M. F. (1855) The Physical Geography of the Sea and its Meteorology. London, Sampson Low. First published (1855) New York, Harper. Reprinted (1963), J. Leighly ed. Cambridge, Mass., Harvard University Press.

Merriman, D. (1948) A posse ad esse. *Jour. Marine Research,* **7,** 139–146.

Merriman, D. (1965) *Edward Forbes—Manxman. Progress in Oceanography.* Edited by M. Sears. London and New York, Pergamon Press, Vol. 3, pp. 191–206.

Merriman, D. (1968) Speculations on life at the depths: a XIXth-century prelude. *Inst. Océanog. Monaco Bull.,* spec. no. 2, **2,** 377–384.

Merriman, D. (1972a) *Challengers of Neptune: the "Philosphers". Royal Soc. Edinburgh, Proc.,* ser. B, **72,** 15–45.

Merriman, D. (1972b) William Carmichael M'Intosh, nonagenarian. *Royal Soc. Edinburgh Proc.,* ser. B, **72,** 99–105.

Merriman, D. and M. (1958) Sir C. Wyville Thomson's letters to Staff Commander Thomas H. Tizard, 1877–1881. *Jour. Marine Research* **17,** 347–374.

Milet-Mureau, L. A. (1799) *A Voyage round the World, performed in the years 1785–88 by the* Boussole *and* Astrolabe, *under the command of J. F. G. de la Pérouse.* London, G. G. & J. Robinson; J. Edwards; T. Payne. Originally published in Paris (1798) as *Voyage de La Pérouse autour de Monde.*

Mill, H. R. (1893) The permanance of ocean basins. *Geog. Jour.* **1**(3), 230–234.

Milne, John (1897) Sub-oceanic changes *Geog. Jour.* **10**:129–146, 259–285.

Moray, Sir Robert (1666) Patternes of the tables proposed to be made for observing of tides, promised in the next foregoing Transactions. *Royal Soc. Philos. London Trans.,* **1**(18), 311–314.

Multhauf, R. P. (1960) The line-less sounder: an episode in the history of scientific instruments. *Jour. Hist. Med.,* **15,** 390–398.

Munk, W. H., and D. E. Cartwright (1966) Tidal spectroscopy and prediction. *Royal Soc. London Philos. Trans.,* ser. A, **259**(1105), 533–581.

Murray, John (1818a) An analysis of sea-water; with observations on the analysis of salt-brines. *Royal Soc. Edinburgh Trans.,* **8,** pt. 1, 205–244.

Murray, John (1818b) A general formula for the analysis of mineral waters. *Royal Soc. Edinburgh Trans.,* **8,** pt. 1, 259–279.

Murray, Sir John (1880) On the structure and origin of coral islands. *Royal Soc. Edinburgh Proc.,* **10**(107), 505–518.

Murray, Sir John, and Johan Hjort (1912) *The Depths of the Ocean.* A general account of the modern science of oceanography based largely on the scientific researches of the Norwegian steamer *Michael Sars* in the North Atlantic. London, Macmillan.

Murray, Sir John, and A. F. Renard (1884) On the microscopic characters of volcanic ashes and cosmic dust, and their distribution in the deep-sea deposits. *Royal Soc. Edinburgh Proc.* **12,** 474–495.

Murray, Sir John, and A. F. Renard (1891) Report on Deep Sea Deposits based on the specimens collected during the voyage of HMS *Challenger* in the years 1872 to 1876. *Challenger* Report (q.v.).

Nansen, Fridtjof (1897) *Farthest North, being the record of a voyage of exploration of the ship* Fram *1893–96 and of a fifteen months' sledge journey by Dr. Nansen and Lieut. Johansen.* London, Archibald Constable, 2 vols.

Nares, G. S. (1872) Investigations of the currents in the Strait of Gibraltar, made in August 1871, by Captain G. S. Nares, R.N., of HMS *Shearwater,* under instructions from Admiral Richards, FRS, Hydrographer of the Admiralty. *Royal Soc. [London] Proc.*, **20**(131), 97-106.

Negretti, H., and J. W. Zambra (1874) On a new deep-sea thermometer. *Royal Soc. [London] Proc.*, **22**(151), 238-241.

Neumann, C. von (1861) *Ueber das Maximum der Dichtigkeit beim Meerwasser.* Munich, Universitatsbuchdrucker.

Newton, Sir Isaac (1687) *Philosophiae Naturalis Principia Mathematica.* London, Sam Smith. Translated by A. Motte, edited by F. Cajori (1934), *Sir Isaac Newton's Mathematical Principles of Natural Philosophy and his System of the World.* Berkeley, University of California Press.

Nicholas of Cusa (1650) *The Idiot in Four Books.* The first and second of Wisdome. The third of the Minde. The fourth of statick Experiments, Or experiments of the Ballance. London, William Leake.

Olson, F. C. W., and M. A. (1958) Luigi Ferdinando Marsigli, the lost father of oceanography. *Florida Acad. Sci. Quart. Jour.*, **21**(3), 227-234.

Palmer, H. R. (1831) Description of a graphical registrer of tides and winds. *Royal Soc. London Philos. Trans.*, **121**, pt. 1, 209-213.

Parrot, G. F. (1833) Expériences de forte compression sur divers corps. *Mémoires de l'Académie Impériale des Sciences.* ser. 6, **2**, 595-630.

Parry, W. E. (1821) *Journal of a Voyage for the Discovery of a North-West Passage from the Atlantic to the Pacific;* performed in the years 1819-20, in HMS *Hecla* and *Griper.* London, John Murray.

Pérès, J. M. (1968) Un précurseur de l'étude du benthos de la Méditerranée: Louis-Ferdinand, comte de Marsilli. *Inst. Océanog. Monaco Bull.*, spec. no. 2, **2**, 369-375.

Pettersson, Otto (1894) Proposed scheme for an international hydrographic survey of the North Atlantic, the North Sea, and the Baltic. *Scottish Geog. Mag.*, **10**(12), 631-635.

Pettersson, Otto (1904, 1907) On the influence of ice-melting upon oceanic circulation. *Geog. Jour.*, **24**(3), 285-333; **30**(3), 273-297.

Philips, Henry (1668) A letter written to Dr. John Wallis by Mr. Henry Philips containing his observations about the true time of the tides. *Royal Soc. London Philos. Trans.* **3**(34), 656-659.

Phipps, C. J. (1774) *A Voyage toward the North Pole undertaken by his Majesty's Command, 1773.* London, J. Nourse.

Pillsbury, J. E. (1891) *The Gulf Stream. Methods of the investigation and results of the research.* Report of the United States Coast and Geodetic Survey for 1890, Washington, appendix 10.

Pliny (1855-1857) *The Natural History of Pliny.* Translated by J. Bostock and H. T. Riley. London, Bohn's Classical Library, 6 vols.

Prestwich, Joseph (1875) Tables of the temperature of the sea at different depths beneath the surface, reduced and collated from the various observations made between the years 1749 and 1868, discussed. *Royal Soc. London Philos. Trans.*, **165**, pt. 2, 587-674.

Proceedings of the First International Congress on the History of Oceanography, 1966 (1968) *Inst. Océanogr. Monaco Bull.*, spec. no. 2, 2 vols.

Proceedings of the Second International Congress on the History of Oceanography (1972) *Royal Soc. Edinburgh Proc.*, ser. B, vols. 72 and 73.

Proudman, J. (1968) Arthur Thomas Doodson. *Biographical Memoirs of Fellows of the Royal Society*, **14**, 189-205.

Riley, J. P., and G. Skirrow (eds) (1965) *Chemical Oceanography.* London and New York, Academic Press, 2 vols.

Rondelet, G. (1558) *L. Histoire Entière des Poissons.* Lion, 2 vols.

Ross, Sir James Clark (1847) *A Voyage of Discovery and Research in the Southern and Antarctic Regions, during the years 1839–1843.* London, John Murray, 2 vols.

Ross, Sir John (1819) *A Voyage of Discovery made under the Orders of the Admiralty, in HMS* Isabella *and* Alexander, *for the purpose of exploring Baffin's Bay, and inquiring into the probability of a North-West Passage.* London, John Murray.

Rumford, Count Benjamin Thompson (1800) Of the propagation of heat in fluids. *Essays, Political, Economical and Philosophical,* T. Cadell, Jr., and W. Davies. London, Vol. 2, pp. 197–386. First published (1797). Reprinted (1968) in *Collected Works of Count Rumford.* Edited by S. C. Brown. Cambridge, Mass. Harvard University Press, (1968–1970), Vol. 1, pp. 117–284.

Sars, G. O. (1872) *On some Remarkable Forms of Animal Life from the Great Deep off the Norwegian Coast:* 1. Partly from the posthumous manuscripts of the late Professor Dr. Michael Sars. Christiania, Kongelist Norsk Frederiks Universitet.

Sars, Michael (1869) Remarks on the distribution of animal life in the depths of the sea. *Annals Mag. Nat. Hist.,* ser. 4 **3**(18), 423–441.

Sarton, George (1927–1948) *Introduction to the History of Science.* Washington, Carnegie Institution, 3 vols.

Sarton, George (1934) Simon Stevin of Bruges (1548–1620). *Isis,* **21**(61), 241–303.

Saunders, William (1805) *A Treatise on the Chemical History and Medical Powers of some of the most celebrated Mineral Waters;* with practical remarks on the aqueous regimen, to which are added, Observations on the use of cold and warm bathing. London, 2nd edition.

Saussure, H. B. de (1779–1796) *Voyages dans les Alpes, précédés d'un essai sur l'histoire naturelle des environs de Genève.* Neuchâtel, 4 vols.

Scheltema, R. S., and A. H. (1972) Deep-sea biological studies in America, 1846 to 1872—their contribution to the *Challenger* expedition. *Royal Soc. Edinburgh Proc.,* ser. B., **72,** 133–144.

Schlee, Susan (1973) *The Edge of an Unfamiliar World.* A History of Oceanography. New York, E. P. Dutton.

Shea, W. R. J. (1970) Galileo's claim to fame: the proof that the earth moves from the evidence of the tides. *British Jour. Hist. Sci.,* **5**(18), 111–127.

Sheeres, Sir Henry (1703) Discourse of the Mediteranian Sea, and the Streights of Gibraltar. In *Miscellanies Historical and Philological:* being a Curious Collection of Private Papers found in the Study of a Noble-man, lately Deceas'd. London, pp. 1–42.

Sibbald, Sir Robert (1692) *Phalainologia nova sive Observationes de rarioribus quibusdam Balaenis in Scotiae Littus nuper ejectis.* Edinburgh, Robert Edward.

Sigsbee, C. D. (1880) *Deep-Sea Sounding and Dredging.* Washington, U.S. Coast and Geodetic Survey.

Sivertsen, Erling (1968) Michael Sars, a pioneer in marine biology, with some aspects from the early history of biological oceanography in Norway. *Inst. Océanogr. Monaco Bull.,* spec. no. 2, **2,** 439–451.

Six, James (1794) *The Construction and Use of a Thermometer, for showing the extremes of Temperature in the Atmosphere, during the Observer's Absence.* Maidstone.

Smith, Thomas (1684) A conjecture about an under-current at the Streights-Mouth. Read before the Oxford Society, Dec. 21, 1683. *Royal Soc. London Philos. Trans.,* **14**(158), 564–566.

Solhaug, T., and G. Saetersdal (1972) The development of fishery research in Norway in the 19th and 20th centuries in the light of the history of fisheries. *Royal Soc. Edinburgh Proc.*, ser. B, **73**, 399–412.

Sorby, H. C. (1861) On the organic origin of the so-called "Crystalloids" of the chalk. *Annals Mag. Nat. Hist.*, ser. 3, **8**(45), 193–200.

Spratt, T. A. B. (1848) On the influence of temperature upon the distribution of the fauna of the Aegean Sea. *Philos. Mag.*, ser. 3, **33**(221), 169–174.

Spratt, T. A. B. (1865) *Travels and Researches in Crete.* London, Jan van Voorst, 2 vols.

Stanley, Owen (1848) On the lengths and velocities of waves. *British Assoc. Adv. Sci. Rept.*, **18**, pt. 2, pp. 38–39.

Stevenson, Robert (1824) *An Account of the Bell Rock Light-House.* Edinburgh, Archibald Constable.

Stommel, H. (1958) *The Gulf Stream. A Physical and Dynamical Description.* Berkeley and Los Angeles, University of California Press; London, Cambridge University Press.

Suess, Eduard (1904–1924) *The Face of the Earth.* Translated by H. Sollas from *Das Antlitz der Erde.* Originally published (1883–1909) Prag, Wien, Leipzig. Oxford, Clarendon Press, 5 vols.

Talbot, W. H. Fox (1833) Proposed method of ascertaining the greatest depth of the ocean. *Philos. Mag.*, ser. 3, **3**(14), 82.

Théodoridès, J. (1968) Les débuts de la biologie marine en France: Jean-Victor Audouin et Henri–Milne Edwards, 1826–1829. *Inst. Océanog. Monaco Bull.*, spec. no. 2, **2**, 417–437.

Thomson, Sir Charles Wyville (1873) *The Depths of the Sea.* London, Macmillan.

Thomson, Sir Charles Wyville (1877) *The Voyage of the Challenger.* The Atlantic. New York, Macmillan, 2 vols.

Thomson, Sir Charles Wyville (1880) General introduction to the zoological series of reports. *Challenger* Report (q.v.) Vol. 1. Zoology.

Thomson, Sir Charles Wyville, and J. Murray, eds. (1880–1895) *Report on the Scientific Results of the Voyage of HMS* Challenger *During the Years 1872–76:* London, Her Majesty's Stationery Office, 50 vols.

Tizard, T. H. (1875) Remarks on the temperatures of the China, Sulu, Celebes and Banda Seas. *HMS* Challenger: *No. 4 Report on ocean soundings and temperatures, Pacific Ocean, China and adjacent seas.* London, Admiralty.

Trégouboff, G. (1968) Les précurseurs dans le domaine de la biologie marine dans les eaux des baies de Nice et de Villefranche-sur-Mer. *Inst. Océanog. Monaco Bull.*, spec. no. 2, **2**, 467–479.

Tricker, R. A. R. (1964) *Bores, Breakers, Waves and Wakes.* London, Mills & Boon.

Vaughan, T. Wayland (1937) *International Aspects of Oceanography, Oceanographic Data and Provisions for Oceanographic Research.* Washington, National Academy of Sciences.

Viglieri, A. (1968) La carte générale bathymétrique des océans établie par S.A.S. le Prince Albert Ier. *Inst. Océanog Monaco Bull.*, spec. no. 2, **1**, 243–253.

Vine, F. J., and D. H. Matthews (1963) Magnetic anomalies over ocean ridges. *Nature*, **199**(4897), 947–949.

Wales, W., and W. Bayly (1777) *The Original Astronomical Observations, made in the course of a Voyage towards the South Pole, and round the World in HMS* Resolution *and* Adventure, *in the years 1772–1775.* London, Board of Longitude.

Wallace, W. J. (1974) *The Development of the Chlorinity/Salinity Concept in Oceanography:* Oceanography Series, no. 7. Amsterdam, London, New York, Elsevier.

Wallich, G. C. (1861) Remarks on some novel phases of organic life, and on the boring powers of minute Annelids, at great depths in the sea. *Annals Mag. Nat. Hist.*, ser. 3, **8**(43), 52-58.

Wallich, G. C. (1862) *The North-Atlantic Sea-Bed:* comprising a diary of the voyage on board HMS *Bulldog,* in 1860; and observations on the presence of animal life, and the formation and nature of organic deposits, at great depths in the ocean. London, John van Voorst.

Wallis, John (1666) An essay of Dr. John Wallis, exhibiting his Hypothesis about the Flux and Reflux of the Sea. *Royal Soc. Philos. Trans.,* **1**(16), 263-281.

Wallis, John (ed.) (1673) *Jeremiae Horroccii, Liverpoliensis Angli, ex Palatinatu Lancastriae, Opera Posthuma.* London, J. Martyn.

Warren, B. A. (1966) Medieval Arab references to the seasonally reversing currents of the North Indian Ocean. *Deep-Sea Research,* **13**(2), 167-171.

Waters, D. W. (1958) *The Art of Navigation in England in Elizabethan and early Stuart Times.* London, Hollis & Carter.

Welander, P. (1968) Theoretical oceanography in Sweden, 1900-1910. *Inst. Océanog. Monaco Bull.,* spec. no. 2, **1**, 169-173.

Went, A. E. J. (1972a) The history of the International Council for the Exploration of the Sea. *Royal Soc. Edinburgh Proc.,* ser. B, **73**, 351-360.

Went. A. E. J. (1972b) Seventy Years Agrowing. A History of the International Council for the Exploration of the Sea, 1902-1972. *Rapp. P.-v. Réun. Cons. int. Explor. Mer,* **165**.

Wilkes, Charles (1845) *Narrative of the United States Exploring Expedition during the years 1838-42, under the command of Charles Wilkes.* Philadelphia, Lea and Blanchard; London, Wiley and Putnam, 5 vols and atlas.

Williams, F. L. (1963) *Matthew Fontaine Maury, Scientist of the Sea.* New Brunswick, New Jersey, Rutgers University Press.

Wilson, George (1846) On the solubility of fluoride of calcium in water, and its relation to the occurrence of fluorine in minerals, and in recent and fossil plants and animals. *Royal Soc. Edinburgh Trans.* **16**, 145-164.

Wilson G., and A. Geikie (1861) *Memoir of Edward Forbes, FRS.* Cambridge and London, Macmillan; Edinburgh, Edmonston and Douglas.

Winthrop, Robert C. (1878) Correspondence of Hartlib, Haak, Oldenburg, and others of the founders of the Royal Society, with Governor Winthrop of Connecticut, 1661-1673. *Mass. Hist. Soc. Proc.,* **16**, 206-251.

Wolff, T. (1967) *Danske Ekspeditioner På Verdenshavene.* Copenhagen, Rhodos.

Wüst, G. (1964) The major deep-sea expeditions and research vessels, 1873-1960. In *Progress in Oceanography,* edited by M. Sears: London and New York, Pergamon, vol. 2, pp. 1-52.

Wüst, G. (1968) History of investigations of the longitudinal deep-sea circulation, 1800-1922. *Inst. Océanog. Monaco Bull.,* spec. no. 2, **1**, 109-120.

Yonge, C. M. (1972) John Graham Dalyell and some predecessors in Scottish marine biology. *Royal Soc. Edinburgh Proc.,* ser. B, **72**, 89-97.

AUTHOR CITATION INDEX

SUBJECT INDEX

About the Editor

MARGARET DEACON was born in West Kilbride, Ayrshire, and educated at Guildford County School and Somerville College, Oxford. From 1966 to 1969 she studied the history of oceanography at the National Institute of Oceanography, Wormley, Surrey, with the aid of grants from the Royal Society of London and the Natural Environment Research Council. From 1969 to 1972 she was a research assistant at the University of Edinburgh, working primarily at the Science Studies Unit. Between 1973 and 1978 she acted as part-time research assistant in the Department of Printed Books and Manuscripts at the National Maritime Museum, Greenwich, while continuing her research. Ms. Deacon has published *Scientists and the Sea, 1650 to 1900: A Study of Marine Science* (1971) and several papers on the early history of oceanography.